Basic Techniques in Biochemistry and Molecular Biology

Basic Techniques in Biochemistry and Molecular Biology

Lajja Das

RANDOM

RANDOM PUBLICATIONS

NEW DELHI (INDIA)

Basic Techniques in Biochemistry and Molecular Biology

ISBN 978-93-5111-204-4

Published in 2014 in India by

Reprint 2020

RANDOM PUBLICATIONS
4376-A/4B, Gali Murari Lal, Ansari Road
New Delhi-110 002
Phone : +91-11-43580356, +91-11-23289044
e-mail: randomexports@gmail.com, sales@randompublications.com, info@randompublications.com

REPRINT-2025

Type Setting by : Keystoneprintads, Delhi-110051
Printed at : Mehra Printers, Delhi-110 092

Preface

Biochemistry and molecular biology is an important component of all disciplines of biology. It uniquely integrates the theories and practices that drive the fields of biology and medicine, comprehensively covering both the methods students will encounter in lab classes and those that underpin recent advances and discoveries.

At its most basic, biochemistry is the study of the chemical processes occurring in living matter. However, this simple definition encompasses an incredibly diverse field of research that touches nearly all aspects of our lives. One of the most pressing issues in our society, environmental degradation, is being addressed by biochemists.

A few examples of work currently being performed include improvements in the efficiency of photosynthesis to increase crop yields, bioremediation of polluted soils, development of new feed-stocks, chemistries for the production of biofuels, genetic mapping of ecosystems to monitor biodiversity, and methodologies for boosting biological capture of carbon. These and other biochemical technologies may play a crucial role in our efforts to find a sustainable means of living.

Biochemistry is the application of chemistry to the study of biological processes at the cellular and molecular level. It emerged as a distinct discipline around the beginning of the 20th century when scientists combined chemistry, physiology and biology to investigate the chemistry of living systems.

Molecular biology explores cells, their characteristics, parts, and chemical processes, and pays special attention to how molecules control a cell's activities and growth.

Looking at the molecular machinery of life began in the early 1930s, but truly modern molecular biology emerged with the uncovering of the structure of DNA in the 1960s. As a science that studies interactions between the molecular components that carry out the various biological processes in living cells, an important idea in molecular biology states that information flow in organisms follows a one-way street: Genes are transcribed into RNA, and RNA is translated into proteins. The book will be essential reading for all bioscience students and pre-clinical medical students.

I thank all members of my team who have helped in the preparation of the book. My special thanks go to "Random Publications" who have published the book.

– Lajja Das

Contents

1

pH and Buffers

INTRODUCTION TO PH

IONIZATION OF WATER

Sometimes the hydrogen of one water molecule will "jump" to another water molecule:

$$H_2O + H_2O = H_3O^+ + OH^-$$

This proton hopping is called the ionization of water (an ion is a positively or negatively charged atom or molecule). This ionization creates a H_3O^+ and a OH^- molecule. The H_3O^+ is often written as simply H^+. This is because a H_3O^+ is just a H^+ that jumps from one water molecule to another: $H_2O = H^+ + OH^-$ So remember, $H_3O^+ = H^+$.

Looking at either of the two chemical equations above, it is important to note that the reverse reaction is also occurring. How much H^+ and OH^- exist in water? Very, very little! The ratio of either H^+ or OH^- to H_2O in neutral water is 1:1,000,000,000! Since this is such a small amount of either H^+ or OH^- , they rarely meet and neutralize each other.

The equilibrium constant, K_{eq} describes the ionization equilibrium of water:

$$K_{eq} = [H^+][OH^-]$$

Because of this relationship it is important to note that if the $[H^+]$ goes up then the $[OH^-]$ must go down, and vice-versa, for the value for the K_{eq} of water must remain constant. For neutral water, the K_{eq} is 1×10^{-14} M and the concentrations of $[H^+]$ and $[OH^-]$ are each 1×10^{-7} M.

Let's look at that last number without the exponent:

$$0.0000001 \text{ M}$$

This is obviously a very small number. A more manageable way to discuss small numbers such as this is to take the negative logarithm. For the concentration of $[H^+]$, this is called the pH. In this case:

$$-\log(0.0000001 \text{ M}) = 7$$

The pH of a solution is simply the negative logarithm of $[H^+]$. The pH of a solution describes the acidity of a solution. Acidic solutions are those with a pH of less than 7 and basic solutions have a pH greater than 7. A solution, like H_2O, with a pH = 7 is neutral. Similarly, the pH could be used to describe a solution in terms of its OH^- concentration. pOH is the negative logarithm of the OH^- concentration.

One useful thing to remember is:

$$pH + pOH = 14.$$

In the body, the pH of blood is 7.4. This corresponds to a $[H^+]$ of about 40 nM. This value can only vary from 37 nM to 43 nM without serious metabolic consequences. pKa In living systems, much of the chemistry involves interactions between acids and bases. Acids are H^+ donors andbases are H^+ acceptors.

$$HA=H^+ + A^-$$

Acid and base reactions are made up of conjugate acid-base pairs. Strong acids are those that readily give up a H^+. Strong bases readily accept a H^+. Weak acids do not readily give up a H^+, but will under the right conditions. A weak base is one that does not easily take up a H^+. The conjugate base for a strong acid is a weak base. In contrast, a strong base has a weak acid as its conjugate. Think of a strong acid as a person with an ugly wig. This person can't wait to get rid of this! So now we have a wig (H^+) and a person without the wig (the weak base). You can understand that the person without the wig doesn't really want to take it back!

However, what about strong base-weak acid conjugate pairs?

$$A^- + H^+ HA$$

Think of it in similar terms: a person in dire need of a wig will grab just about anything. This is our strong base. Now on the right side of the equation we have someone with a wig who really doesn't want to give it up, just as a weak acid does not want to part with its H^+. We can think of a strong acid or base as having a "flip side", which is its weak base or weak acid conjugate, respectively.

How readily an acid gives up its H^+ is expressed by the acid dissociation constant, or K_a:

$$K_a = \frac{[H+][A-]}{[HA]}$$

The pK_a is the negative logarithm of the K_a. Strong acids have small pK_as. Looking at the above equation, it can be seen that when $[A^-] = [HA]$, then $K_a = [H^+]$. Then the pK_a = pH. This is the basis for the Henderson-Hasselbalch equation:

$$pH = pK_a + \log\frac{[H+][A-]}{[HA]}$$

This is Henderson-Hasselbalch Equation.

TITRATION CURVES

Acid with a Strong base: When acids or base are added to a solution, the pH changes. The controlled addition of an acid or base to a solution is called a titration. A titration curve can be made by plotting the pH changes against the volume of acid or base added to a solution. Many titration curves are made by plotting pH on the y axis and the volume of base added to the solution on the x axis. For example, pH changes to acetic acid, CH_3COOH, and hydrochloric acid, HCl, with 0.1M sodium hydroxide, NaOH, is shown to the left. Notice the shape of the curves. The CH_3COOH curve starts out at a higher pH than the HCl curve.

This is because CH_3COOH is a weaker acid than HCl. As NaOH is added to the solution of CH_3COOH, the pH rises. However, the addition of NaOH to the HCl solution does not significantly change the pH. It takes less NaOH to change the pH of a weak acid solution than it does to change the pH of a strong acid solution. Eventually all of the acid in both solutions will have reacted with base, such that adding more base quickly increases the pH. This is called the equivalence point.

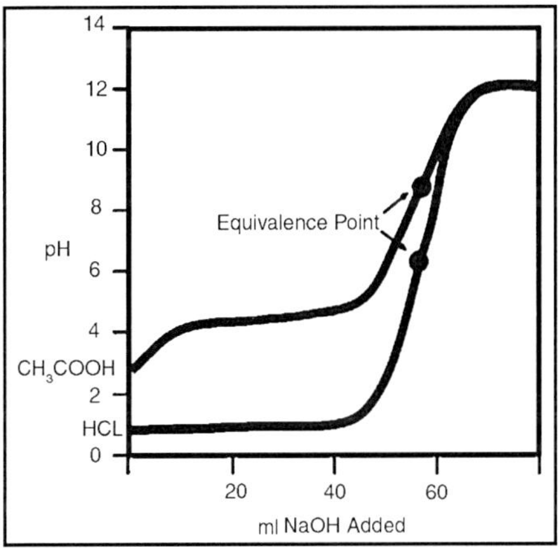

Fig. Titration Curves of Both a Strong and Weak.

This is the point where all the acid has reacted with the base, such that further additions of base quickly raise the pH. The equivalence point for a strong acid-strong base titration is pH = 7. For weak acids-strong base titrations, the equivalence point is pH >7.

BUFFERS

The pH of a solution is dependent on the concentration of H^+ ions. Addition or removal of H^+ ions, then, can greatly affect the pH of a solution. In the body, the pH of cells and extracellular fluids can vary from pH 8 in pancreatic fluid to pH 1 in stomach acids. The average pH of blood is 7.4, and of cells is 7 – 7.3. Although there is great variation in pH between the fluids in the body, there is little variation within each system. For example, blood pH only varies between 7.35 – 7.45 in a healthy individual. Large changes in pH can be life threatening. How does the body maintain a constant blood pH? The body uses a buffer system to withstand changes in pH. Buffers are made up of a mixture of a weak acid with its conjugate base or a weak base with its conjugate acid. Remember that an acid donates a H^+. A weak acid does not donate its H^+ as easily. Similarly, a weak base will not accept a H^+ as well as a strong base.

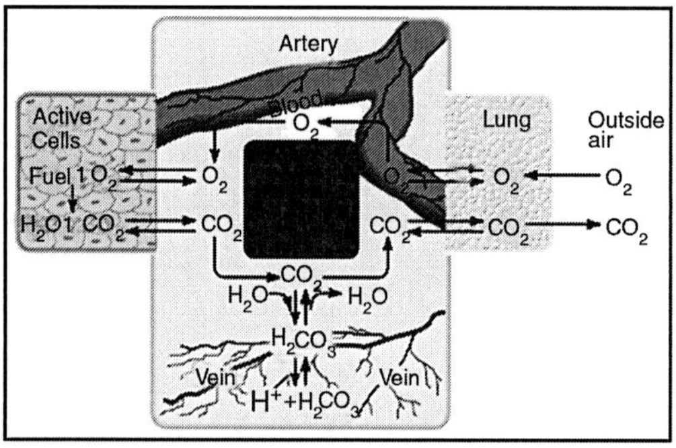

Fig. The Bicarbonate Blood Buffering System

Buffers maintain pH by binding H^+ or OH^- ions. This stabilizes changes in pH. The bicarbonate buffer system maintains blood pH near pH 7.4. The carbonic acid, H_2CO_3, in the blood is in equilibrium with the carbon dioxide (CO_2), in the air. Buffers are most effective in a pH range near its pK_a. This is where the titration curve is most shallow, and where the pH is least affected by added acid or base. For example, in the titration curve of phosphoric acid:

$$H_3PO_4 \underset{}{\overset{pK_a1}{\rightleftharpoons}} H_2PO_4^- + H^+ \overset{pK_a2}{\rightleftharpoons} HPO_4^{-2} + H^+ \overset{pK_a3}{\rightleftharpoons} PO_4^{-3} + H^+$$

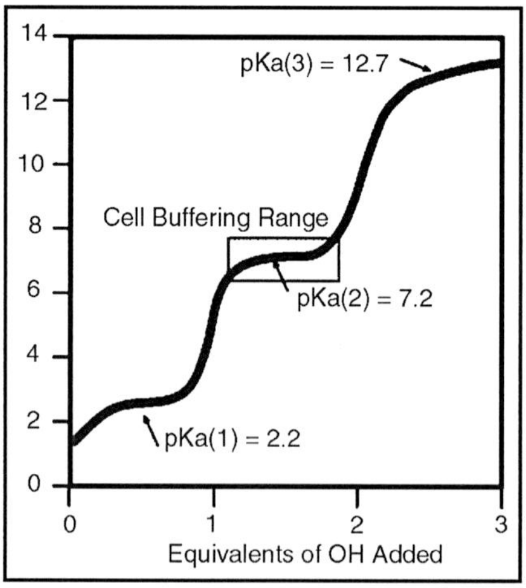

Fig. The Titration of a Polyprotic Weak Acid

The region in the rectangle is the buffering region of biological importance. Note that the slope of the curve is shallow within one pH unit of the pK_a value. This is where the buffering range is most effective, meaning that the buffer is able to resist changes in pH.

In chemistry, the degrees of acidity or alkalinity of a substance are expressed in pH values. The neutral point, where a solution would be neither acid nor alkaline is pH 7. Increasing acidity is expressed as a number less than 7, and increasing alkalinity is a number great than 7. Maximum acidity is pH 0 and maximum alkalinity is pH 14. Because each unit on the scale represents a logarithm, there is 10-fold difference between each unit (pH 5 is 10 times as acid as pH 6 and pH 4 is 100 times as acid as pH 6. Expressed mathematically, pH is logarithm of the hydrogen ion concentration divided into one - Log 1/H+

An acid is a substance that releases hydrogen into a solution. An alkaline substance is one that removes hydrogen from a solution. Most body functions occur at only certain levels of acidity or alkalinity, as well as, many enzymes and chemical reactions work best at a specific pH. A minute change in pH can have a profound effect on body function.* *Some of the pH values of a number of biological materials are given below:*

Table A - pH Values of Representative Biological Materials

Material	pH value
Blood, normal limits	7.3 - 7.5
Blood, extreme limits	7.0 - 7.8
Enzymes, Activity Range of	
Amylopsin, optimum	7.0
Erepsin, optimum	7.8
Invertase, optimum	5.5
Lipase, optimum	7.0 - 8.0
Maltase, optimum	6.1 - 6.8
Pepsin, optimum	1.5 - 2.4
Trypsin, optimum	8.0 - 9.0
Fruit Juices	
Apple	3.8
Banana	4.6
Grapefruit	3.0 - 3.3
Orange	3.1 -4.1
Tomato	4.2
Gastric juice (adult)	0.9 - 1.6
Milk (cows)	6.2 - 7.3

Plants (Extracted Juice)

Alfalfa tops	5.9
Carrot	5.2
Cucumber	5.2
Peas, field	6.8
Potato	6.1
Rhubarb, stalks	3.4
String beans	5.2
Sweat	4.5 - 7.1
Saliva	6.2 - 7.6
Urine (human)	4.2 - 8.0
Tears	7.2

PH - POTENTIAL OF HYDROGEN DESCRIPTION

WHAT IS HYDROGEN

Hydrogen (H) is the lightest and simplest element. A Hydrogen atom is made up of one proton and one electron. At room temperature, Hydrogen is gas - as it cools it becomes liquid. Hydrogen is one of 9 essential macronutrients and is a major component of organic molecules. Hydrogen supplies building blocks of all organisms whether plant, animal or human. Each water molecule contains two Hydrogen molecules and one Oxygen molecules (H20).

Ideal pH Values

When pH = 8: There are 6 billion H+ ions in solution.
When pH = 7: There are 60 billion H+ ions in solution.
When pH = 6: There is 600 billion H+ ions in solution.

Ideal pH Values Of Blood, Urine and Saliva

- Arterial Blood – 7.40 – 7.45
- Capillary – 7.35 – 7.45
- Venous – 7.30 – 7.35
- Saliva – 6.5
- Urine – 6.8

pH Values Of Body

Saliva – (6.0 – 7.4)	Skeletal Muscle - (6.9 - 7.2)
Gastric Secretion - (1.0 - 3.5)	Heart - (7.0 - 7.4)
Pancreatic Secretion - (8.0 - 8.3)	Liver - (7.2)
Bile - (7.8)	Brain (7.1)
Small Intestine Secretion - (7.5 - 8.0)	Urine - (4.5 - 8.0)
Arterial Blood - (7.4 - 7.45)	Capillary Blood - (7.35 - 7.4)
Venous Blood - (7.3 - 7.35)	Feces - (4.6-8.4)

CLINICAL ACIDOSIS AND ALKALOSIS

The blood pH is supposed to remain "practically constant" according to our best text books. The physician who starts to check his patients, however, cannot occur. Instead of the normal figure of pH 7.3 to 7.5 he finds it varying from 7.2 to 7.8 (1). The measurement of pH is attended with difficulty because the blood loses carbon dioxide (CO_2) and thereby rises the pH rapidly after it has been withdrawn. In two minutes a rise of.2 may occur (1). pH metres are available where the electrodes are built into a syringe so the pH may be read directly before the blood is exposed to air. (National Technical Labouratories, Pasadena, California).

The importance of clinical pH testing was illustrated by the following report of a pharmacist. He had been supplying for a number of months a capsule formula of aspirin, caffeine, phenacetin and ammonium chloride to an arthritic patent where the remedy had been successful in controlling arthritic pains. The capsule dispensed was a standard formulas of a pharmaceutical house, with a specific catalog number. The patient returned two days after receiving a renewed supply of capsules, complaining that they were ineffective, and different from the previously used kind. On checking the records, it was found that the original formulas had been revised, and the supposedly unnecessary ammonium chloride had been omitted. On supplying capsules of ammonium chloride alone, the patient reported the usual successful results.

What was the biochemistry behind this: Simply that a high pH tends to throw calcium out of solution in the body fluids, and bursitis, arthritis, neuritis, lumbago, sciatica, and a host of other painful syndromes develop as a consequence. Cheilosis (lips redden, fissures at the angles) may occur and herpes simplex, and other virus types of diseases become active. The low calcium bicarbonate in the body fluids seems responsible, Vitamin F and Vitamin D as well as the Vitamin C complex are also factors, their deficiency aggravates the situation.

Allergies become acute, the calcium deficiency aspect of allergic sensitivity is well known. Normal rates of wound healing are greatly reduced, ulcers then too become static. Ammonium chloride is a natural constituent of gastric juice, calcium chloride is the form of calcium found in arrowroot starch, so both of these materials are physiological therapeutic agents, factors common to body biochemistry and not new and foreign substances to the physiological economy (as are synthetics and antibiotics).

It is described that Ammonium Chloride is useful in the treatment of bronchial affections, hepatic congestion, pelvic cellulitis, muscular rheumatism, gout, sciatica, chronic glandular enlargement, hemicrania, senile gangrene, dysmenorrhea, leucorrhea. Calcium Chloride is listed as useful in the treatment of hemorrhage, hemorrhagic endometritis, menorrhagia, erytherma nodosum, tetanus, spasmophillia, blackwater fever, hay fever, asthma, hemophilia, albuminuria, nephritis, typhoid, coryza, tuberculosis, osteomalacia, scrofula, rachitis, arthritis, spasm of glottis, infant convulsions, urticaria, pruritis, among others. All of these effects are no doubt in the main simply accomplished by correcting the unbalanced state of the buffers and mineral salts in the tissue fluids. It will be noted that Vitamin C complex is a desirable synergist in most of the conditions listed. Gout is mentioned as one situation where ammonium chloride may help.

Recently it has been discovered that the juice of cherries contains some active agent effective in gout, 4 to 6 ounces of the juice (preferably unsweetened, but may be canned or cooked form) promoting relief with in a few days with consistent improvement. This is the first time that the possibility of gout being a deficiency disease has been suspected. We can say that the ingestion of sodium compounds particularly sodium phosphate - is definitely aggravative. Gout patients should see that their diet contains a preponderance of potassium.

Sodium promotes the precipitation of gouty concretions of sodium urate in the tissues. A very important aspect of alkalosis is the fact the habitual use of milk of magnesia promotes the condition. It also tends to cause epistaxis (nosebleed - can be secondary to nasal and sinus issues; scarlet fever or typhoid; arteriosclerosis; hypertension and anemias), very specifically. You will be surprised how many of the patients who ask for a remedy for nosebleed are using milk of magnesia as a laxative. Citrus fruits being of a highly alkaline ash, with a high content of organic acid (citric), need special attention. The first and immediate effect of ingestion of a few ounces of grapefruit or lemon juice is to lower (acidity) the blood pH. The patient with alkalosis feels temporarily better.

Later in the day, after the citric acid has been destroyed by oxidation as a fuel (it is classed as carbohydrate) the aggravation of the alkaline state become apparent. The temporary effect of the citric acid is to cause calcium to be picked up - no doubt from some bone reserves - and after the pH change this calcium is deposited elsewhere - in physiologically undesirable spots. The treatment for bursitis obviously becomes simple. No longer can we recommend X ray treatment with equanimity. Let the physiological treatment be the preferred method.

Guanidine, a fatigue and tissue poison, and end product of the breakdown of creatine (combated by Vitamin E complex which stops the loss of creatine from the tissues), is the most potent organic alkaline substance know. It specifically precipitates calcium (recall our comment that it is diffusion from tired heart muscle into coronary vessels is the cause of precipitation therein of calcium), is normally reconverted into creatine by the influence of the parathyroid hormones, and with the assistance of the thyroid (2). Blood guanidine levels rise eight fold after parathyroidectomy (3).* We can see where the guanidine effect as a fatigue poison is contributory to the arthritis, sciatica, etc., the patient often tell us how fatigue aggravates his state of misfortune. Local inflammation also can release guanidine, its presence insure a spastic state of blood vessels, contributing to gangrene, of necessity. There seems no doubt that the effect of oxygen metabolizing vitamins - the Vitamin C and Vitamin E complexes - is vital in combating these morbid reactions by preventing the degradation of tissue elements into guanidine. Where Vitamin E deficiency has caused the tissue demand for oxygen to rise up to 250 per cent, of normal (4), the augmented release of end products certainly includes guanidine, and we see here how the find of Dr. Shute of London, Ontario, that Vitamin E reduces capillary hemorrhage may be rationalized.*

A vicious cycle is set up, the more oxygen demand the less supplied by reason of the constrictive effect of guanidine on blood vessels. Degeneration, rupture of capillaries is inevitable. As Vitamin E promotes more oxygen supply, that is why it also is good in gangrene (especially the diabetic type) and in "capillary fragility". (The true antifragility vitamin is Vitamin P, which provides a special calcium to promote collagen formation.) Vitamin F opposes the toxic effect of guanidine by providing more of the diffusible calcium (from the colloidal blood reserves) that is precipitated out of the body fluids by guanidine, thereby ameliorating the spastic, nervous and irritative (allergic) reactions. The thyroid hormone physiologically promotes the resorption and dissolution of protein structures that have reached the end of their physiological cycle (the theory of the dynamic state of living tissue, replaced and rebuilt at specific time intervals). That is why in children, thyroid deficiency is the cause of delayed development. In the adult, a hyperactive thyroid may again cause toxicosis by promoting more tissue poisons than the eliminative system can tolerate. In old people, this may be a critical situation, thyroid sometimes becoming a violent poison.

Only because the normal synergists are not available to maintain the desirable complete cycle of activity. We can see that Vitamin F is vital to the prevention of this toxicosis. It tells why the thyroid puts its secretion into the blood after ingestion of Vitamin F to the extent of a doubling of the iodine content of the blood (5). The normal activity of the gland had been blocked by reason of Vitamin F deficiency. In this blocking, we have apparently the explanation for prostate hypertrophy.

Once the thyroid is permitted to secrete, its hormone makes quick work of the fibrous tissue collected in the prostate. This also explains why we consider the Vitamin F complex, and Vitamin F2 the most friendly vitamins for the older person. The Vitamin F2 is a more highly developed compound of the basic F (a sensitized fatty acid), in which this fatty acid is combined into a phospholipid molecule that appears to have a specific function of catalyzing protective (Insulating) layers in nerve tissue and cell nuclear structure. Thereby protecting the basic controls of metabolic activity. Loss of appetite and wasting disease suggests the possible need for this vitamin.

Appetite restoration is immediate where the deficiency exists (In adults and children alike.) There seems to be as much clinical acidosis as alkalosis (1). The effects of acidosis may be aggravated by a diabetic state where organic acids cannot be oxidized normally, instead must be combined with reserve alkaline salts and excreted (5). Where sodium bicarbonate is administered to relieve diabetic acidosis, the urine should be watched and dosage stopped when a neutral is reached, according to this authority. Normally, blood glutamine provides ammonia for this purpose. Beet root and leaf are best vegetable sources of glutamine, is very high if grown on high ammonia content soils. (Plant cells form glutamine to dispose of excess ammonia). When abnormal blood pH is found, it must be considered only as a sign that something is wrong. The symptoms may be entirely different with the same abnormal pH figure in different patients, in view of the fact that there is an infinite number of combination of chemical situations that could alter pH - as many as there are different acids and alkalies. Just as a plump patient may be bloated with gas, water logged, fat, or hyper muscular. It is a preliminary classification, not a final. The pH of the blood is a resultant of the forces acting upon it. The regulation of the blood CO_2 (carbon dioxide) by respiration control Centre of the brain is a major factor. The parathyroid by eliminating the alkaline guanidine, and the kidney, eliminating mineral salt (both acid and alkaline as may be required) no doubt exercises a supplementary control. The clinical value of vinegar is more or less well known as an alternative of merit. Creatine is methylguanidine acetic acid, and is often craved by the patient, very logically it would seem. Cider vinegar being the preferred form, no doubt the malic acid of the apple is a factor, calcium malate being a catalyzer of polyphenol detoxification.

ACIDOSIS AND ALKALOSIS INFORMATION

Regulatory Factors

The following factors tend to promote normalization of pH through physiological mechanisms. They do not supply mineral elements which directly influence body chemistry. To directly influence pH values through acid or alkaline mineral administration, see: Alkalosis or Acidosis. Normalization of the function of the glands and kidneys which regulates pH balance should be the primary consideration. However, in most cases specific support of either acid or alkaline mineral therapy is necessary where immediate results are to be obtained, particularly in acute situations.

Primary

- *Renatrophin PMG*: Kidneys are thought to be the most important organ in acid-base balance regulation.
- *Drenatrophin PMG*: Adrenal support regulation of sodium-potassium-chloride balance.
- *Thytrophin PMG*: Thyroid supports regulation of calcium-

magnesium-phosphorus balance.

Normally the endocrine glands, along with the kidneys regulate the pH of the blood, just as the pancreas with it insulin regulated the blood sugar levels. The sex glands, along with the adrenals, seem to be the main endocrines involved. Particularly after menopause or male climacteric there may be serious changes in the body's economy whereby acid-base disorders develop, which may be aggravated, if not caused by dietary circumstance - just as diabetes is aggravated by a high sugar intake.*

Synergistic Supplements

- *Pituitrophin PMG: Pituitary is the hormonal control of* endocrine system
- *Ovatrophin PMG*: Specific cell activator
- *Utrophin PMG*: Specific cell activator
- *Prostate PMG*: Specific cell activator
- *Orchic PMG*: Specific cell activator

WATER, PH, AND NON-COVALENT BONDING

WATER

Water is essential for life. It covers 2/3 of the earth's surface and every living thing is dependent upon it. The human body is comprised of over 70 per cent water, and it is a major component of many bodily fluids including blood, urine, and saliva. What accounts for the ubiquitous use of water in living systems? If we take a step back and consider the structure of water and compare it to another substance, methane, we can understand the unique properties water possesses that make it well suited for biological systems. Water (H_2O) is made up of 2 hydrogen atoms and one oxygen atom, with a total atomic weight of 18 daltons.

The structure of the electrons surrounding water is tetrahedral, resembling a pyramid. For comparison, Methane (CH_4) is made up of one carbon and 4 hydrogens. Note that methane is similar to water in that it weighs 16 daltons and also has a tetrahedral structure, yet has very different physical properties.

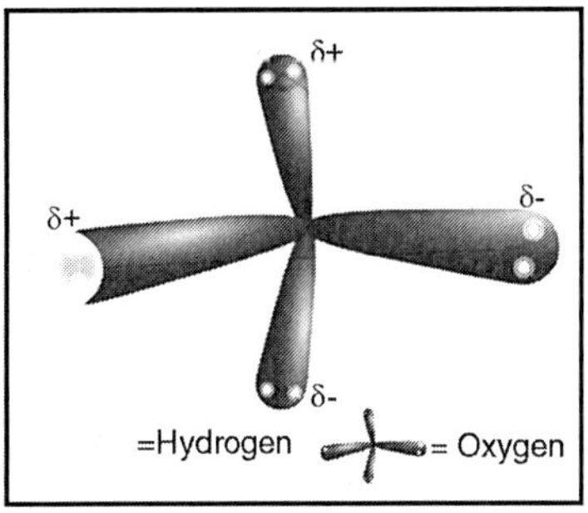

Fig. Molecular Geometry of Water: The Orbitals are Pyramidal, Yet the Atoms form a Bent Shape.

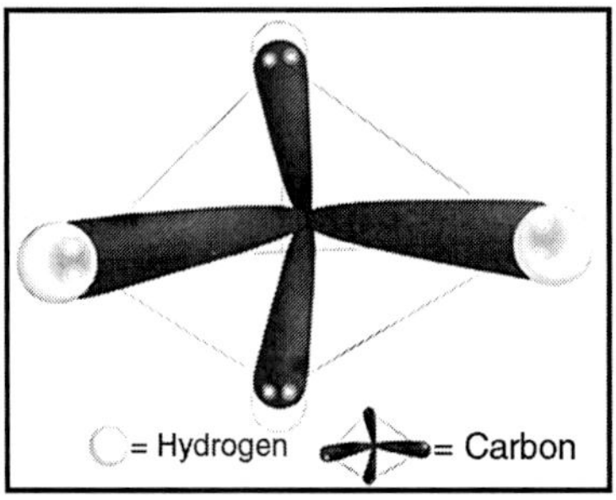

Fig. Molecular Geometry of Methane: Pyramidal

All the vertices on the methane that form the points of a pyramid are occupied by a hydrogen. However, only two vertices in water are occupied with hydrogens. The other two vertices are each occupied by a lone pair of electrons. This fact causes the water molecule to have a bent molecular shape. Oxygen is highly electrophilic (electron loving). This means that even though the oxygen in water is bound to each of the hydrogens by a covalent bond, the oxygen "pulls" the shared electrons closer to itself. This unequal sharing of the electrons in the O-H bond in water causes the hydrogens to have a partial positive charge (positive dipole), and the oxygen has a partial negative charge (negative dipole). Water is called a polar molecule because it has a positive side and a negative side, called a dipole moment. In contrast, the carbon in methane shares electrons equally with the 4 hydrogens. Methane's does not have a dipole moment, and in contrast with water is non-polar.

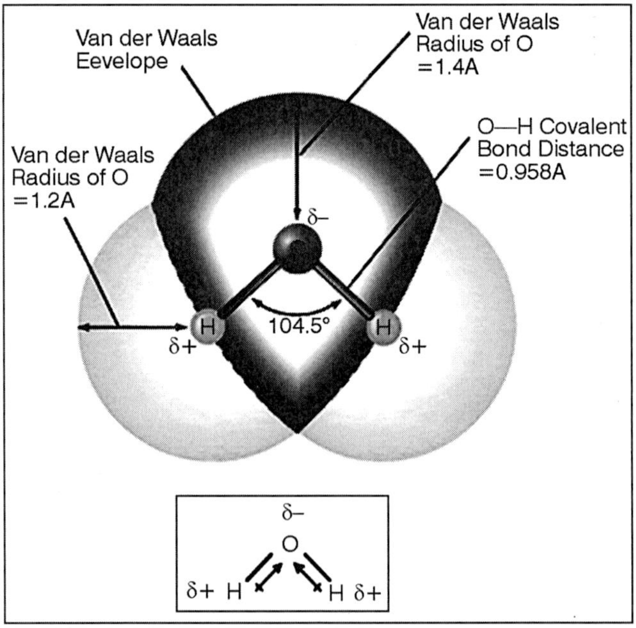

Fig. The "Magnet-like" Dipole Moment in Water

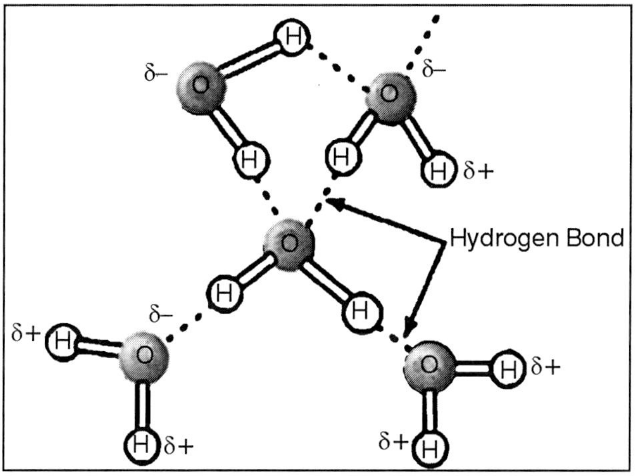

Water molecules, then, have both a partial positive and a partial negative charge. Since opposites attract, this means that water molecules are attracted to each other, like the positive and negative ends of a magnet. Because of these attractive forces, water molecules are in close proximity to one another, making water very dense. Methane molecules do not have a dipole attraction for one another, and therefore and are spaced farther apart.

Therefore, despite its similar size and mass to water, methane is much less dense than water. This is the reason that under room temperature situations, water exists as a liquid while methane is a gas. We know that water can be converted to its gaseous phase, steam, but only by applying a lot of energy in the form of heat to disrupt the large attraction of the water molecules for one another.

Hydrogen Bonding

Water molecules are bound together through hydrogen bonds. Hydrogen bonds arise when a hydrogen that is covalently bound to either an oxygen, sulfur, or nitrogen atom nears an electrophilic atom, such as oxygen. The electrophilic atom "pulls" the hydrogen closer to itself. The end result is that the hydrogen is now shared (unequally) between the atom to which it is covalently bound and the electrophilic atom to which it is attracted (O-H...O). Since water is made up solely of hydrogens covalently bound to oxygens and electrophilic oxygens, it is easy to see why there are so many hydrogen bonds in water.

The molecules in water are in constant motion and the hydrogen bonds are constantly being broken and reformed. In fact, the average hydrogen bond lasts only a one-trillionth of a second (10^{-12} s)! Each hydrogen bond has an average energy of 20 kJ/mol. This is much less than an O-H covalent bond, which is 460 kJ/mol. Even though an individual hydrogen bond is relatively weak, the large number of hydrogen bonds that exist in water which pull the molecules together give water its special density and phase transition properties. This is much like the story of Gulliver's Travels, when Gulliver awoke on the island of Lilliput to find himself tied down with many tiny strings.

Each string was individually weak, but together they were enough to bind Gulliver. On average, a liquid water molecule will have 3 out of 4 possible hydrogen bonds. This is because the molecules in liquid are in constant motion. In ice, water will form all 4 possible hydrogen bonds because the molecules in a solid are essentially locked in place. Water molecules are further apart from one another in ice than they are in the liquid state. Ice is therefore not as dense as liquid water.

This is why icebergs float in the ocean and why ice floats in a beverage. This contrasts greatly from the normal relationship between a compound's solid and liquid forms, with the solid form usually being of a higher density. While the unusual phase density properties of water might not seem too important in biological reactions, it is very important for life on earth. If ice did not float, then in very cold bodies of water ice would sink, removing its insulating effects from the water's surface. Lakes, rivers, and major portions of the oceans thus would freeze completely from the bottom up. The world would be a much different place, and much less suited for life.

Water is a Good Solvent

The polar nature of water, with its partial positive and partial negative dipole, allows it to dissolve charged molecules (ions) easily. Water is thus an excellent solvent for charged compounds. The positive side of water surrounds negatively charged molecules, and the negatively charged side of water surrounds positively charged molecules.

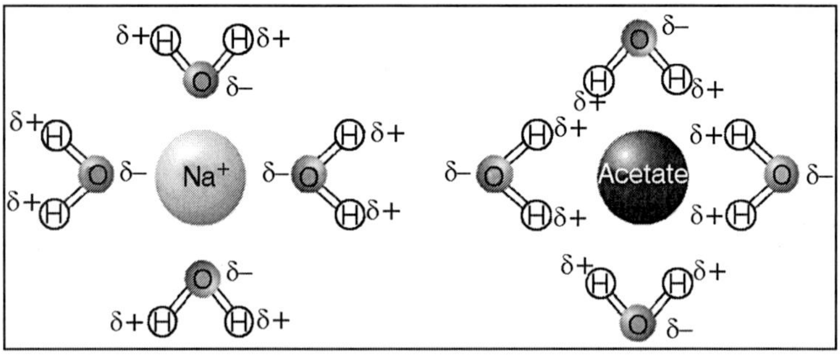

In this way, water makes "solvation shells" around ions. Water can also readily dissolve other polar molecules, even if they are not positively or negatively charged. The solvent properties of water allow for dissolved metals and buffering systems that are very important for the workhorses of life, enzymes. However, the saying "oil and water don't mix" is true—water cannot dissolve oil.

This is because oily substances are non-polar. Non-polar substances (which lack dipoles) are also called *hydrophobic* (water fearing). Hydrophobic substances gather together to exclude water as best they can. This is why you see oil droplets in water. This is also important for the stability and structure of enzymes.

NON-COVALENT BONDING

Non-covalent bonds are not at strong as covalent bonds, but they are important in the stabilization of molecules. In contrast to covalent bonds, non-covalent bonds do not share electrons.

Noncovalent bonds include:

- Electrostatic interactions
- van der Waals forces
- Hydrophobic interactions
- Hydrogen bonds

Electrostatic interactions are formed between positive and negative ions. The bond is non-directional, meaning that the pull of the electrons does not favour one atom over another. An example is NaCl, which is formed between the positively charged Na^+ ion and the negatively charged Cl^- ion. The average strength of an electrostatic interaction is 15 kilojoules/mol (kJ/mol). The bond strength lessens when the distance between the two ions increases.

Van der Waals forces are weak forces between temporary dipoles. These forces may be attractive or repulsive. They are also non-directional. The average bond energy for van der Waals forces is on the order of 10 kJ/mol. Hydrophobic interactions result when non-polar molecules are in a polar solvent, e.g. H_2O. The non-polar molecules group together to exclude water (hydrophobic means water fearing). By doing so they minimize the surface area in contact with the polar solvent.

A hydrogen bond results when a hydrogen atom that is covalently bound to an electronegative atom (e.g. O, N, S) is shared with another electronegative atom. A hydrogen bond is directional towards the electronegative atom. An example of this is the hydrogen bonds formed in water. Hydrogen bonds are constantly being made and remade. The half-life of a hydrogen bond is about 10 seconds. A hydrogen bond has an energy of 21 kJ/mol. As stated above, non-covalent bonds are not as strong as covalent bonds, but the additive effect of many non-covalent bonds can stabilize a molecule. Non-covalent bonds are very important in the structure of proteins.

Bond type	Energy (kJ/mol)
Covalent, e.g. C-C	350
Electrostatic	15
van der Waal's	10
Hydrogen	21

TEST AND PROBLEMS

pH Questions

Below are a few sample questions to elucidate the concepts discussed in the book.

- A simple pH problem
- pH of a weak acid
- A buffer problem
- A more involved buffer problem

A Simple pH Problem

What is the pH 0.1 M HCl?

Answer: The pH of a solution is the negative logarithm of the concentration of H^+ ions. In this case, HCl is a very strong acid. HCl completely dissociates to H^+ and Cl^-.

$$HCl \rightarrow H^+ + Cl^-$$

The concentration of HCl is 0.1 M, if all of it dissociates to H^+, then

$$[H^+] = 0.1$$
$$-\log(0.1) = 1.26$$
$$pH = 1.26$$

Remember that a pH of 7 is neutral, and that the pH scale is logarithmic, meaning each pH unit represents a factor of 10. This HCl solution has a difference of 5.74 pH units from water (7 - 1.26). Taking the inverse log of 5.74 means that this HCl solution is more than 500,000 times as acidic as water.

pH of a Weak Acid Solution

What is the pH of a 0.05 M solution of acetic acid? (The K_a of acetic acid $= 1.8 \times 10^{-5}$)

Answer: Determining the pH of a weak acid solution is more complicated than for a strong acid. This is because a weak acid does not completely dissociate.

$$CH_3COOH \square H^+ + CH_3COO^-$$

To simplify matters, it is helpful to construct a grid:

Reaction	CH_3COOH	H^+	CH_3COO^-
Initial	0.05 M	0 M	0 M
Change	(−x) M	+ x M	+x M
Equilibrium	(0.05−x) M	x M	x M

To determine the $[H^+]$, we must employ the K_a of acetic acid, which is 1.8×10^{-5}.

$$Ka = \frac{[H+][CH_3COO-]}{[CH_3COOH]} = 1.8 \times 10-5$$

Substituting the equilibrium values in the table above, we get:

$$1.8 \times 10\text{–}5 = \frac{(x\ M)(x\ M)}{(0.05 - x)\ M}$$

Since acetic acid is a weak acid, we can assume that very little of it will dissociate. *This simplifies the above equation to:*

$$1.8 \times 10^{-5} = \frac{(x\ M)(x\ M)}{0.05}$$

This can be easily solved:

9×10^{-7} M = x^2 M

x = 9.5×10^{-4} M

x = 0.00095 M

Now that we have the [H+], we can solve for pH:

pH = –log[0.00095]

pH = 3

Buffer Problem

Question 1: What is the ratio of [lactate]/[lactic acid] at pH = 5? (The pK_a of lactic acid = 3.85)

Answer: To solve this problem, we need to utilize the Henderson-Hasselbalch equation:

$$pH = pKa + \log \frac{[H+][A-]}{[HA]}$$

Table in the book lists the pKa of lactic acid as 3.85. Now just substitute into the equation:

$$5 = 3.85 + \log \frac{[\text{lactate}]}{[\text{lactic acid}]}$$

$$1.15 = \log \frac{[\text{lactate}]}{[\text{lactic acid}]}$$

$$14 = \frac{[\text{lactate}]}{[\text{lactic acid}]}$$

There is 14 times as much lactate as lactic acid at pH 5.

PROCESS OF BIOLOGICAL BUFFERS

Many biochemical processes are markedly impaired by even small changes in the concentrations of free H+ ions. It is therefore usually necessary to stabilise the H+ concentration *in vitro* by adding a suitable buffer to the medium, without, however, affecting the functioning of the system under investigation. A buffer keeps the pH of a solution constant by taking up protons that are released during reactions, or by releasing protons when they are consumed by reactions. The observation that partially neutralised solutions of weak acids or bases are resistant to changes in pH when small amounts of strong acids or bases are added led to the concept of the'buffer'. The Biological Buffers imparts basic knowledge on biological buffer substances and support in terms of selecting the best buffer substance for your assay system.

Be aware of the fact that some buffers serve as substrate/ inhibitor of enzymes (e.g. phosphate) or the temperature sensitivity of the pH of some buffers, especially Tris, probably the most frequently employed buffer substance.

BUFFER CONCEPT

Many biochemical processes are markedly impaired by even small ch anges in the concentrations of free H+ ions. It is Therefore usually necessary to stabilise the H+ concentration *in vitro* by adding a suitable buffer to the medium, without, however, affecting the functioning of the system under investigation.

A buffer keeps the pH of a solution constant by taking up protons that are released during reactions, or by releasing protons when they are consumed by reactions. The observation that partially neutralised solutions of weak acids or bases are resistant to changes in pH when small amounts of strong acids or bases are added led to the concept of the'buffer'.

BUFFER CAPACITY

Buffers consist of an acid and its conjugated base. The quality of a buffer is determined by its buffer capacity, i.e. its resistance to changes in pH when strong acids or bases are added. In other words: the buffer capacity corresponds to the amount of H+ or OH- ions that can be neutralised by the buffer.

The buffer capacity is related to the buffer concentration. The graph described by the relation of the pH to the addition of H+/OH– ions is called the titration curve. The point of inflection of the curve corresponds to the pKa value. At this point, the buffer capacity is at its maximum at the pKa value.

This point therefore corresponds to the mid-point of the pH range covered by the buffer and is where the concentration of acid and base is the same. In the area of this pH range, therefore, relatively large amounts of H+/OH– ions result in only small changes in pH. A basic principle is that a buffer that has a pH value of one pH unit above or below the pKa value loses so much buffer capacity that it no longer has any real buffer function.

Based on the Henderson-Hasselbalch equation pH = pKa + log [A–]/[HA] for the calculation of the pH of a weak acid or alkaline solution, the ions in the water must also be taken into account when working in pH ranges below 3.0 and above 11.0. Most biochemical reactions, however, take place in the pH range between 6.0 and 10.0.

THE PH VALUE

Conductivity can also be detected in highly purified water due to the OH– and H3O+ ions resulting from the autoprotolysis of water. This intrinsic dissociation of the water is an equilibrium reaction and the product of the concentrations of the two ions represents a constant:

$$K = [H3O+] \times [OH-].$$

This value is depends on the temperature only and is 10-14 for purified water at 22°C. Depending on which of the two ions is present in a higher concentration in a solution, the solution is called to be acidic or alkaline. To express this fact in terms of a simple number, the negative exponent of the easily measurable hydronium ion concentration [H3O+] was chosen.

This dimensionless number is called the pH value. The pH can also be described as the negative, decadic logarithm of the hydronium ion concentration of a solution:

$$pH = -\log [H3O+]$$

The hydronium ion concentration of pure water is 10–7 mol/L, as can be derived from the above equation for the ion product. Therefore, the pH value is 7. In an acidic solution, the concentration of H3O+ ions is increased (e.g. from 10–7 mol/L to 10–2 mol/L) and the pH is therefore < 7. In an alkaline solution, the hydronium ion concentration is decreased and this results in a pH > 7.

THE PKA VALUE

To change the pH value of a solution, substances are dissolved in the solution which release H+ ions into the water (acids), thus raising the H3O+ ion concentration and lowering the pH value, or which decrease the H3O+ ion concentration and thus increase the pH value by taking up H+ ions (bases). As always in chemistry, this reaction is an equilibrium reaction, and the capacity of a substance to shift this equilibrium in one direction or another is determined by the potency of the acid. It is calculated from the equilibrium constants using the following equation $K = [A] \times [H3O+]/ [HA]$, requirements.

DILUTION AND RINSE BUFFERS

Rowmanowsky stains (Giemsa, Leishmans, Wrights etc) are pH sensitive. Tinctorial (stain picture) quality can be altered by changes in the final pH of staining reagent, this can be caused by differences in the quality or source of the water used to dilute such stains. Tinctorial quality and consistency can be maintained by the use of dilution buffers, which when added to the diluent water will produce stain reagents of the required pH.

- Acid pH (<7.0) will emphasize eosinophilic staining.
- Basic pH (>7.0) will emphasize azureophilic (basophilic) staining.
- In general, blood stains are diluted to produce solutions of slightly acid pH, 6.50-6.80.

We offer a range of buffer concentrates covering pH's 6.50/ 6.80/ 7.00/ 7.20/ 7.40/ 7.60, supplied in 10 vial or 25 vial packs each vial makes 5 litres of dilution buffer. They offer great value for money and the certainty of preparing blood stains of the required pH time and again.

To maintain the tinctorial qualities of the final stain slides should be rinsed in the same buffer.

- Concentrates - ready-to-use.
- High quality, accurate, time saving and good value for money.

BUFFERS AND IONIZATION OF AMINO ACIDS

- Buffers resist change. pH buffers resist change in pH when acid (H+) or base (OH–) is added.
- pH = –log [H+] –where bracket indicates molar concentration. If [H+] = 10^–7 M, then pH = 7. This is neutral pH since [H+] = [OH–] and it is middle of pH scale. Draw a pH scale by calculating the pH of 1 M HCl and 1 M NaOH. Use the equation Kw = 10^–14 = [H+] × [OH–], to find the pH of 1 M NaOH.
- pH of a buffered solution can be found by using the Henderson-Hasselbach equation:

pH = pK + log([A–]/[HA])

All acids and bases have a conjugate acid (HA) and a conjugate base (A–). At [HA] = [A–], pH = pK (prove this with the above equation).

At the pK, a buffer maximally resists change in pH and...the buffering zone is considered to be 1 pH unit above and below the pK.

CONCEPTS

Buffering zone = pK +/–1 pH unit:

- At pH = pK, [HA] = [A–].
- At pH below pK, [HA] > [A–].
- At pH above pK, [HA] <[A–].
- Amino acids (AAs) are more complex than simple buffers and...have at least two pK values because AAs have at least two ionizing groups:

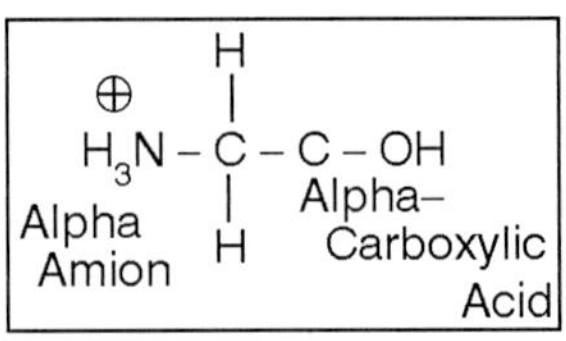

Glycine Structure

Example: Glycine has an alpha-amino group and alpha-carboxylic acid group. Each ionizable group has a pK (called pK–1 and pK–2). Glycine pK–1

= 2.3 (alpha-carboxylic acid); pK–2 = 9.6 (alpha amine). Draw a titration curve for Glycine for practice. In this class, we treat all ionizing groups with in a molecule, as acting independently of other ionizing groups. The pH on the titration curve where an AA has no net charge is called the pI or isoelectric point. The pI is calculated by averaging the two pK values on either side of the neutral form (ie form with no net charge).

All AAs must have at least 3 ionic forms:

- "AA+1","AA0", and"AA-1".
- Where"AA+1" = AA form with net charge of +1,
- "AA0" = neutral form with no net charge, and
- "AA-1" = AA form with net charge of –1.

While an AA may have other ionic forms, only these 3 forms count when finding pI:

- AA+1 goes to AA0 via pK–below.
- AA0 goes to AA-1 via pK–above.

Idea

pI = average of (pK-below) + (pK–above).
where pK-below is the pK between AA+1 and AA0.
and pK-above is the pK between AA0 and AA–1.

For more complex AAs, like Asp, Glu, Lys, Arg and His - with 3 ionizing groups, Other forms exist with charges other than +1, zero, and -1. For these AAs, the best approach is to start with the fully protonated form (ie. One with all the protons it can take on its ionizable groups).

Calculate the net charge on this group (it must be positively charged), then titrate the AA to remove the first proton and find net charge again. Continue doing this until all the protons have been removed. Keep track of this by making a simple model. Then, find the form with no net charge (ie. AA0) and use the pK values, which govern the transition from AA+1 to AA0 and AA0 to AA-1, as the ones to calculate the pI.

His has 3 ionizing groups, alpha-carboxylic acid (pK 1.8), side-chain amino (pK 6.0) and alpha-amino (pK 9.2):

- His+2 goes to His+1 via pK 1.8;
- His+1 goes to His0 via pK 6.0;
- His0 goes to His-1 via pK 9.2.

Therefore, pI = (6.0 + 9.2)/2 = 7.6:

- Some of you have trouble figuring out which group ionizes first. Here are some rules to apply to help you..
  - Carboxylic acids ionize at acidic pH; ie. Carboxylic acids give up their protons at acid pHs.
  - Amino groups ionize at basic pH; ie. Amines give up their protons at basic or alkaline pHs.

– When groups with a similar chemical nature are present:

(a) Carboxylic acids near an amino group have a more acidic pK than isolated carboxylic acids.

(b) Amino groups near a carboxylic acid have a more acidic pK than isolated amines.

(b) Aromatic amines (like in His side chain) have a pK near neutral.

Apply the above rules and concepts to calculate the pI for all the other complex AAs.

Asp, Glu, Lys, and Arg. Their pK values and structures are shown below:

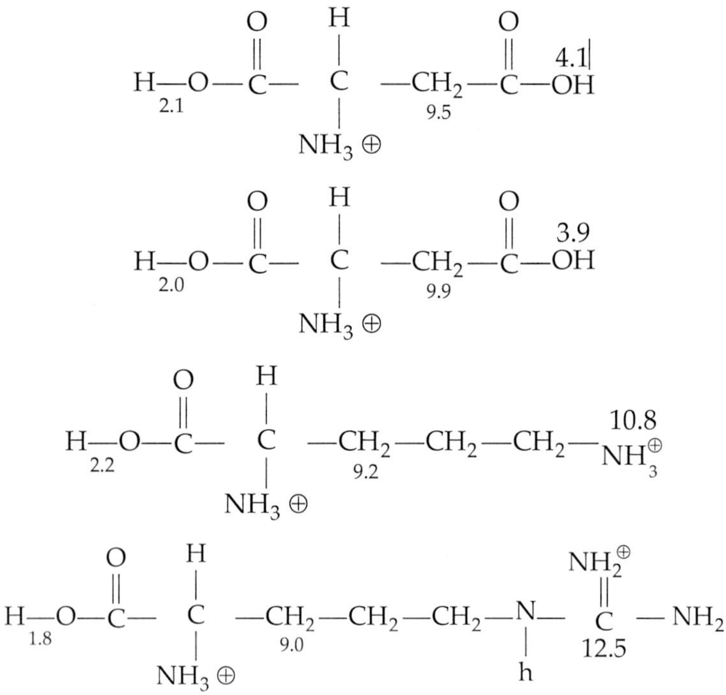

BIOCHEMISTRY AS CHEMISTRY OF WATER

We immediately see the evident truth in this statement when we consider the complex thermodynamics of protein and peptide structural stability and the prominent role of solvent in influencing this equilibrium. Proteins unfold from their native fold upon the addition of urea or other denaturants, peptide helicity increases upon addition of TFE, and alcohol is even known to induce helical structure in regions that otherwise adopt beta-sheet conformations. While the helix inducing effects of alcohol upon peptide conformation have long been used to probe and enhance latent structure, thereby making it measurable by CD or NMR, a recent paper illustrates how this dependence of structure on solvent conditions may be harnessed to make a sort of'conformational switch'. Successful molecular design or peptide engineering applications will necessarily include an assessment of solvation effects for conformational stability. Turning from a macroscopic picture of the role of water and aqueous solutions in establishing the equilibria of proteinaceous substances to a more microscopic view, we can consider the surprising role played by discrete water molecules in antigen-antibody recognition. In a 1993 paper in Current Opinion in Immunology, Mariuzza and Poljak report a case in which interfacial water plays a critical role in determining the complementarity of the antigen and antibody surfaces.

This work was illuminating, since previous work was leading to the conclusion that the packing density was lower at protein interfaces than in protein interiors. It also appeared that polar atoms at these interfaces were apparently tolerating the absence of neutralizing partners - far more than they do in protein interiors. In this work, a 1.8 angstrom crystal structure of an immune complex was studied and revealed that the water network was an integral part of the interface. As pointed out by these authors, the analysis of low resolution crystal structures may in fact lead to erroneous conclusions on the nature of the interface in these complexes. Furthermore, the evidence reported in this work points to the inescapable conclusion that the intermolecular complementarity essential for molecular recognition is not necessarily imparted solely by the large molecules.

And, contrary to what our notions of solvent entropy might lead us to expect, bound water may play a role in stabilizing the structure of single proteins as well as that of complexes. And, this bound water may, apparently, serve this function at the surface of the protein, and not solely while confined to the interior. That's just the tip of the iceberg as far as far as water and its impact for biological systems at what is considered physiological temperature and pressure; the realm normally explored by biochemists and pharmacologists.

Much, however is being learned about the so-called"hot life", life found at thermal vents deep in the ocean floor. Reports have documented the existence of organisms at temperatures in excess of 110 deg. C (see Shock for a list of references). These organisms are referred to collectively as thermophiles. It is important to note the enormous temperature variation of sea water at these depths, with a documented range from 2 deg. C to 900 deg. C at depths of a few km where the high pressure precludes boiling. Hot life, or the existence of thermophiles, has been noted at temperature in excess of 100 deg. C at these elevated pressures, and it remains a matter of some dispute whether or not bacteria can exist at even higher temperatures still. So, what are the properies of ambient sea water for organisms at deep sea thermal vents? We can expect these properties to influence the kinetics and thermodynamics of the biochemistry pertinent to this life, just as they influence the systems with which we are, perhaps, more familiar. We can begin to understand the fluid environment presented to thermophiles by considering the temperature dependence of a few properties of pure water. A quick look at the following data from Shock show the dramatic effect of temperature on water's dielectric. Water at 25 deg. C and 1 atm pressure has the familiar dielectric near 80; however near water's critical point the dielectric has fallen to less than 10. For comparison, ethanol has a dielectric of roughly 25 at STP.

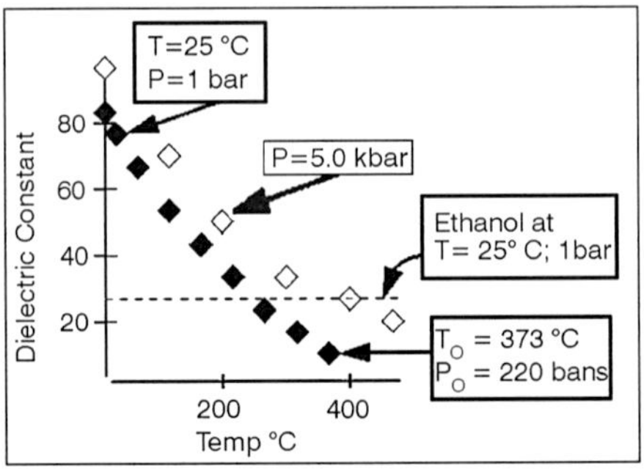

Fig. From Shock.

In this graph all white and blue diamonds pertain to water data points. The blue diamonds are for pressures below the critical point where water exists in two phases, while the white diamonds pertain to 5 kbar data. For the low pressure data, two key data points are indicated. The grey dashed line indicates the dielectric of ethanol at standard conditions. The weak pressure dependence of this data should be noted.

In order to understand the relevance of this sort of analysis to a complete understanding of ambient conditions at deep sea thermal vents, it should be understood that the pressure at these depths ranges from a few hundred bars to several kilobars. Shock should be consulted for additional details, data and a general groovy time. Well, if all this is true, then water must have a pretty different structure at these T&P's than we are used to. Right? Not surprisingly, this is what is being found. Some of these results are shown in this graph:

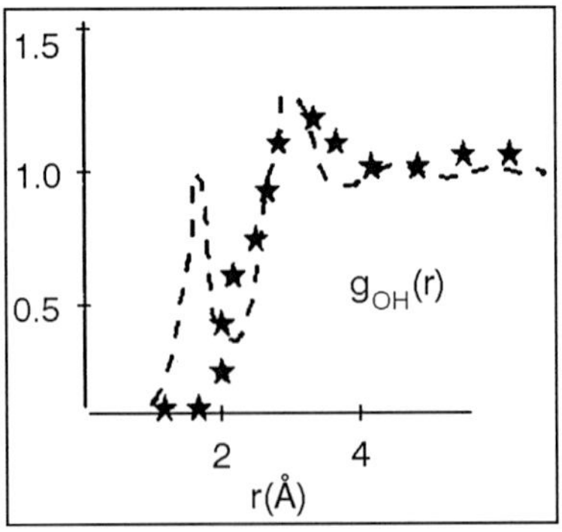

Fig. From Postorino

The pair distribution function for intermolecular OH atom pairs is shown as a function of interatomic distance for water at 25 deg. C (red dashed line) and 400 deg. C. (This graph is drawn to approximate that shown in the original paper which should be consulted for quantitative work.) This particular pair distribution function tells us the likelihood with which a H atom will be found at any distance from an O atom (The intramolecular data isn't shown here, but the water molecule does have the same structure at 400 deg. C as at 25 deg. C. This graph considers only intermolecular associations.) Immediately, then, we see that at room temperature it is relatively likely to find these atom pairs separated by 1.95 angstroms. This is the hydrogen bond. And then there are strong secondary structural phenomena, manifest as the damped oscillations that we see here out to beyond 5 angstroms, all due to the overwhelming hydrogen bonding Behaviour.

Looking at the high temperature data, it's clear that the hydrogen bonding peak, which dominates the room temperature data, is completely gone. The OO correlation function, not reproduced here, shows a slight broadening of the nearest-Neighbour peak with temperature. The high temperature OO pair correlation function also yields a larger coordination number than is seen at room temperature. In other words, each water molecule is surrounded by more water molecules in its first solvation shell at 400 deg. C than at 25 deg. C. At 400 deg. C the packing is not restricted by the severe hydrogen-bonding geometrical constraints of the the room temperature fluid. This result is in line with the close packing Behaviour of most liquids. These results taken together with those for the dielectric phenomena described above begin to paint a picture of high temperature water. Water at high temperature exhibits many of the properties we associate with less polar liquids under standard conditions. In order to understand something of the biochemistry that might pertain under extreme aqueous conditions, Shock has been estimating the stability of peptidic species at elevated temperatures and pressures.

Specifically he has been asking questions like"How does the free energy of the peptide bond change with temperature and pressure?" A snapshot view of some of this work will serve to further the discussion of aqueous properties under extreme conditions while at the same time giving a hint for some of the very different biochemistry that will undoubtedly be revealed by further study in this area. Considering only Glycine here, the free energy of interest here is that which determines the following equilibrium:

$$\text{Glycine} + \text{Glycine} = \text{Glycylglycine} + H_2O$$

In other words, what is the free energy cost to form the peptide bond between two Glycine molecules? This equilibrium can be calculated from the free energies of formation for the individual species involved in this reaction, Glycine and Glycylglycine, and the temperature dependence of this free energy in the following way:

$$\Delta G^\circ_{PT} = \Delta G^\circ_f - S_{Pr,Tr}(T - Tr) + \int C^\circ_P$$

$$-T\int C^\circ_P d\ln T + \int V^\circ dP$$

This equation has the usual host of characters. G represents free energy, S is entropy, Cp is heat capacity, V is volume and P is pressure. If any of this is murky please consult the original paper by Shock. What we're basically doing is getting the free energy (OK, so it's really the standard partial molal Gibbs free energy) at any T&P from that at STP by integrating over the temperature and pressure dependence of the heat capacity and volume. In order to do this, we need to know something about the way heat capacity and volume depend on the thermodynamic variables of T and P, respectively. We'll look at this first:

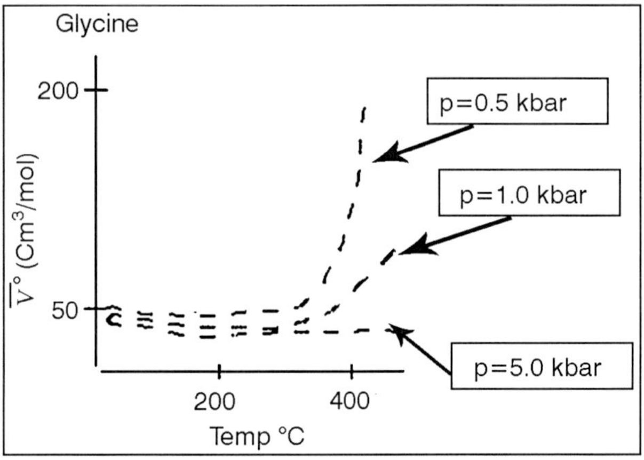

Standard partial molal volume for Glycine. From Shock. (This graph is drawn to approximate that shown in the original paper which should be consulted for quantitative work.) These results show the very large dependence of the molal volume on temperature at lower pressures, and the much weaker temperature dependence at extremely high pressures. Here are the results for the heat capacity:

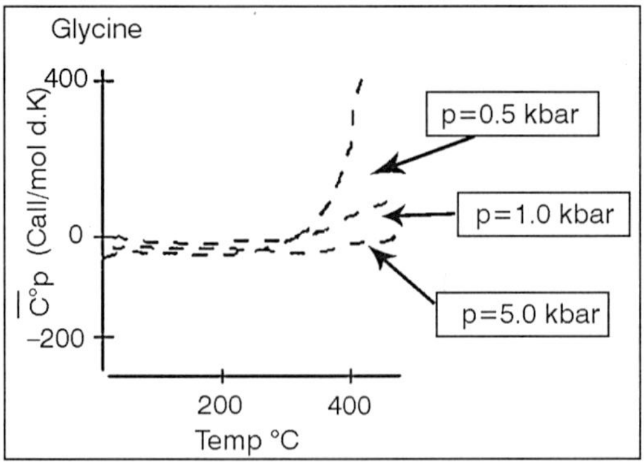

Standard partial molal heat capacity (constant pressure variety) for Glycine. From Shock. (This graph is drawn to approximate that shown in the original paper which should be consulted for quantitative work.) Pretty much the same thing. A large T dependence is witnessed at lower temperature, but this Behaviour is convincingly squashed at high pressure. The results for Glycine have been shown here.

Analogous results for Glycylglycine are obtained from these by calculating a regression equation that relates, for instance, the partial molal volume of Glycine to that of Glycylglycine. Putting it all together with some other necessary data (like the free energy of formation at STP, and the entropy at STP; you got it, Shock) the free energies of formation for both Glycine and Glycylglycine can be obtained at any temperature and pressure, and from these the free energy governing the equilibria may be directly obtained.

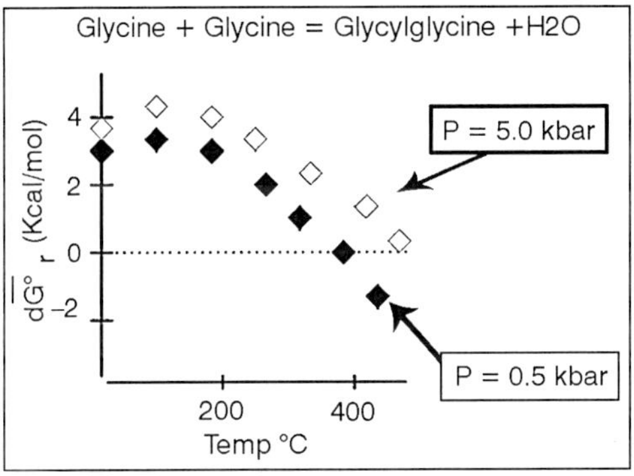

Free energy required to form the Glycylglycine dipeptide from the constituent Glycine moieties. From Shock. (This graph is drawn to approximate that shown in the original paper which should be consulted for quantitative work.) The results indicate that while it costs free energy to form the dipeptide at lower temperatures and pressures (at STP the free energy of this reaction is roughly 3.5 kcal/mol), at extremes of temperature and pressure the dipeptide is the thermodynamically favoured species. In any event, these results are consistent with those for many other dipeptides not covered here and predict that peptide bonds become more stable at elevated temperatures. This Behaviour just begins to hint at the interesting biochemical phenomena that will be revealed for life in water under extreme conditions where water has very different properties from those to which we are accustomed.

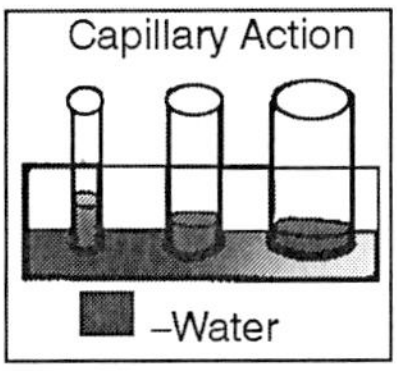

Water has a high specific heat capacity. It takes a lot of energy to change the temperature of water. This characteristic makes water a good insulator and a good coolant. When you sweat, your body is using water as a coolant. The eVapourating water removes heat with it. At night, oceans are a good insulator. The energy that the sun spent in heating the water all day is slowly released into the night.

Finally, water is the most dense at 4 degrees Celsius. Hydrogen bonding arranges water molecules into hollow"cells" when water freezes, making it less dense than liquid water. This characteristic of water is what allows fish in lakes and ponds to survive in the winter. When the water freezes, it becomes less dense and floats to the surface -- leaving the bottom of the lake or pond unfrozen.

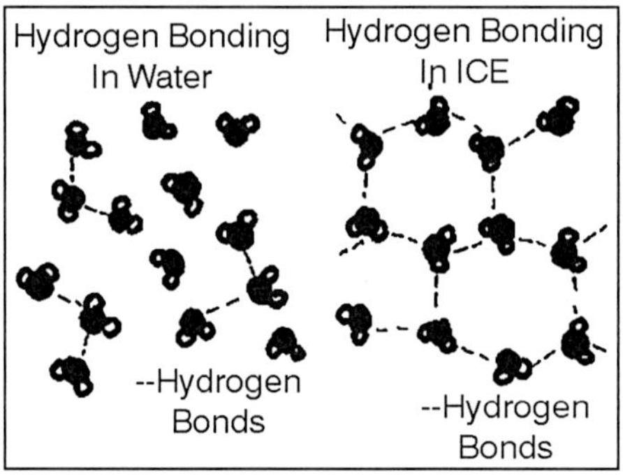

WEAK INTERACTIONS IN AQUEOUS SYSTEMS

Water is a good solvent for polar solutes with which it forms hydrogen bonds. Non-polar compounds dissolve poorly in water. They can not form a hydrogen bond with the solvent. Numerous weak non covalent interactions influence the folding of macromolecules. The physical properties of aqueous solutions are strongly influenced by the concentration of solutes. Ex: The tendency for water to move across a semi permeable membrane is the osmotic pressure.

IONIZATION OF WATER

Pure water is slightly ionized. pH is defined as the negative logarithm of the molar hydronium ion concentration. An acid that ionizes completely in water is a strong acid. An acid which is partially ionized in water is a weak acid. pk provides the measure of the ease of the dissociation of the proton. The lower the number the stronger is the acid.

BUFFERING AGAINST PH

Buffers are aqueous systems that tend to resist in changes in pH when small amounts or acids or base are added. It consists of a weak acid and its conjugate base. As an example a mixture of equal concentrations of acetic acid and acetate ion is a buffer system. Every life form is extremely sensitive to slight pH changes. Human blood for example needs to remain within the pH range of 7.38 to 7.42. The relationship between pH and pk is described by Henderson - Hasselbach equation. It is used to calculate the pH of biological fluids. Two especially important biological buffers are the phosphate and bicarbonate systems. The property of a solution that depends on the concentration of a dissolved solute but not on its chemical identity is known as colligative properties. Substances that do not ionize in water are non electrolytes. Vapour pressure, boiling point, freezing point and osmosis are colligative properties.

WATER AS A REACTANT

Water is very often a direct participant in the chemical reactions of living cells. Formation of ATP from ADP and inorganic phosphate is an example of a condensation reaction in which the elements of water are eliminated. Two amino acid molecules are covalently joined through a peptide bond with a loss of water molecule.

2

Proteins

INTRODUCTION

Proteins are any of a group of complex organic compounds containing carbon, hydrogen, nitrogen and sulfur. Proteins, the principle constituents of the protoplasm of all cells, are of high molecular weight and consist of alpha-amino acids joined by peptide linkages. Different amino acids are commonly found in proteins, each protein having a unique, genetically defined amino-acid sequence, which determines its specific shape and function. They serve as enzymes, structural elements, hormones, immunoglobulins, etc. and are involved in oxygen transport, muscle contraction, electron transport, and other activities. The importance of adequate protein intake to proper immune function has been extensively studied. The most severe effects of"Protein-Calorie Malnutrition" (PCM) are on cell-mediated immunity, although all facets of immune function are ultimately affected. PCM is not, however, a single nutrient deficiency. It is normally associated with multiple nutrient deficiencies, and some immune dysfunctions attributed to PCM are most likely due to these other factors. Partial deficiencies of dietary vitamins produce a comparatively greater depression on immune functions than do partial protein deficiencies. Non-the less, adequate protein is essential for optimal immune function. High protein diets are not recommended for individuals with kidney or liver disease. There are no other known side effects.

A macromolecule made from amino acids; some consist of one, long polypeptide chain while others are made from several intertwined polypeptides (proteios = first place) Proteins are about 50 per cent of the dry weight of most cells, and are the most structurally complex macromolecules known. Each type of protein has its own unique structure and function.

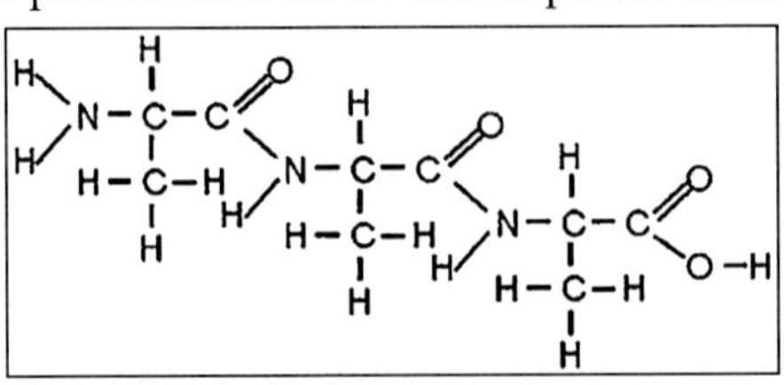

POLYMER

A large molecule made of many, smaller, repeating, identical or similar subunits (poly = many; mer = a part) Polymers are any kind of large molecules made of repeating identical or similar subunits called monomers. The starch and cellulose we previously discussed are polymers of glucose, which in that case, is the monomer. Proteins are polymers of about 20 amino acids (the monomer).

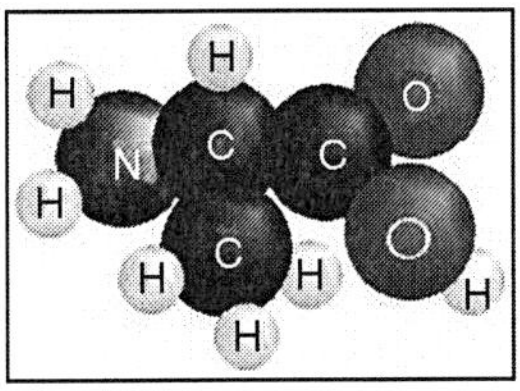

Amino acids are built from a central carbon bonded to four different groups. (run your mouse over the amino acid to see the names of the parts).

- Hydrogen (–H),
- Amino group (–NH_2),
- Carboxyl group (–COOH), and some side chain symbolized by"R".

In the approximately-20 amino acids found in our bodies, what varies is the side chain. Some side chains are hydrophilic while others are hydrophobic Since these side chains stick out from the backbone of the molecule, they help determine the properties of the protein made from them.

Most naturally-occurring amino acids are the l- form, whereas synthetically-produced amino acids give a 50:50 mixture. Notice that these molecules are mirror images of each other, thus there is no way you can rotate one molecule to make it look like the other.

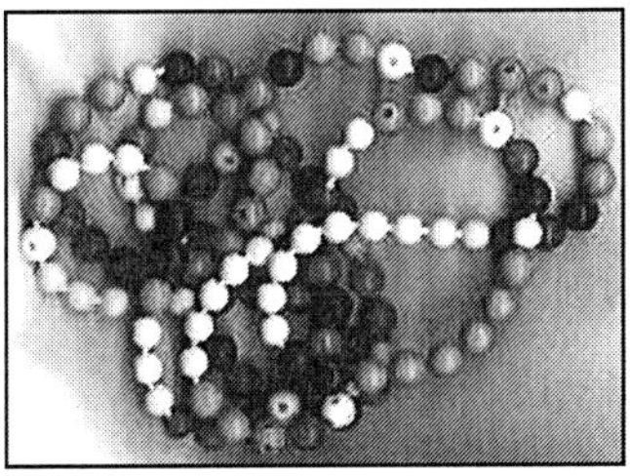

To form protein, the amino acids are linked by dehydration synthesis to form peptide bonds. The chain of amino acids is also known as a polypeptide. Some proteins contain only one polypeptide chain while others, such as hemoglobin, contain several polypeptide chains all twisted together. The sequence of amino acids in each polypeptide or protein is unique to that protein, so each protein has its own, unique 3-D shape or native conformation. If even one amino acid in the sequence is changed, that can potentially change the protein's ability to function. For example, sickle cell anemia is caused by a change in only one nucleotide in the DNA sequence that causes just one amino acid in one of the hemoglobin polypeptide molecules to be different.

Because of this, the whole red blood cell ends up being deformed and unable to carry oxygen properly.

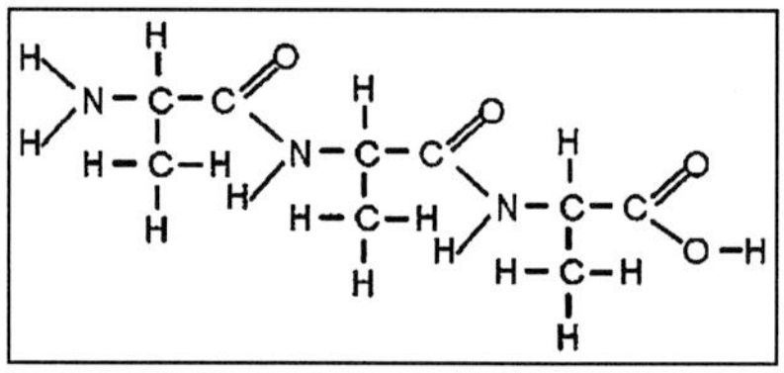

As a polypeptide chain forms, it naturally twists and bends into its native conformation. One of the things that helps determine the native conformation of a protein is the side chains of all the amino acids involved. Remember some amino acid side chains are hydrophobic while others are hydrophilic. In this case, the"likes" attract: all the hydrophobic side chains (here represented by yellow beads) try to"get together" in the Centre of the molecule, away from the watery environment, while the hydrophilic side chains are attracted to the outside of the molecule, near the watery environment. Additionally, some of the hydrophilic side chains have groups of atoms attached that make them acidic, while others have groups attached that make them basic. Side chains with acidic ends are attracted to side chains with basic ends, and can form ionic bonds. Thus, the side chains interacting with each other help to hold the protein in its native conformation.

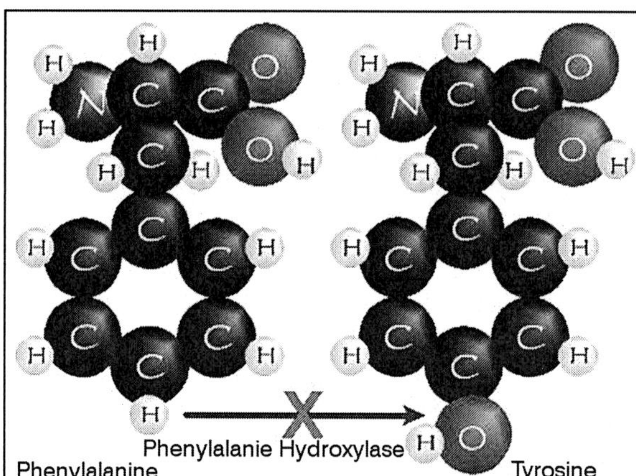

Denaturation is when a protein loses its native conformation. For example, egg white (also called albumen) contains a protein called albumin which is water-soluble. However, if heated, albumin becomes denatured and loses its ability to be water-soluble. There are several possible things that can denature proteins including changing the temperature (adding heat), changing the pH or salt concentration of the solution, or putting the protein into a hydrophobic solvent. In a hydrophobic solvent, the amino acids with hydrophobic side chains (the yellow beads) would all try to go to the outside of the molecule, and all those with hydrophilic side chains would cluster in the Centre of the molecule. If a protein remains water-soluble when denatured (unlike albumin), it can return to its native conformation if/when placed back into a"normal" environment.

PKU or phenylketonuria is a genetic disorder for which, by law, all newborns must be tested. People with this disorder are lacking an enzyme called phenylalanine hydroxylase needed to turn the amino acid, phenylalanine into tyrosine so it can be eliminated via the urine. Since phenylalanine cannot be eliminated, it builds up and in children whose nervous systems are still developing, causes mental retardation. According to the Merck Manual, the body is slowly able to turn it into several other compounds which can be eliminated via the urine, but these chemical reactions are too inefficient to get rid of the excess phenylalanine. If PKU is detected shortly after birth, the child is put on a special diet low in phenylalanine and has normal mental development.

Since phenylalanine is an amino acid that is needed by the body for normal growth and development, the goal is to limit intake to just what the child needs without any excess that could build up in his/her system. The Merck Manual says that, unfortunately, most natural protein sources, including milk, are too high in phenylalanine for children with PKU, so these babies/children must be put on a special formula from which most, if not all, of the phenylalanine has been removed. As the child grows, some low-protein, natural foods such as fruits, vegetables, and some kinds of grain are OK to eat. The needed phenylalanine is supplied via carefully-measured amounts of natural protein plus the residual in the formula. One area of"debate" is how long the special diet must be continued. Some people think the diet may be discontinued after the brain finishes developing at about age 5. Others have noted some Behaviour problems and learning difficulties that appear to be related to going off the diet, and recommend that people with PKU stay on the diet for life.

Staying on the special diet is especially important if a woman who has PKU is even thinking about trying to become pregnant. Because any excess phenylalanine in her system could cross the placenta and could cause mental retardation in the baby, especially in the first few weeks while the nervous system is forming, she has to make sure that she adheres to the special diet from before conception. Note that this is not a major concern for a normal woman carrying a baby with PKU (she wouldn't know that until after the baby was born and tested) because the woman's body can digest phenylalanine, including any excess from the baby's body that is sent back, via the placenta, to her body.

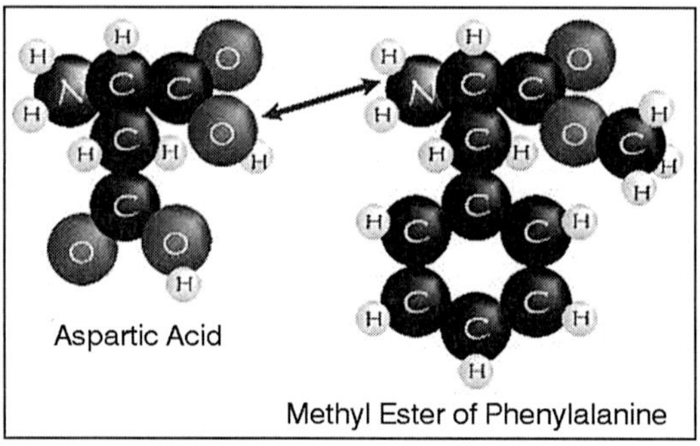

Nutrasweet has the chemical name aspartame and is made of two amino acids: l-aspartic acid and the methyl ester of phenylalanine, thus people with PKU can't have it. Look at a can of diet soft drink that is sweetened with aspartame. Somewhere on the can, there's always a cautionary label warning people with PKU not to drink it. But what about the rest of us? That methyl group on the phenylalanine can break off pretty easily, and at temperatures as"low" as 86° F (= 30° C - lower than body temperature or a summer day), aspartame starts to disintegrate, releasing methanol (which in larger quantity causes blindness, etc. - it's toxic). As the liver tries to deal with the methanol using normal metabolic pathways, it ends up turning the methanol into formaldehyde, which is a known carcinogen. Interestingly, because of this and several other concerns, originally there was controversy over FDA approval of aspartame. Even still, its safety is controversial. More recently, some people have suggested a possible link between aspartame consumption and an increase in brain tumors in this country.

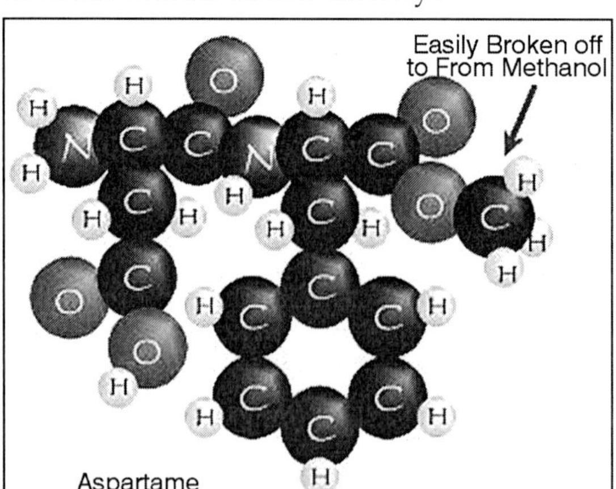

Equal® is not the same as Nutrasweet, but does contain some of it. There's an interesting psychological thing here: in our culture, people think they need sweetener by the teaspoon, but Nutrasweet is so much sweeter than sugar that only a tiny pinch would be needed to gain the equivalent sweetness. Artificial sweetner manufacturers know that if they don't make a product such that one teaspoon is as sweet as one teaspoon of the sugar people are used to using, people will use one teaspoon anyway (a tiny pinch just doesn't look like enough), then say it was too sweet and stop buying it. Thus, artificial sweetners are designed to take up about a teaspoon of space, and Equal® is mostly glucose (which diabetics and hypoglycemics shouldn't have) with a little Nutrasweet added, so in theory, not quite as much of the sugar is needed (also remember that glucose isn't as sweet as sucrose).

To make the sweetness of one teaspoon of sweetner powder equivalent to that of one teaspoon of sucrose, various fillers (things like cellulose) are added. These products also contain a number of other chemicals, including silicon dioxide (SiO2)"to prevent caking." This is the chemical name for sand or glass (although, from what I've heard, a bit of silicon is needed for healthy fingernails)! Pre-mixed, pre-packaged items, such as artificially-sweetened soft drinks, can contain all aspartame as their sweetener because the manufacturers of these products do not have to be concerned with the"one teaspoon" psychological factor. Remember, because aspartame degrades into methanol at warm temperatures, it should not be used in anything hot or anything that will be cooked (coffee sweetener, cake mixes, cocoa, etc.), and bottles of artificially-sweetened soft drinks that have been sitting out in the hot sun at the local gas station are suspect.

FORCES CONTROLLING PROTEIN STRUCTURE

HYDROGEN BONDING

Polypeptides contain numerous proton donors and acceptors both in their backbone and in the R-groups of the amino acids. The environment in which proteins are found also contains the ample H-bond donors and acceptors of the water molecule. H-bonding, therefore, occurs not only within and between polypeptide chains but with the surrounding aqueous medium.

Hydrophobic Forces

Proteins are composed of amino acids that contain either hydrophilic or hydrophobic R-groups. It is the nature of the interaction of the different R-groups with the aqueous environment that plays the major role in shaping protein structure. The spontaneous folded state of globular proteins is a reflection of a balance between the opposing energetics of H-bonding between hydrophilic R-groups and the aqueous environment and the repulsion from the aqueous environment by the hydrophobic R-groups. The hydrophobicity of certain amino acid R-groups tends to drive them away from the exterior of proteins and into the interior. This driving force restricts the available conformations into which a protein may fold.

Electrostatic Forces

Electrostatic forces are mainly of three types; charge-charge, charge-dipole and dipole-dipole. Typical charge-charge interactions that favor protein folding are those between oppositely charged R-groups such as K or R and D or E. A substantial component of the energy involved in protein folding is charge-dipole interactions. This refers to the interaction of ionized R-groups of amino acids with the dipole of the water molecule. The slight dipole moment that exist in the polar R-groups of amino acid also influences their interaction with water. It is, therefore, understandable that the majority of the amino acids found on the exterior surfaces of globular proteins contain charged or polar R-groups.

Van der Waals Forces

There are both attractive and repulsive van der Waals forces that control protein folding. Attractive van der Waals forces involve the interactions among induced dipoles that arise from fluctuations in the charge densities that occur between adjacent uncharged non-bonded atoms. Repulsive van der Waals forces involve the interactions that occur when uncharged non-bonded atoms come very close together but do not induce dipoles. The repulsion is the result of the electron-electron repulsion that occurs as two clouds of electrons begin to overlap. Although van der Waals forces are extremely weak, relative to other forces governing conformation, it is the huge number of such interactions that occur in large protein molecules that make them significant to the folding of proteins.

BUILDING BLOCKS - AMINO ACIDS

An amino acid is defined as the molecule containing an amino group (NH_2), a carboxyl group (COOH) and an R group. It has the following general formula R-CH(NH_2)-COOH. The R group differs among various amino acids. In a protein, the R group is also called a sidechain.

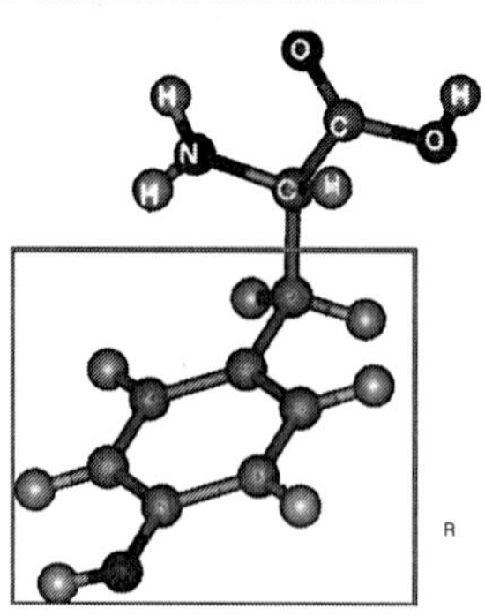

Fig. The Structure of an Amino Acid (Tyrosine).

There are over 300 naturally occurring amino acids on earth, but the number of different amino acids in proteins is only 20.

THE TWENTY AMINO ACIDS OF PROTEINS

Based on the physical and chemical properties of R groups, the 20 amino acids of proteins may be classified as follows.

1. *Acidic*: including aspartic acid (aspartate) and glutamatic acid (glutamate). In a neutral solution, the R group of an acidic amino acid may lose a proton and become negatively charged.
2. *Basic*: including lysine, arginine and histidine. In a neutral solution, the R group of a basic amino acid may gain a proton and become positively charged. Interaction between positive and negative R groups may form a salt bridge, which is an important stabilizing force in proteins.
3. *Aromatic*: including tyrosine, tryptophan and phenylalanine. Their R groups contain an aromatic ring.
4. *Sulfur:* including cysteine and methionine. Their R groups contain a sulfur atom (S). The disulfide bond formed between two cysteine residues provides a strong force for stabilizing the globular structure. A unique feature about methionine is that the synthesis of all peptide chains starts from methionine.
5. *Uncharged hydrophilic:* including serine, threonine, asparagine and glutamine. Their R groups are hydrophilic and capable of forming hydrogen bonds.
6. *Inactive hydrophobic:* including glycine, alanine, valine, leucine and isoleucine. These amino acids are more likely to be buried in the protein interior. Their R groups do not form hydrogen bonds and rarely participate in chemical reactions.
7. *Special structure*: including proline. In most amino acids, the R group and the amino group are not directly connected. Proline is the only exception among 20 amino acids found in protein. Due to this special feature, proline is often located at the turn of a peptide chain in the three-dimensional structure of a protein.

PH AND PKA

A molecule, or an atom group in a molecule, may lose or gain a proton when the molecule is placed in an aqueous solution. The exact probability that a molecule will be protonated or deprotonated depends on the pK_a of the molecule and the pH of the solution. Let AH be an atom group in a molecule. AH could be neutral or charged. After AH loses a proton, it is denoted by A^-. The protonation/deprotonation reaction may be written as

$$AH \rightleftharpoons H^+ + A^-$$

Define its equilibrium constant,

$$K_a = [H^+][A^-]/[AH]$$

where [..] represents the concentration.
Taking logarithm on both sides,

$$\log K_a = \log [H^+] + \log [A^-]/[AH]$$

Define

$$pK_a = -\log K_a; \quad pH = -\log [H^+]$$

we obtain

$$pK_a = pH - \log [A^-]/[AH]$$

or

$$[A^-]/[AH] = 10^{(pH - pK_a)}$$

pH	$[H^+]$ (mol/l)	
0	10^{-0}	
1	10^{-1}	↑
2	10^{-2}	
3	10^{-3}	Increasing acidity
4	10^{-4}	
5	10^{-5}	
6	10^{-6}	
7	10^{-7}	Neutral
8	10^{-8}	
9	10^{-9}	
10	10^{-10}	Increasing alkalinity
11	10^{-11}	
12	10^{-12}	
13	10^{-13}	↓
14	10^{-14}	

From the last equation, we can see that half of molecules will lose protons if they are in a solution with pH = pK_a. The higher the pH value, the more likely a molecule will lose a proton. On the other hand, from the definition of pH, we find that the proton concentration of a solution becomes lower at higher pH, thereby capable of accepting more protons. This physical concept is consistent with the result we just obtained from the equation.

- The carboxyl group is most likely negatively charged.
- The amino group is most likely positively charged
- The R group of aspartate and glutamate are most likely negatively charged.
- The R group of lysine and arginine are most likely positively charged.
- The R group of tyrosine at pH = 7 is most likely neutral.
- The R group of histidine has 10 per cent probability to become positively charged at pH = 7, but the probability increases to 50 per cent at pH = 6. Thus, histidine is very sensitive to pH change in the physiological range.

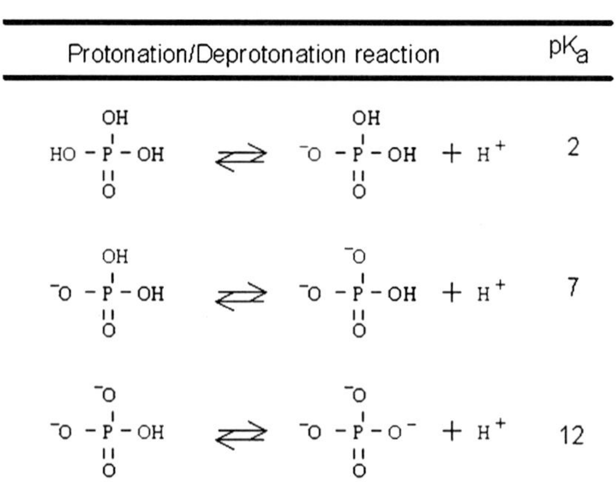

Fig. pK_a Values of Phosphate Groups.

This information tells us that in a neutral solution H_3PO_4 is very likely to lose a proton and become H_2PO4^-. The latter has a 50per cent chance of losing another proton, becoming HPO_4^{2-}. The chance for HPO_4^{2-} to lose an additional proton is very small. Hence, the average charge of a phosphate group is about 1.5 electron charges.

MAMMALIAN MDR TRANSPORT PROTEINS

A series of rather unexpected observations led to discovery of the first eukaryotic ABC protein. Oncologists noted that tumor cells often became simultaneously resistant to several chemotherapeutic drugs with unrelated chemical structures; similarly, cell biologists observed that cultured cells selected for resistance to one toxic substance (e.g., colchicine, a microtubule inhibitor) frequently became resistant to several other drugs, including the anticancer drug adriamycin. Subsequent studies showed that this resistance is due to enhanced expression of a*multidrug-resistance (MDR) transport protein* known as *MDR1*. In this member of the ABC superfamily, all four domains are "fused" into a single 170,000-MW protein. This protein uses the energy derived from ATP hydrolysis to*export* a large variety of drugs from the cytosol to the extracellular medium. The *Mdr1* gene is frequently amplified in multidrug-resistant cells, resulting in a large overproduction of the MDR1 protein.

Most drugs transported by MDR1 are small hydrophobic molecules, which diffuse from the culture medium across the plasma membrane into the cell. The ATP-powered export of such drugs from the cytosol by MDR1 means a much higher extracellular drug concentration is required to kill cells. That MDR1 is an ATP-powered small-molecule pump has been demonstrated with liposomes containing the purified protein. The ATPase activity of these liposomes is enhanced by different drugs in a dose-dependent manner corresponding to their ability to be transported by MDR1. Not only does MDR1 transport a varied group of molecules, but all these substrates compete with one another for transport by MDR1. Although the mechanism of action of MDR1-assisted transport has not been definitively demonstrated, the *flippase model*, is a likely candidate.

Substrates of MDR1 are primarily planar, lipid-soluble molecules with one or more positive charges, and they move spontaneously from the cytosol into the cytosolic-facing leaflet of the plasma membrane. The hydrophobic portion of a substrate molecule is oriented toward the hydrophobic core of the membrane, and the charged portion toward the polar cytosolic face of the membrane and is still in the cytosol. The substrate diffuses laterally until encountering and binding to a site on the MDR1 protein that is within the bilayer. The protein then "flips" the charged substrate molecule into the exoplasmic leaflet, an energetically unfavorable reaction powered by the coupled ATPase activity of MDR1. Once in the exoplasmic face, the substrate diffuses into the aqueous phase on the outside of the cell. Support for the flippase model of transport by MDR1 comes from MDR2, a homologous protein present in the region of the liver cell plasma membrane that faces the bile duct. MDR2 has been shown to flip phospholipids from the cytosolic-facing leaflet of the plasma membrane to the exoplasmic leaflet, thereby generating an excess of phospholipids in the exoplasmic leaflet; these phospholipids peel off into the bile duct and form an essential part of the bile. An alternative *pump model* also has been proposed for MDR1. According to this model, drug molecules in the cytosol bind directly to a single small-molecule binding site on the cytosolic face of the MDR1 protein; subsequent ATP hydrolysis powers movement of the bound drug through the protein to the aqueous phase on the outside of the cell by a mechanism similar to that of other ATP-powered pumps.

MDR1 protein is expressed in abundance in the liver, intestines, and kidney — sites from which natural toxic products are removed from the body. Thus the natural function of MDR1 may be to transport a variety of natural and metabolic toxins into the bile, intestinal lumen, or forming urine. During the course of its evolution, MDR1 appears to have coincidentally acquired the ability to transport drugs whose structures are similar to those of these toxins. Tumors derived from these cell types, such as hepatomas (liver cancers), frequently are resistant to virtually all chemotherapeutic agents and thus difficult to treat, presumably because the tumors exhibit increased expression of the MDR1 or MDR2 proteins.

Cystic Fibrosis Transmembrane Regulator (CFTR) Protein

Discovery of another ABC transport protein came from studies of cystic fibrosis (CF), the most common lethal autosomal recessive genetic disease of Caucasians. This disease is caused by a mutation in the *CFTR* gene, which encodes a chloride-channel protein that is regulated by cyclic AMP (cAMP), an intracellular second messenger. These Cl^{-} channels are present in the apical plasma membranes of epithelial cells in the lung, sweat glands, pancreas, and other tissues. An increase in cAMP stimulates Cl^{-} transport by such cells from normal individuals, but not from CF individuals who have a defective CFTR protein.

The sequence and predicted structure of the encoded CFTR protein, based on analysis of the cloned gene, are very similar to those of MDR1 protein except for the presence of an additional domain, the regulatory (R) domain, on the cytosolic face. The Cl^{-}-channel activity of CFTR protein clearly is enhanced by binding of ATP. Moreover, cAMP activates a protein kinase that phosphorylates, and thereby activates, CFTR. When purified CFTR protein is incorporated into liposomes, it forms Cl^{-} channels with properties similar to those in normal epithelial cells. And when the wild-type CFTR protein is expressed by recombinant techniques in cultured epithelial cells from CF patients, the cells recover normal Cl″-channel activity. This latter result raises the possibility that gene therapy might reverse the course of cystic fibrosis. Since CFTR protein is similar to MDR1 in structure, it may also function as an ATP-powered pump of some as-yet unidentified molecule. In any case, much remains to be learned about this fascinating class of ABC transport proteins.

MEMBRANE-ATTACHED PROTEINS

Membrane-attached proteins can be divided into three groups:

- Bound to the cytosolic face of the plasma membrane by myristate;
- Bound to the cytosolic face of the plasma membrane by farnesyl;
- Bound to the extracelluar face of the plasma membrane by glycosylphosphatidylinositol.

TRANSMEMBRANE PROTEINS

A transmembrane protein is a protein which spans the entire length of the cell membrane. It is embedded between the phospholipids, providing a channel through which molecules and ions can pass into the cell. Transmembrane proteins also facilitate communication between cells by interacting with chemical messengers. Many biological processes, such as the metabolism of glucose and the production of fatty acids, are triggered after a particular transmembrane protein is activated.

The insulin receptor is an example of a transmembrane protein which interacts with a chemical messenger, namely insulin. The receptor acts as a target on the surface of the cell for the insulin molecule. After the molecule docks with the receptor, the receptor typically releases chemicals that cause the movement of a glucose transporter to the surface of the cell. This allows the cell to absorb large amounts of glucose from the external environment, leading to glucose metabolism and eventually energy production.

Another job of the transmembrane protein is to shuttle ions, such as sodium and potassium, across the cell membrane to maintain the chemical environment. Some cells cannot perform specific tasks if the ion channels are not working properly. An important example of this is the voltage-gated ion channels of nerve cells. In the resting state, the ion channel is typically closed, preventing ions from crossing the membrane. As soon as a stimulus is detected, such as a cut or burn, a nerve impulse is sent from one end of the nerve cell to the other end. This can only occur if the ion channels open and allow the ions to flow through the cell membrane.

In order to organize properly, cells also require transmembrane proteins to examine the environment within which the cell resides. For example, muscle cells generally organize around other muscle cells, while skin cells organize around other skin cells. Integrins are a broad category of transmembrane proteins that perform this organizing function. Integrins also anchor cells to substrates, facilitating cell migration and wound healing. A cell's growth, division and death are generally dependent on the signals received by integrins. A transmembrane protein can be classified as alpha helical or beta barrel, depending on how the protein chain is organized. Alpha helical proteins consist of a single chain while the beta barrel proteins have several protein chains organized side by side. The alpha helical protein is generally coiled, and the beta barrel protein is twisted into a closed structure that resembles a barrel.

G-PROTEIN-COUPLED RECEPTORS

The major role of G-protein-coupled receptors is to transmit signals into the cell. They are characterized by seven transmembrane segments. This class of membrane proteins can respond to a wide range of agonists, including photon, amines, hormones, neurotransmitters and proteins. Some agonists bind to the extracellular loops of the receptor, others may penetrate into the transmembrane region.

Immunoglobulin Family

This protein family transmit signals for immune responses. Each of them contains one or two transmembrane segments. The immunoglobulin (i.e., antibody) consists of four peptide chains: two heavy chains and two light chains. V_H and C_H represent the variable and conserved domains of the heavy chain, respectively; V_L and C_L denote those for the light chain. Variable domains are the binding site of antigens. To send signals into the cell, this class of proteins are typically coupled to non-receptor protein tyrosine kinase located on the cytosolic face of the plasma membrane.

Transport Proteins

According to the Transport Commission (TC) system, all transport proteins may be divided into the following classes:

a. Channels/Pores

   Examples: ion channels, porins

b. Electrochemical Potential-driven transporters

   Examples: uniporters, symporters and antiporters

c. Primary Active Transporters

   Examples: Na^+-K^+ ATPase, H^+ ATPase, ABC Transporter Superfamily and CFTR channel.

d. Group Translocators

e. Transport Electron Carriers

f. Auxiliary Transport Proteins

g. Incompletely Characterized

CHANNELS/PORES

Ion Channels [I]

Ion channels are a special class of proteins that conduct small ions such as Na^+, K^+, Ca^{2+} or Cl^-. The opening and closing of ion channels may depend on membrane voltage or the binding of molecules (the ligand). Its mechanism remains unclear. However, the mystery of their selectivity has been solved.

Ion Selectivity

Potassium Channels

By definition, a potassium channel conducts mainly K^+ ions while excluding Na^+ and Ca^{2+} ions. We note that a Na^+ ion is smaller than a K^+ ion. How could a channel conduct K^+ ions while excluding smaller ions? The answer lies in the ion's hydration energy. The diameter of the potassium channel's selectivity filter (the narrowest region of its channel pore) should be about the same as the size of a Cs^+ ion (diameter = 3.3Å) because Cs^+ is the largest ion which can barely pass through the potassium channel. In a solution, ions are surrounded by water molecules. The first hydration shell of a K^+ or Na^+ ion contains six water molecules. The diameters of unhydrated Na^+ and K^+ ions are 1.96Å and 2.66Å, respectively. The effective diameter of a water molecule in a hydrated ion is larger than 2Å. Therefore, to pass through the selectivity filter, the K^+ or Na^+ ion must remove four water molecules from its first hydration shell, leaving only two (one at the front and anther at the back).

The smaller Na^+ ion requires greater dehydration energy than the K^+ ion, because its nucleus has shorter distance with surrounding water molecules and thus interacting more strongly. As a result, the Na^+ ion is harder to pass through the potassium channel than the K^+ ion. The observation that lithium (Li^+) is excluded from the potassium channel can also be explained by the same energetic consideration. Divalent cations such as Ca^{2+} should also be excluded because their dehydration energies are much higher than monovalent cations.

Sodium Channels

In the sodium channel, the Na^+ ion is more permeable than the K^+ ion. This is because the selectivity filter of the sodium channel is slightly larger than that of the potassium channel. It is large enough to accommodate a Na^+ ion attached with three water molecules, but not enough for a K^+ ion attached with three water molecules. Therefore, to pass through the sodium channel, the Na^+ ion needs to remove only three, but the K^+ ion has to remove four, water molecules from its first hydration shell. The required dehydration energy for the K^+ ion is greater than the Na^+ ion.

Calcium Channels

In calcium channels, the permeability of monovalent cations (Na^+ and K^+) is about three orders of magnitude smaller than the Ca^{2+} permeability. This ion selectivity does not seem to involve hydration, because Ca^{2+} is more heavily hydrated than Na^+, and the unhydrated diameters of Ca^{2+} and Na^+ are almost identical. Then, how could calcium channels select Ca^{2+} over Na^+? Although the permeability of monovalent cations in the calcium channel is quite small at normal ionic concentrations, large monovalent cationic current can be observed in the absence of Ca^{2+} and other divalent cations. This suggests that the calcium channel is basically permeable to both divalent and monovalent cations, but the selectivity arises from competition between ions. The calcium channel may contain a negatively charged binding site to facilitate ion conduction. The monovalent cations simply cannot compete with Ca^{2+} for this binding site. This idea has been demonstrated experimentally. In the calcium channel, if a negatively charged glutamate residue in the pore-lining region is mutated into a positively charged lysine, the calcium channel becomes more permeable to Na^+ than Ba^{2+}. Conversely, in the sodium channel, mutation of a pore-lining lysine residue into glutamate transforms the channel from a Na^+-selective to a Ca^{2+}-selective channel.

Channel Structures

The detailed 3D structures of most ion channels are not known. However, from the hydrophobicity profile of their amino acid sequences, it is possible to obtain the domain structures of most ion channels.

K Channels

There are many types of potassium channels. Most commonly observed K channels are composed of four subunits, each is homologous to the Shaker protein. The hydrophobicity profile indicates that it contains six hydrophobic segments, designated as S1 - S6. These segments are likely to be the transmembrane domains. Other experimental results suggests that the P-region is lining the channel pore.

Na and Ca Channels

A Na or Ca channel consists of a major pore-forming subunit and possibly other small auxiliary subunits. The major pore-forming subunit is called the α subunit which can be divided into four similar domains. Each domain is analogous to a Shaker protein with six transmembrane segments and a P-region. Thus, an α subunit is sufficient to form an ion channel.

Synaptic Channels

A large class of ion channels are specifically located at the synapses (the junction between nerve cells).

Porin

The nonpolar interior of membranes limits the free diffusion of polar molecules. In most membranes various proteins serve as channels and pumps which regulate the movement of ions, metabolites and even water across the membrane.

Due to the specificity of these proteins there is a high degree of control over what passes through the membrane.Several membranes such as the outer membranes of mitochondria, chloroplasts and bacteria, however, do not need the same high degree of control over what gets across. Rather than synthesizing many selective proteins to carry polar molecules across, it appears that cells make only a few types of transporters, like the porin pictured below, which let almost anything below a certain size across. It may simply be more cost effective to put a few porins in these membranes then many more selective proteins. These "leaky" membranes are still effective barriers to the movement of large molecules such as proteins.

Below are two views of a model of porin from the bacterium *Rhodopseudomonas blastica.* In the left image the model is shown from the side as if you are in the middle of the membrane. The height of the protein is about the same as the thickness of the membrane in which it sits. The right image shows the model from the end where the central "hole" through which ions and metabolites pass is clearly visible. Regions of secondary structure are shown in yellow (beta sheet), pink (alpha helix) and blue (turn). The strands of colour follow the backbone of the molecule. If all of the atoms and bonds in the protein were visualized the image would be much too complicated.Why is the tertiary structure of porin described as a beta barrel?

Electrochemical Potential-driven Transporters

Uniporter-Catalyzed Transport

We begin our discussion of membrane transport proteins with the simplest type, which catalyze uniport transport. The plasma membrane of most cells contains several uniporters that enable amino acids, nucleosides, sugars, and other small molecules to enter and leave cells down their concentration gradients. Similar to enzymes, uniporters accelerate a reaction that is already thermodynamically favored, and the movement of a substance across a membrane down its concentration gradient will have the same negative ΔG value whether or not a protein transporter is involved. This type of movement sometimes is referred to as facilitated transport (or facilitated diffusion).

Many chemical reactions that are thermodynamically favored will not occur unless an appropriate enzyme is present; such is also the case with movement of hydrophilic molecules across biological membranes. Unlike the substrates of enzymatic reactions, however, transported substances undergo no chemical change during movement across a membrane.

Three Main Features Distinguish Uniport Transport from Passive Diffusion Three properties of uniporter-catalyzed movement of glucose and other small hydrophilic molecules across a membrane distinguish this type of transport from passive diffusion:

1. The rate of facilitated transport by uniporters is far higher than predicted by Fick's equation describing passive diffusion. Because the transported molecules never enter the hydrophobic core of the phospholipid bilayer, the partition coefficient K is irrelevant.
2. Transport is specific. Each uniporter transports only a single species of molecule or a single group of closely related molecules.
3. Transport occurs via a limited number of uniporter molecules, rather *than throughout the phospholipid bilayer.* Consequently, there is a maximum transport rate V_{max} that is achieved when the concentration gradient across the membrane is very large and each uniporter is working at its maximal rate.

Cotransport by Symporters and Antiporters

Besides ATP-powered pumps, cells have a second, discrete class of proteins that import or export ions and small molecules, such as glucose and amino acids, against a concentration gradient. These proteins use the energy stored in the electrochemical gradient of Na^+ or H^+ ions to power the uphill movement of another substance, a process called cotransport. For instance, the energetically favored movement of a Na^+ ion (the "cotransported" ion) into a cell across the plasma membrane, driven both by its concentration gradient and by the transmembrane voltage gradient can be coupled obligatorily to movement of the "transported" molecule (e.g., glucose) against its concentration gradient. When the transported molecule and cotransported ion move in the same direction, the process is called symport; when they move in opposite directions, the process is called antiport.

STRUCTURE OF PROTEINS

This explains how amino acids combine to make proteins and what is meant by the primary, secondary and tertiary structures of proteins. Quaternary structure isn't covered. It only applies to proteins consisting of more than one polypeptide chain. There is a mention of quaternary structure on the IB chemistry syllabus, but on no other UK-based syllabus at this level. Quaternary structure can be very complicated, and I don't know exactly what depth the IB syllabus wants for this. I suspect what is wanted is fairly trivial. IB students should ask the advice of their teacher or lecturer.

PRIMARY STRUCTURE OF PROTEINS

Drawing the Amino Acids

In chemistry, if you were to draw the structure of a general 2-amino acid, you would probably draw it like this:

NH2
|
R – CH – COOH

However, for drawing the structures of proteins, we usually twist it so that the"R" group sticks out at the side. It is much easier to see what is happening if you do that.

R
|
NH_2 – CH – COOH

That means that the two simplest amino acids, glycine and alanine, would be shown as:

H
|
NH_2 – CH – COOH
Glycine

CH_3
|
NH_2 – CH – COOH
alanine

PEPTIDES AND POLYPEPTIDES

Glycine and alanine can combine together with the elimination of a molecule of water to produce a dipeptide. It is possible for this to happen in one of two different ways - so you might get two different dipeptides.

Either:

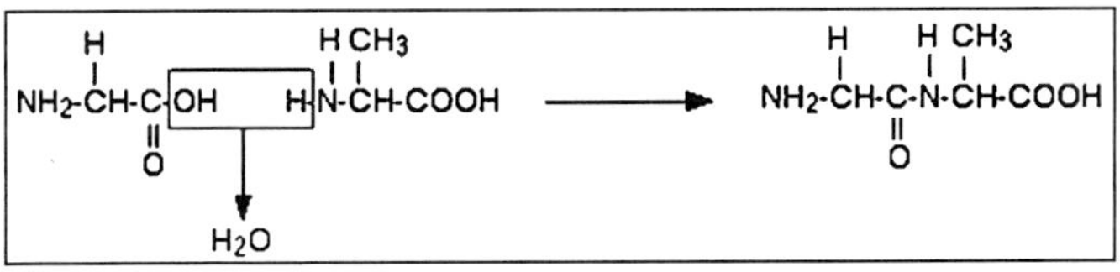

Or:

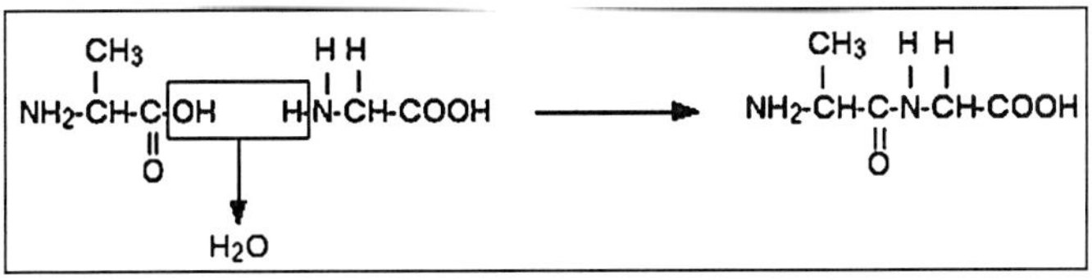

In each case, the linkage shown in blue in the structure of the dipeptide is known as a peptide link. In chemistry, this would also be known as an amide link, but since we are now in the realms of biochemistry and biology, we'll use their terms.

If you joined three amino acids together, you would get a tripeptide. If you joined lots and lots together (as in a protein chain), you get a polypeptide. A protein chain will have somewhere in the range of 50 to 2000amino acid residues. You have to use this term because strictly speaking a peptide chain isn't made up of amino acids. When the amino acids combine together, a water molecule is lost.

The peptide chain is made up from what is left after the water is lost - in other words, is made up of amino acid residues. By convention, when you are drawing peptide chains, the -NH_2group which hasn't been converted into a peptide link is written at the left-hand end. The unchanged –COOH group is written at the right-hand end. The end of the peptide chain with the –NH_2 group is known as theN-terminal, and the end with the –COOH group is the C–.

Terminal

A protein chain (with the N-terminal on the left) will therefore look like this:

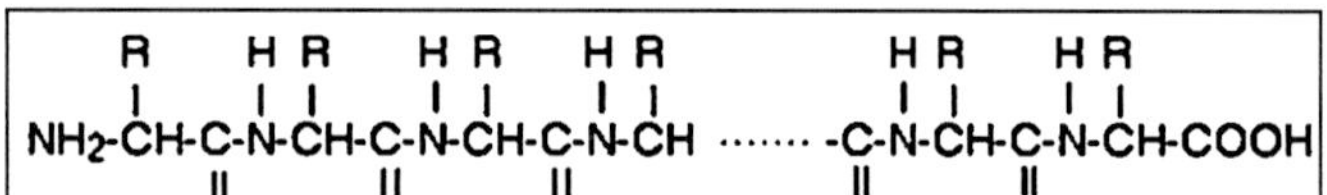

The"R" groups come from the 20 amino acids which occur in proteins. The peptide chain is known as the backbone, and the"R" groups are known as side chains.

Note: In the case where the"R" group comes from the amino acid proline, the pattern is broken. In this case, the hydrogen on the nitrogen nearest the"R" group is missing, and the"R" group loops around and is attached to that nitrogen as well as to the carbon atom in the chain.

I mention this for the sake of completeness - not because you would be expected to know about it in chemistry at this introductory level. At its simplest, the term is used to describe the order of the amino acids joined together to make the protein. In other words, if you replaced the"R" groups in the last diagram by real groups you would have the primary structure of a particular protein. This primary structure is usually shown using abbreviations for the amino acid residues. These abbreviations commonly consist of three letters or one letter.

Using three letter abbreviations, a bit of a protein chain might be represented by, for example:

Try.Gly.Lys.pro.Val.Gly. Lys.Lys.Arg.Arg.Pro.Val.Lys.
Tyr.Pro.Ala.Gly.Glu...

If you look carefully, you will spot the abbreviations for glycine (Gly) and alanine (Ala) amongst the others. If you followed the protein chain all the way to its left-hand end, you would find an amino acid residue with an unattached –NH_2 group. The N-terminal is always written on the left of a diagram for a protein's primary structure - whether you draw it in full or use these abbreviations.

The wider definition of primary structure includes all the features of a protein which are a result of covalent bonds. Obviously, all the peptide links are made of covalent bonds, so that isn't a problem. But there is an additional feature in proteins which is also covalently bound. It involves the amino acid cysteine.

$$\begin{array}{c} CH_2SH \\ | \\ NH_2 - CH - COOH \end{array}$$

Cysteine

If two cysteine side chains end up next to each other because of folding in the peptide chain, they can react to form a sulphur bridge. This is another covalent link and so some people count it as a part of the primary structure of the protein.

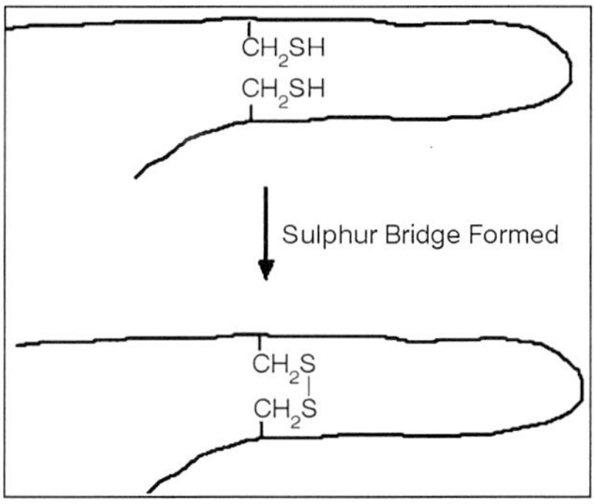

Because of the way sulphur bridges affect the way the protein folds, other people count this as a part of the tertiary structure (see below). This is obviously a potential source of confusion!

Important: You need to know where your particular examiners are going to include sulphur bridges - as a part of the primary structure or as a part of the tertiary structure. You need to check your current syllabus and past papers. If you are studying a UK-based syllabus and haven't got these, follow this link to find out how to get hold of them.

Secondary Structure of Proteins

Within the long protein chains there are regions in which the chains are organised into regular structures known as alpha-helices (alpha-helixes) and beta-pleated sheets.

These are the secondary structures in proteins. These secondary structures are held together by hydrogen bonds. These form as shown in the diagram between one of the lone pairs on an oxygen atom and the hydrogen attached to a nitrogen atom:

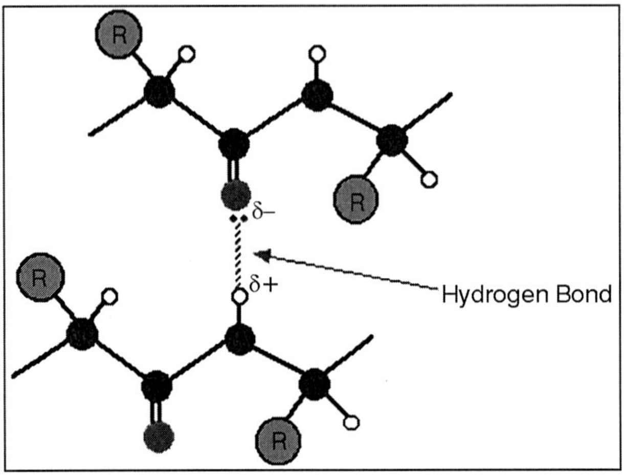

Important: If you aren't happy about hydrogen bonding and are unsure about what this diagram means, follow this link before you go on. What follows is difficult enough to visualise anyway without having to worry about what hydrogen bonds are as well! You must also find out exactly how much detail you need to know about this next bit. It may well be that all you need is to have heard of an alpha-helix and know that it is held together by hydrogen bonds between the C=O and N-H groups. Once again, you need to check your syllabus and past papers - particularly mark schemes for the past papers.

The Alpha-helix

In an alpha-helix, the protein chain is coiled like a loosely-coiled spring. The"alpha" means that if you look down the length of the spring, the coiling is happening in a clockwise direction as it goes away from you.

Note: If your visual imagination is as hopeless as mine, the only way to really understand this is to get a bit of wire and coil it into a spring shape. The lead on your computer mouse is fine for doing this! The next diagram shows how the alpha-helix is held together by hydrogen bonds. This is a very simplified diagram, missing out lots of atoms. We'll talk it through in some detail after you have had a look at it.

What's wrong with the diagram? Two things: First of all, only the atoms on the parts of the coils facing you are shown. If you try to show all the atoms, the whole thing gets so complicated that it is virtually impossible to understand what is going on.

So, what do you need to notice? Notice that all the"R" groups are sticking out sideways from the main helix. Notice the regular arrangement of the hydrogen bonds. All the N-H groups are pointing upwards, and all the C=O groups pointing downwards.

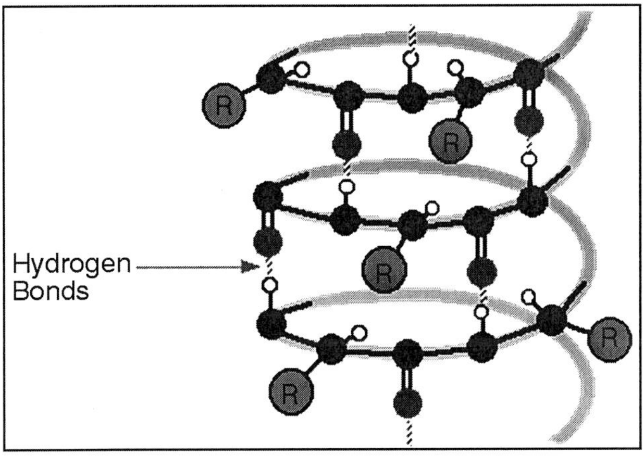

Each of them is involved in a hydrogen bond. And finally, although you can't see it from this incomplete diagram, each complete turn of the spiral has 3.6 (approximately) amino acid residues in it. If you had a whole number of amino acid residues per turn, each group would have an identical group underneath it on the turn below. Hydrogen bonding can't happen under those circumstances.

Each turn has 3 complete amino acid residues and two atoms from the next one. That means that each turn is offset from the ones above and below, such that the N-H and C=O groups are brought into line with each other.

Beta-pleated Sheets

In a beta-pleated sheet, the chains are folded so that they lie alongside each other. The next diagram shows what is known as an"anti-parallel" sheet. All that means is that next-door chains are heading in opposite directions. Given the way this particular folding happens, that would seem to be inevitable.

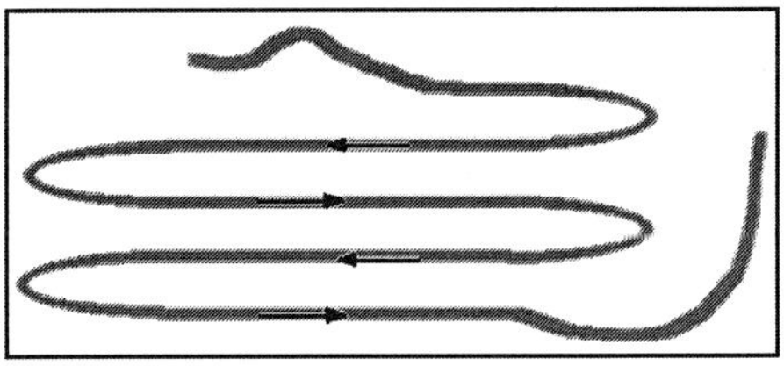

It isn't, in fact, inevitable! It is possible to have some much more complicated folding so that next-door chains are actually heading in the same direction. We are getting well beyond the demands of UK A level chemistry (and its equivalents) now. The folded chains are again held together by hydrogen bonds involving exactly the same groups as in the alpha-helix.

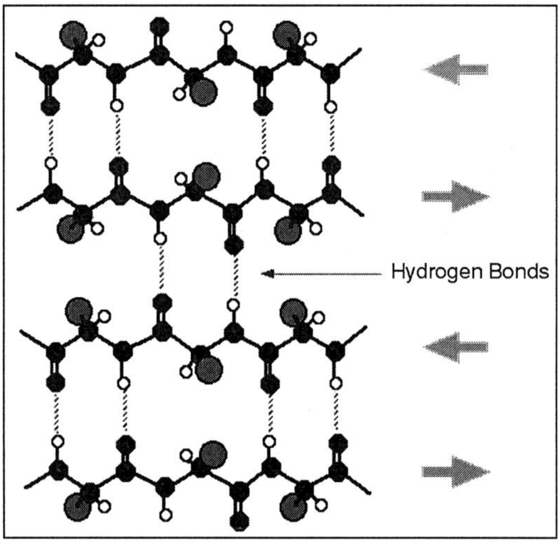

Note: Note that there is no reason why these sheets have to be made from four bits of folded chain alongside each other as shown in this diagram. That was an arbitrary choice which produced a diagram which fitted nicely on the screen!

TERTIARY STRUCTURE OF PROTEINS

Definition

The tertiary structure of a protein is a description of the way the whole chain (including the secondary structures) folds itself into its final 3-dimensional shape. This is often simplified into models like the following one for the enzyme dihydrofolate reductase. Enzymes are, of course, based on proteins.

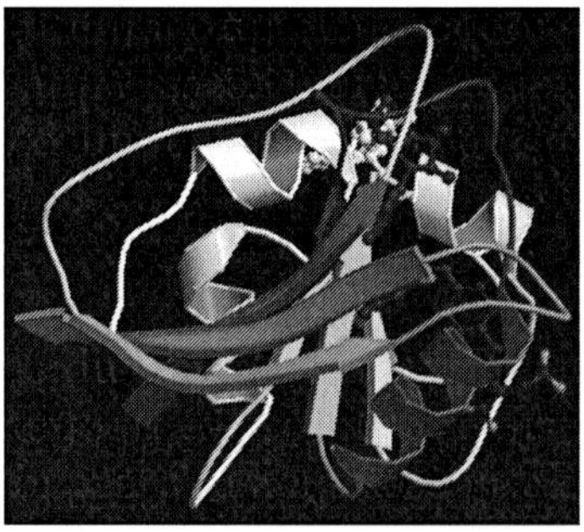

Note: This diagram was obtained from the RCSB Protein Data Bank. If you want to find more information about dihydrofolate reductase, their reference number for it is 7DFR. There is nothing particularly special about this enzyme in terms of structure. The model shows the alpha-helices in the secondary structure as coils of"ribbon". The beta-pleated sheets are shown as flat bits of ribbon ending in an arrow head. The bits of the protein chain which are just random coils and loops are shown as bits of"string". The colour coding in the model helps you to track your way around the structure - going through the spectrum from dark blue to end up at red. You will also notice that this particular model has two other molecules locked into it (shown as ordinary molecular models). These are the two molecules whose reaction this enzyme catalyses.

Holding Protein into its Tertiary Structure

The tertiary structure of a protein is held together by interactions between the the side chains - the"R" groups. There are several ways this can happen.

Ionic Interactions

Some amino acids (such as aspartic acid and glutamic acid) contain an extra -COOH group. Some amino acids (such as lysine) contain an extra $-NH_2$ group.

You can get a transfer of a hydrogen ion from the -COOH to the -NH2 group to form zwitterions just as in simple amino acids. You could obviously get an ionic bond between the negative and the positive group if the chains folded in such a way that they were close to each other.

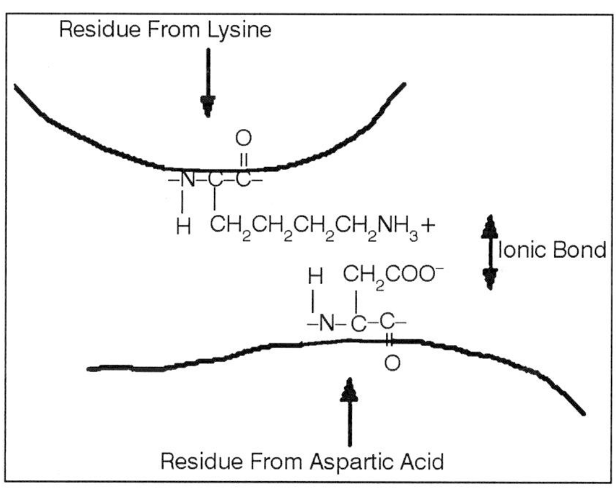

Hydrogen Bonds

Notice that we are now talking about hydrogen bonds between side groups - not between groups actually in the backbone of the chain. Lots of amino acids contain groups in the side chains which have a hydrogen atom attached to either an oxygen or a nitrogen atom. This is a classic situation where hydrogen bonding can occur.

For example, the amino acid serine contains an -OH group in the side chain. You could have a hydrogen bond set up between two serine residues in different parts of a folded chain.

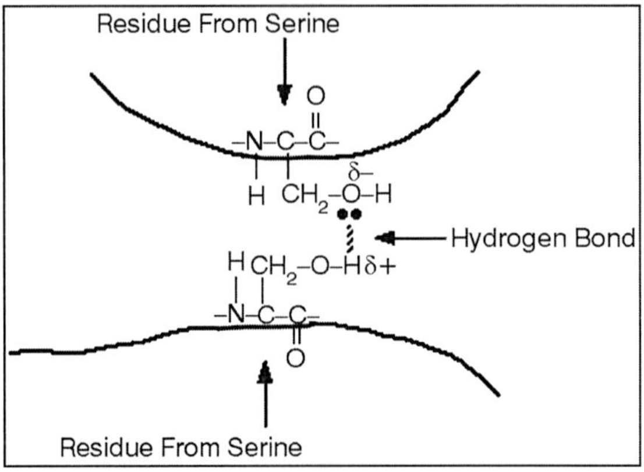

You could easily imagine similar hydrogen bonding involving -OH groups, or -COOH groups, or -$CONH_2$ groups, or -NH_2 groups in various combinations - although you would have to be careful to remember that a -COOH group and an -NH_2 group would form a zwitterion and produce stronger ionic bonding instead of hydrogen bonds.

Van der Waals Dispersion Forces

Several amino acids have quite large hydrocarbon groups in their side chains. A few examples are shown below. Temporary fluctuating dipoles in one of these groups could induce opposite dipoles in another group on a nearby folded chain.

The dispersion forces set up would be enough to hold the folded structure together.

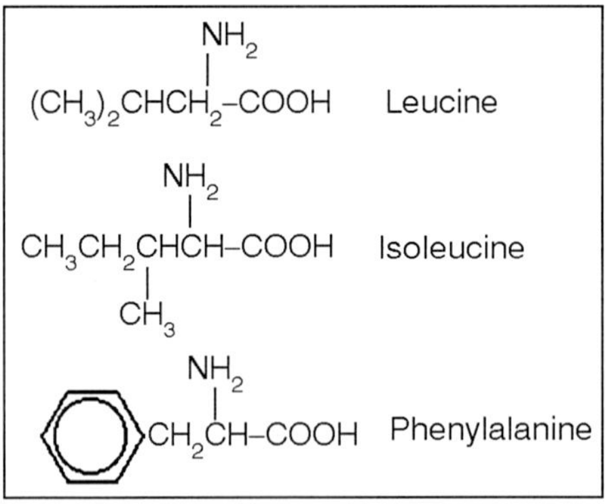

Sulphur Bridges

Sulphur bridges which form between two cysteine residues have already been discussed under primary structures. Wherever you choose to place them doesn't affect how they are formed.

PRIMARY ACTIVE TRANSPORTERS

ACTIVE TRANSPORT BY ATP-POWERED PUMPS

We turn now to the ATP-powered pumps that transport ions and various small molecules against their concentration gradients. The general structures of the four principal classes of these transport proteins are depicted in figure and their properties are summarized in table. Note that the P, F, and V classes transport ions only, whereas the ABC superfamily class transports small molecules as well as ions.

P-class ion pumps containa transmembrane catalytic á subunit, which contains an ATP-binding site, and usually a smaller β subunit, which may have regulatory functions. Many of these pumps are tetramers composed of two and two β subunits. During the transport process, at least one of the á subunits is phosphorylated (hence the label "P"), and the transported ions are thought to move through the phosphorylated subunit. This class includes the Na^+/K^+ ATPase in the plasma membrane, which maintains the Na^+ and K^+ gradients typical of animal cells, and several Ca^{2+} ATPases, which pump Ca^{2+} ions out of the cytosol into the external medium or into the lumen of the sarcoplasmic reticulum (SR) of muscle cells. Another member of the P class, found in acid-secreting cells of the mammalian stomach, transports protons (H^+ ions) out of and K^+ ions into the cell. The H^+ pump that maintains the membrane electric potential in plant, fungal, and bacterial cells also belongs to this class.

The structures of *F-class* and *V-class ion pumps* are similar to each other but unrelated to and more complicated than P-class pumps. F- and V-class pumps contain at least three kinds of transmembrane proteins and five kinds of extrinsic polypeptides that form the cytosolic domain. Several of the transmembrane and extrinsic subunits in F-class and V-class pumps exhibit sequence homology, and each pair of homologous subunits is thought to have evolved from a common polypeptide. All known V and F pumps transport only protons in a process that does not involve a phosphoprotein intermediate. V-class pumps generally function to maintain the low pH of plant vacuoles and of lysosomes and other acidic vesicles in animal cells by using the energy released by ATP hydrolysis to pump protons from the cytosolic to the exoplasmic face of the membrane against the proton electrochemical gradient.

F-class pumps are found in bacterial plasma membranes and in mitochondria and chloroplasts. In contrast to V pumps, they generally function to power the synthesis of ATP from ADP and P_i by movement of protons from the exoplasmic to the cytosolic face of the membrane down the proton electrochemical gradient. The final class of ATP-powered transport proteins is larger and more diverse than the other classes. Referred to as the *ABC (ATP-binding cassette) superfamily,* this class includes more than 100 different transport proteins found in organisms ranging from bacteria to humans. Each

ABC protein is specific for a single substrate or group of related substrates including ions, sugars, peptides, polysaccharides, and even proteins. All ABC transport proteins share a common organization consisting of four "core" domains: two transmembrane (T) domains, forming the passageway through which transported molecules cross the membrane, and two cytosolic ATP-binding (A) domains. In some ABC proteins, the core domains are present in four separate polypeptides; in others, the core domains are fused into one or two multidomain polypeptides.

All classes of ATP-powered pumps have one or more binding sites for ATP, and these are always on the cytosolic face of the membrane. Although these proteins are often called ATPases, they normally do not hydrolyze ATP into ADP and P_i unless ions or other molecules are simultaneously transported. Because of the tight coupling between ATP hydrolysis and transport, the energy stored in the phosphoanhydride bond is not dissipated. Thus ATP-powered transport proteins are able to collect the free energy released during ATP hydrolysis and use it to move ions or other molecules uphill against a potential or concentration gradient.

The energy expended by cells to maintain the concentration gradients of Na^+, K^+, H^+, and Ca^{2+} across the plasma and intracellular membranes is considerable. In nerve and kidney cells, for example, up to 25 percent of the ATP produced by the cell is used for ion transport; in human erythrocytes, up to 50 percent of the available ATP is used for this purpose. In cells treated with poisons that inhibit the aerobic production of ATP (e.g., 2,4-dinitrophenol), the ion concentration inside the cell gradually approaches that of the exterior environment as the ions move through plasma membrane channels down their electric and concentration gradients. Eventually treated cells die: partly because protein synthesis requires a high concentration of K^+ ions and partly because in the absence of a Na^+ gradient across the cell membrane, a cell cannot import certain nutrients such as amino acids. Studies on the effects of such poisons provided early evidence for the existence of ion pumps. In this section, we discuss in some detail examples of the P, V, and ABC classes of ATP-powered pumps.

Plasma-Membrane Ca^{2+} ATPase Exports Ca^{2+} Ions from Cells. In order for Ca^{2+} to function in intracellular signaling, its cytosolic concentration usually must be kept below 0.1 – 0.2 μM. (Although some cytosolic Ca^{2+} is bound to negatively charged groups, it is the concentration of *free,* unbound Ca^{2+} that is critical to its signaling function.) The plasma membranes of animal, yeast, and probably plant cells contain Ca^{2+} ATPases that transport Ca^{2+} out of the cell against its electrochemical gradient. These P-class ion pumps help maintain the concentration of free Ca^{2+} ions in the cytosol at a low level. In addition to a catalytic subunit containing an ATP-binding site, as found in other P-class pumps, plasmamembrane Ca^{2+} ATPases also contain the Ca^{2+}-binding regulatory protein calmodulin. A rise in cytosolic Ca^{2+} induces the binding of Ca^{2+} ions to calmodulin, which triggers an allosteric activation of the Ca^{2+} ATPase; as a result, the export of Ca^{2+} ions from the cell accelerates, and the original low cytosolic concentration of free Ca^{2+} is restored rapidly.

Muscle Ca^{2+} ATPase Pumps Ca^{2+} Ions from the Cytosol into the Sarcoplasmic Reticulum. Besides the plasma-membrane Ca^{2+} ATPase, muscle cells contain a second, different Ca^{2+} ATPase that transports Ca^{2+} from the cytosol into the lumen of the sarcoplasmic reticulum (SR), an internal organelle that concentrates and stores Ca^{2+} ions. As discussed the SR and its calcium pump (referred to as the *muscle calcium pump*) are critical in muscle contraction and relaxation: release of Ca^{2+} ions from the SR into the muscle cytosol causes contraction, and the rapid removal of Ca^{2+} ions from the cytosol by the muscle calcium pump induces relaxation.

Because the muscle calcium pump constitutes more than 80 percent of the integral protein in SR membranes, it is easily purified and characterized. Each transmembrane catalytic subunit has a molecular weight of 100,000 and transports two Ca^{2+} ions per ATP hydrolyzed. In the cytosol of muscle cells, the free Ca^{2+} concentration ranges from 10^{-7} M (resting cells) to more than 10^{-6} M (contracting cells), whereas the *total* Ca^{2+} concentration in the SR lumen can be as high as 10^{-2} M. Sites on the cytosolic surface of the muscle calcium pump have a very high affinity for Ca^{2+} ($K_m = 10^{-7}$ M), allowing the pump to transport Ca^{2+} efficiently from the cytosol into the SR against the steep concentration gradient.

The concentration of free Ca^{2+} within the sarcoplasmic reticulum is actually much less than the total concentration of 10^{-2} M. Two soluble proteins in the lumen of SR vesicles bind Ca^{2+} and serve as a reservoir for intracellular Ca^{2+}, thereby reducing the concentration of free Ca^{2+} ions in the SR vesicles, and consequently decreasing the energy needed to pump Ca^{2+} ions into them from the cytosol. The activity of the muscle Ca^{2+} ATPase is so regulated that if the free Ca^{2+} concentration in the cytosol becomes too high, the rate of calcium pumping increases until the cytosolic Ca^{2+} concentration is reduced to less than 1 μM. Thus in muscle cells, the calcium pump in the SR membrane can supplement the activity of the plasma-membrane pump, assuring that the cytosolic concentration of free Ca^{2+} remains below 1 μM.

The current model of the mechanism of action of the Ca^{2+} ATPase in the SR membrane is outlined in Figure. Coupling of ATP hydrolysis with ion pumping involves several steps that must occur in a defined order.When the protein is in one conformation, termed *E1*, two Ca^{2+} ions bind in sequence to high-affinity sites on the cytosolic surface (step 1). Then an ATP binds to its site on the cytosolic surface; in a reaction requiring that a Mg^{2+} ion be tightly complexed to the ATP, the bound ATP is hydrolyzed to ADP and the liberated phosphate is transferred to a specific aspartate residue in the protein, forming a high-energy acyl phosphate bond, denoted by E1~P (step 2). The protein then changes its conformation to E2 – P, generating two lowaffinity Ca^{2+}-

binding sites on the exoplasmic surface, which faces the SR lumen; this conformational change simultaneously propels the two Ca^{2+} ions through the protein to these sites (step 3) and inactivates the high-affinity Ca^{2+}-binding sites on the cytosolic face. The Ca^{2+} ions then dissociate from the exoplasmic surface of the protein (step 4). Following this, the aspartyl-phosphate bond in E2 – P is hydrolyzed, causing E2 to revert to E1, a change that inactivates the exoplasmic-facing Ca^{2+}-binding sites and regenerates the cytosolicfacing Ca^{2+}-binding sites (step 5).

Thus phosphorylation of the muscle calcium pump by ATP favors conversion of E1 to E2, and dephosphorylation favors the conversion of E2 to E1. While only E2 – P, not E1~P, is actually hydrolyzed, the free energy of hydrolysis of the aspartyl-phosphate bond in E1~P is greater than that for E2 – P. The reduction in free energy of the aspartyl-phosphate bond in E2 – P, relative to E1~P, can be said to power the E1 E2 conformational change. The affinity of Ca^{2+} for the cytosolic-facing binding sites in E1 is a thousandfold greater than the affinity of Ca^{2+} for the exoplasmic-facing sites in E2; this difference enables the protein to transport Ca^{2+} unidirectionally from the cytosol, where it binds tightly to the pump, to the exoplasm, where it is released.

For instance, the muscle calcium pump has been isolated with phosphate linked to an aspartate residue, and spectroscopic studies have detected slight alterations in protein conformation during the E1 E2 conversion. On the basis of the protein's amino acid sequence and various biochemical studies, investigators proposed the structural model for the catalytic subunit shown in Figure. The membrane-spanning helices are thought to form the passageway through which Ca^{2+} ions move. The bulk of the subunit consists of cytosolic globular domains that are involved in ATP binding, phosphorylation of aspartate, and energy transduction. These domains are connected by "stalks" to the membrane-embedded domain.

As noted previously, all P-class ion pumps, regardless of which ion they transport, are phosphorylated during the transport process. The amino acid sequences around the phosphorylated aspartate in the catalytic subunit are highly conserved in all proteins of this type. Thus the mechanistic model in Figure probably is generally applicable to all these ATP-powered ion pumps. In addition, the subunits of all the P pumps examined to date have a similar molecular weight and, as deduced from their amino acid sequences derived from cDNA clones, have a similar arrangement of transmembrane helices. These findings strongly suggest that all these proteins evolved from a common precursor, although they now transport different ions. Na^+/K^+ ATPase Maintains the Intracellular Na^+ and K^+ Concentrations in Animal Cells. A second P-class ion pump that has been studied in considerable detail is the Na^+/K^+ ATPase present in the plasma membrane of all animal cells. This ion pump is a tetramer of subunit composition $_2\beta_2$. (Classic Experiment 15.1

describes the discovery of this enzyme.) The β polypeptide is required for newly synthesized subunits to fold properly in the endoplasmic reticulum but apparently is not involved directly in ion pumping.

The subunit is a 120,000-MW nonglycosylated polypeptide whose amino acid sequence and predicted membrane structure are very similar to those of the muscle Ca^{2+} ATPase. In particular, the Na^+/K^+ ATPase has a stalk on the cytosolic face that links domains containing the ATP-binding site and the phosphorylated aspartate to the membrane-embedded domain. The overall process of transport moves three Na^+ ions out of and two K^+ ions into the cell per ATP molecule split. Several lines of evidence indicate that the Na^+/K^+ ATPase is responsible for the coupled movement of K^+ and Na^+ into and out of the cell, respectively. For example, the drug *ouabain*, which binds to a specific region on the exoplasmic surface of the protein and specifically inhibits its ATPase activity, also prevents cells from maintaining their Na^+/K^+ balance. Any doubt that the Na^+/K^+ ATPase is responsible for ion movement was dispelled by the demonstration that the enzyme, when purified from the membrane and inserted into liposomes, propels K^+ and Na^+ transport in the presence of ATP.

The mechanism of action of the Na^+/K^+ ATPase, outlined in figure is similar to that of the muscle calcium pump, except that ions are pumped in both directions across the membrane. In its E1 conformation, the Na^+/K^+ ATPase has three high-affinity Na^+-binding sites and two low-affinity K^+-binding sites on the cytosolic-facing surface of the protein. The K_m for binding of Na^+ to these cytosolic sites is 0.6 mM, a value considerably lower than the intracellular Na^+ concentration of H"12 mM; as a result, Na^+ ions normally will fill these sites. Conversely, the affinity of the cytosolic K^+-binding sites is low enough that K^+ ions, transported inward through the protein, dissociate from E1 into the cytosol despite the high intracellular K^+ concentration. During the E1 E2 transition, the three bound Na^+ ions move outward through the protein.

Transition to the E2 conformation also generates two high-affinity K^+ sites and three low-affinity Na^+ sites on the exoplasmic face. Because the K_m for K^+ binding to these sites (0.2 mM) is considerably lower than the extracellular K^+ concentration (4 mM), these sites will fill quickly with K^+ ions. In contrast, the three Na^+ ions, transported outward through the protein, will dissociate into the extracellular medium from the low-affinity Na^+ sites on the exoplasmic surface despite the high extracellular Na^+ concentration. Similarly, during the E2 E1 transition, the two bound K^+ ions are transported inward. V-Class H^+ ATPases Pump Protons across Lysosomal and Vacuolar MembranesAll V-class ATPases transport H^+ ions only. These proton pumps, present in the membranes of lysosomes, endosomes, and plant vacuoles, function to acidify the lumen of these organelles. The acidity of the lysosomal lumen, usually H"4.5 – 5.0, can be measured precisely in living cells by use of particles labeled with a pH-sensitive fluorescent dye. Cells phagocytose these particles and transfer them to the lysosomes. The ability of different wavelengths of visible light to excite fluorescence is highly dependent on pH, and the lysosomal pH can be calculated from the spectrum of the fluorescence emitted. Maintenance of the 100-fold or more proton gradient between the lysosomal lumen (pH H"4.5 – 5.0) and the cytosol (pH H"7.0) depends on ATP production by the cell.

The ATP-powered proton pumps in lysosomal and vacuolar membranes have been isolated, purified, and incorporated into liposomes. As illustrated in figure these V-class proton pumps contain two discrete domains: a cytosolic-facing hydrophilic domain (V_1) composed of five different polypeptides and a transmembrane domain (V_0) containing 9 – 12 copies of proteolipid c, one copy of protein b, and one copy of protein a. The subunit composition of the cytosolic domain is $\alpha_3\beta_3\gamma\delta\varepsilon$; the and β subunits contain the sites where ATP binding and hydrolysis occur. Each transmembrane c subunit is thought to span the membrane two times; the c and a subunits together form the proton-conducting channel. Unlike P-class ion pumps, the V-class H^+ ATPases are not phosphorylated and dephosphorylated during proton transport. Similar V-class ATPases are found in the plasma membrane of certain acid-secreting cells. These include osteoclasts, bone-resorbing macrophagelike cells, which bind to a bone and seal off a small segment of extracellular space between the plasma membrane and the surface of the bone. HCl secreted into this space by osteoclasts dissolves the calcium phosphate crystals that give bone its rigidity and strength.

Another example is the mitochondria-rich epithelial cells lining the toad bladder; the apical plasma membrane of these cells contain many V-class H^+ ATPases, which function to acidify the urine. As we discuss later, the membrane of plant vacuoles contains two proton pumps: a typical V-class H^+ ATPase and another one that utilizes the energy released by hydrolysis of inorganic pyrophosphate (PP_i) to pump protons into the vacuole. This PP_i-hydrolyzing proton pump, believed to be unique to plants, has an amino acid sequence different from any other ion-transporting proteins ATP-powered proton pumps cannot acidify the lumen of an organelle (or the extracellular space) by themselves.

The reason for this is that pumping of protons would rapidly cause a buildup of positive charge on the exoplasmic face of the membrane on the inside of the vesicle membrane and a corresponding buildup of negative charges on the cytosolic face. In other words, the pump would generate a voltage across the membrane, exoplasmic face positive, which would prevent movement of protons into the vesicle before a significant H^+ concentration gradient had been established. In fact, this is the way that H^+ pumps generate an insidenegative potential across plant and yeast plasma membranes. In order for an organelle lumen or an extracellular space (e.g., the outside of an osteoclast) to become acidic, movement of H^+ up its concentration gradient must be accompanied by

1. Movement of an equal number of anions in the same direction or
2. Movement of equal numbers of a different cation in the opposite direction. The first process occurs in lysosomes and plant vacuoles whose membranes contain V-class H^+ ATPases and ion channels through which accompanying anions (e.g., Cl″) move. The second occurs in the lining of the stomach, which contains a P-class H^+/K^+ ATPase that pumps one H^+ outward and one K^+ inward.

THE ABC SUPERFAMILY TRANSPORTS A WIDE VARIETY OF SUBSTRATES

As noted earlier, all members of the very large and diverse ABC superfamily of transport proteins contain two transmembrane (T) domains and two cytosolic ATP-binding (A) domains. The T domains, each built of six membrane-spanning α helices, form the pathway through which the transported substance (substrate) crosses the membrane and determine the substrate specificity of each ABC protein.

The sequence of the A domains is H″30 to 40 percent homologous in all members of this superfamily, indicating a common evolutionary origin. Some ABC proteins also contain a substrate-binding subunit or regulatory subunit.

Bacterial Plasma-Membrane Permeases

The plasma membrane of many bacteria contain numerous *permeases* that belong to the ABC superfamily. These proteins use the energy released by hydrolysis of ATP to transport specific amino acids, sugars, vitamins, or even peptides into the cell. Since bacteria frequently grow in soil or pond water where the concentration of nutrients is low, these ABC transport proteins allow the cells to concentrate amino acids and other nutrients in the cell against a substantial concentration gradient. Bacterial permeases generally are *inducible;* that is, the quantity of a transport protein in the cell membrane is regulated by both the concentration of the nutrient in the medium and the metabolic needs of the cell.

In *E. coli* histidine permease, a typical bacterial ABC protein, the two transmembrane domains and two cytosolic ATP-binding domains are formed by four separate subunits. In gram-negative bacteria such as *E. coli,* which have an outer membrane, a soluble histidine-binding protein in the periplasmic space assists in transport. This soluble protein binds histidine tightly and directs it to the T subunits, through which histidine crosses the membrane powered by ATP hydrolysis. Mutant *E. coli* cells that are defective in any of the histidine-permease subunits or the soluble binding protein are unable to transport histidine into the cell, but are able to transport other amino acids whose uptake is facilitated by other transport proteins. Such genetic analyses provide strong evidence that histidine permease and similar ABC proteins function to transport solutes into the cell.

PEPTIDE OR PROTEIN

Once a peptide or protein has been synthesized and released from the ribosome it often undergoes further chemical transformation. This post-translational modification may involve the attachment of other moieties such as acyl groups, alkyl groups, phosphates, sulfates, lipids and carbohydrates. Functional changes such as dehydration, amidation, hydrolysis and oxidation (*e.g.* disulfide bond formation) are also common. In this manner the limited array of twenty amino acids designated by the codons may be expanded in a variety of ways to enable proper functioning of the resulting protein. Since these post-translational reactions are generally catalyzed by enzymes, it may be said: "Virtually every molecule in a cell is made by the ribosome or by enzymes made by the ribosome."

Modifications, like phosphorylation and citrullination, are part of common mechanisms for controlling the behaviour of a protein. As shown on the left below, citrullination is the post-translational modification of the amino acid arginine into the amino acid citrulline. Arginine is positively charged at a neutral pH, whereas citrulline is uncharged, so this change increases the hydrophobicity of a protein.

Phosphorylation of serine, threonine or tyrosine residues renders them more hydrophilic, but such changes are usually transient, serving to regulate the biological activity of the protein. Other important functional changes include iodination of tyrosine residues in the peptide thyroglobulin by action of the enzyme thyroperoxidase. The monoiodotyrosine and diiodotyrosine formed in this manner are then linked to form the thyroid hormones T3 and T4, shown on the right below.

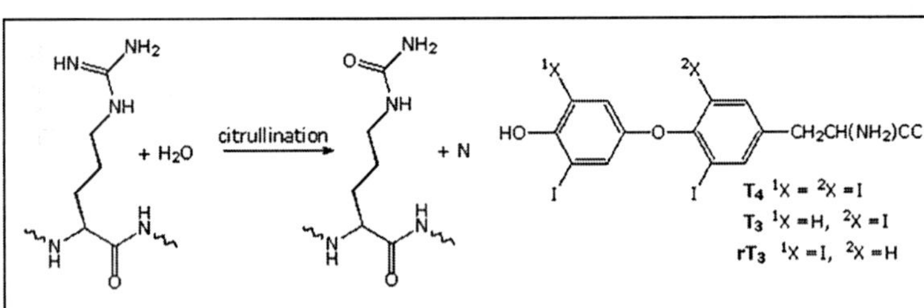

Amino acids may be enzymatically removed from the amino end of the protein. Because the "start" codon on mRNA codes for the amino acid methionine, this amino acid is usually removed from the resulting protein during post-translational modification. Peptide chains may also be cut in the middle to form shorter strands. Thus, insulin is initially synthesized as a 105 residue preprotein. The 24-amino acid signal peptide is removed, yielding a proinsulin peptide. This folds and forms disulfide bonds between cysteines 7 and 67 and between 19 and 80. Such dimeric cysteines, joined by a disulfide bond, are named cystine. A protease then cleaves the peptide at arg31 and arg60, with loss of the 32-60 sequence (chain C). Removal of arg31 yields mature insulin, with the A and B chains held together by disulfide bonds and a third cystine moiety in chain A. The following cartoon illustrates this chain of events.

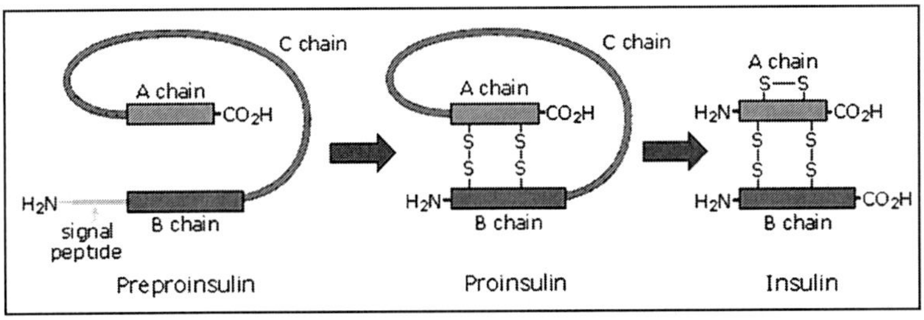

Nisin is a polypeptide (34 amino acids) made by the bacterium Lactococcus lactis. Nisin kills gram positive bacteria by binding to their membranes and targeting lipid II, an essential precursor of cell wall synthesis. Such antimicrobial peptides are a growing family of compounds which have received the name lantibiotics due to the presence of lanthionine, a nonproteinogenic amino acid with the chemical formula $HO_2C\text{-}CH(NH_2)\text{-}CH_2\text{-}S\text{-}CH_2\text{-}CH(NH_2)\text{-}CO_2H$.

Lanthionine is composed of two alanine residues that are crosslinked on their β-carbon atoms by a thioether linkage (*i.e.* it is the monosulfide analog of the disulfide cystine). Lantibiotics are unique in that they are ribosomally synthesized as prepeptides, followed by post-translational processing of a number of amino acids (*e.g.* serine, threonine and cysteine) into dehydro residues and thioether crossbridges.

Nisin is the only bacteriocin that is accepted as a food preservative. Several nisin subtypes that differ in amino acid composition and biological activity are known. A typical structure is drawn below, and a Jmol model will be presented by clicking on the diagram.

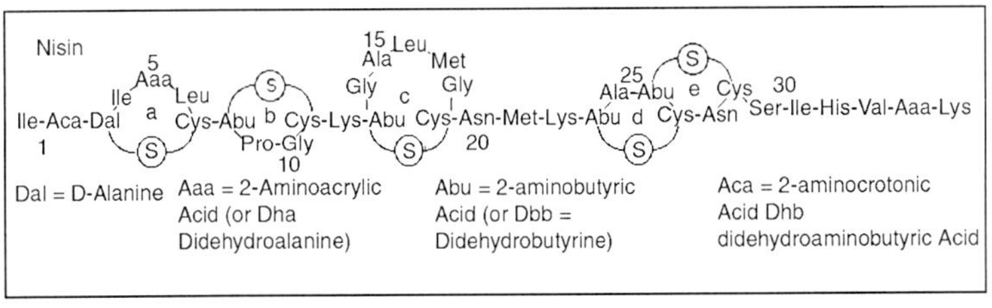

The bacterial cell wall is a cross-linked glycan polymer that surrounds bacterial cells, dictates their cell shape, and prevents them from breaking due to environmental changes in osmotic pressure. This wall consists mainly of peptidoglycan or murein, a three-dimensional polymer of sugars and amino acids located on the exterior of the cytoplasmic membrane.

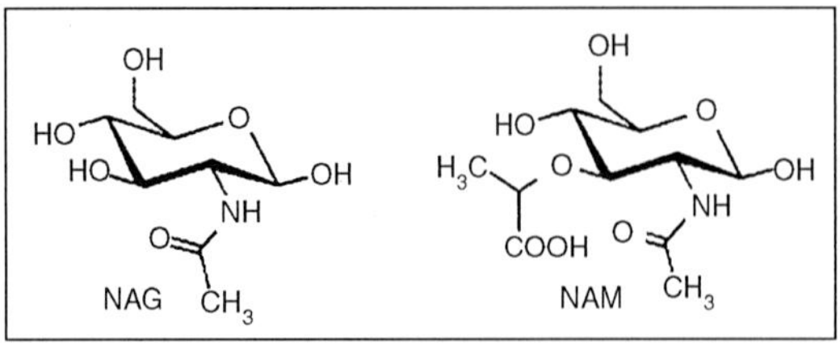

The monomer units are composed of two amino sugars, N-acetylglucosamine (NAG) and N-acetylmuramic acid (NAM), shown on the right. Transglycosidase enzymes join these units by glycoside bonds, and they are further interlinked to each other via peptide cross-links between the pentapeptide moieties that are attached to the NAM residues. Peptidoglycan subunits are assembled on the cytoplasmic side of the bacterial membrane from a polyisoprenoid anchor.

Lipid II, a membrane-anchored cell-wall precursor that is essential for bacterial cell-wall biosynthesis, is one of the key components in the synthesis of peptidoglycan. Peptidoglycan synthesis via polymerization of Lipid II is illustrated in the following diagram. Cross-linking of the peptide side chains is then effected by transpeptidase enzymes. A model of Lipid II complexed with nisin may be examined as part of the previous Jmol display.

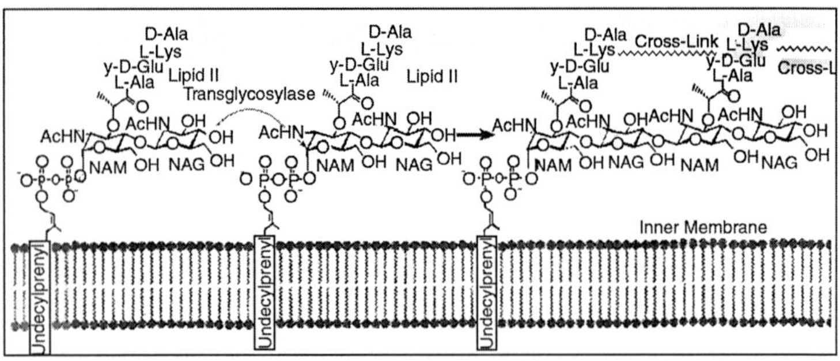

In order for bacteria to divide by binary fission and increase their size following division, links in the peptidoglycan must be broken, new peptidoglycan monomers must be inserted, and the peptide cross links must be resealed. Transglycosidase enzymes catalyze the formation of glycosidic bonds between the NAM and NAG of the peptidoglycan monomers and the NAG and NAM of the existing peptidoglycan.

Finally, transpeptidase enzymes reform the peptide cross-links between the rows and layers of peptidoglycan making the wall strong. Many antibiotic drugs, including penicillin, target the chemistry of cell wall formation. The effectiveness of choosing Lipid II for an antibacterial strategy is highlighted by the fact that it is the target for at least four different classes of antibiotic, including the clinically important glycopeptide antibiotic vancomycin.

The growing problem of bacterial resistance to many current drugs, including vancomycin, has led to increasing interest in the therapeutic potential of other classes of compound that target Lipid II. Lantibiotics such as nisin are part of this interest.

3

Vitamins

INTRODUCTION

Vitamins are a multibillion dollar industry.1 They are readily available to the public and are the focus of the most frequently asked questions to pharmacists.2 Surprisingly, most pharmacy students receive little training on the many roles of vitamins in nutrition.2 Therefore, this lecture attempts to not only reinforce fundamental biochemical and metabolic pathways in the human body, but also to provide pharmacy students with practical information.

The diet is a vast source of important nutrients. These nutrients include several important classes of biomolecules such as:

- Energy yielding components (carbohydrates, lipids, and proteins).
- Essential and nonessential amino acids.
- Essential fatty acids.
- Minerals.
- Vitamins.

Vitamins are organic substances that must be provided by the diet either because they cannot be biosynthesized or the amount that is provided through biosynthesis is inadequate for maintaining normal health. Vitamins are broadly divided into 2 classes based upon their hydrophobicity. The more hydrophilic vitamins are termed the water-soluble vitamins and are composed of the B-complex vitamins and vitamin C. The more hydrophobic vitamins, referred to as the fat-soluble vitamins, are composed of Vitamins A, D, E, and K. This article focuses primarily on the watersoluble vitamins due to their greater role in the major metabolic cycles, which are the primary focus of this course.

The normal North American diet is sufficient to prevent significant vitamin deficiencies and the related diseases associated with these deficiencies. However, there is increasing concern that slight vitamin deficiencies in a number of water-soluble vitamins (B1, B6, B12, folate and vitamin C) are risk factors for diseases such as such as depression, cancer, cardiovascular disease, and neural tube defects. The prevalence of this slight vitamin deficiency is likely related to diet, since most of these vitamins are supplied by fruit and vegetable intake. A recent survey has shown that only 25 per cent of the population meets their daily intake of 5 servings.3 Food preparation is another source of vitamin loss. For example, heating food for more than 2 hours causes more than a 10 per cent loss of most water-soluble vitamins. Refrigeration, freezing, and reheating have all been shown to lead to further loss of vitamins. Exposure to light causes significant loss of riboflavin from foods and the combination of heating and light can almost completely remove this vitamin from food.

Thiamin (Vitamin B1)

Thiamin functions in the body as thiamin pyrophosphate (TPP) an important enzyme co-factor.5 Thiamin reacts with adenosine triphosphate (ATP) to form thiamin pyrophosphate through a reaction mediated by the enzyme thiamin diphosphokinase. Following its production, TPP is incorporated into 2 important enzymes: pyruvate dehydrogenase and ketoglutarate dehydrogenase. Pyruvate dehydrogenase is part of a multi-enzyme complex that acts to convert pyruvate generated in glycolysis into acetyl-CoA for entry into the tricarboxylic acid (TCA) cycle. Thiamin pyrophosphate is also used as a co-factor for the enzyme -ketoglutarate dehydrogenase, which is a key point of regulation in the TCA cycle. -Ketoglutarate dehydrogenase is involved in the conversion of ketoglutarate to succinyl CoA. In the case of both enzymes, TPP assists in decarboxylation of a small ketoacid.

In addition to its uses in metabolism, thiamin may enhance circulation and blood formation. It is required for maintenance of the nervous system and is used in the biosynthesis of the neurotransmitters acetylcholinc and *hydroxybutyrate (GABA). Thiamin also is used in the production of hydrochloric acid in the stomach and, therefore, has a role in digestion.

Thiamin is provided in the diet through most grains. A deficiency of thiamin is called beriberi and it is most often observed in Southeast Asia.6 The main staple of diet in this part of the world is rice, which does not provide dietary thiamin to as great an extent as other grains. The symptoms of beriberi include limb pain, muscle weakness, and low cardiac output. all symptoms are related to the diminished capacity of the major energy producing pathways that are dependent on TPP as a co-factor.

Mild deficiencies of thiamin are sometimes observed in the elderly and in low-income groups on restricted diets. The earliest symptoms of thiamin deficiency are loss of appetite, constipation, and nausea. Other symptoms such as mental depression, peripheral neuropathy, irritability, and fatigue are related to the role of thiamin in maintaining a healthy nervous system. The most common form of thiamin deficiency in the United States is alcoholic neuritis. Because alcoholics normally have a poor appetite, overall food consumption is low. In addition, alcoholics are predisposed to developing nutritional deficiencies since alcohol is their primary calorie source. Under the broad umbrella of alcoholic neuritis come two disorders, Wernicke's Syndrome and Korsakoff s Psychosis.7 The symptoms of Wernicke's Syndrome (ophthalmoplegia, nystagmus, and ataxia) respond quickly to administration of thiamin, whereas the more severe memory and learning disorders associated with Korsakoff s Psychosis are refractory to thiamin treatment.

Riboflavin (Vitamin B2)

The reduced forms of these enzyme co-factors can donate these electrons to return to their previous fully oxidized forms. It is this ability to act as a conduit for electron transfer reactions that makes FAD and FMN such important enzyme cofactors.

Riboflavin is available from a wide variety of dietary sources such as milk, cheese, meat, eggs, and cereal products. Symptoms associated with riboflavin deficiency include sore throat, dermatitis, anemia, neuropathy, and cataract formation. Riboflavin deficiency is not normally observed in the United States; however, deficiency may be observed as part of a general case of malnutrition or in cases of chronic alcoholism.8 The conversion of riboflavin to FMN is required for absorption and transport into many tissues. The conversion has been shown to be inhibited by hypothyroidism and the structurally similar medications chlorpromazine, adriamycin, quinacrine, tetracycline, and tricyclic antidepressants.9-12

Niacin (Vitamin B3)

Niacin can be found in foods such as meats, breads, and beans. Mild niacin deficiencies have similar symptoms to those observed with riboflavin, which is not surprising due to the similar roles both of these vitamins play in biochemical reactions. Niacin deficiencies are occasionally observed in alcoholics, cases of general malnutrition, and in the elderly on restricted diets. A severe deficiency of niacin is known as pellagra, which is derived from the Italian phrase meaning rough skin. Pellagra is marked by dermatitis and also is notable for causing a blackening of the tongue. Early cases of pellagra were first observed in Europe shortly after the introduction of corn from the voyages of Christopher Columbus. In these cases, poor farmers who were raising corn as animal feed were particularly susceptible. Because niacin in corn is not bioavailable unless treated with a strong base such as lye, the farmers developed niacin deficiencies.

Europeans did not know this processing method until revealed to them by Native-Americans during the early colonial period of North America. However, the connection between pellagra and niacin was not known until the early 20th century. Pellagra was a significant health issue in the United States over the period from 1900-1940 resulting in over 100,000 deaths. Today most diets are supplemented with niacin through enriched flour, which receives its name because of the added niacm.13 Niacin can be administered in doses of 2g to 4g to causes a decrease in circulating levels of cholesterol and LDL. While the cholesterol lowering effects of niacin are desirable there are potential side effects from such large doses of this vitamin. The most immediate reaction observed from large doses of niacin is vasodilation resulting in flushing.

Over time there may be a reduction in fatty acid mobilization causing a depletion of glycogen and lipid stores in muscle tissue. Long-term exposure may also elevate blood glucose and uric acid levels, suggesting increased risk for patients who are on the borderline for diseases such as diabetes and gout. Prolonged use of high doses of niacin can lead to elevated levels of the serum enzymes alanine aminotransferase and aspartate aminotransferase, which may suggest liver damage.

Pyridoxine (Vitamin B6)

Pyridoxine is the precursor to the active enzyme cofactor pyridoxal phosphate (PLP). Pyridoxal phosphate is a critical co-factor for enzymes involved in reactions involving many amino acids. The N-terminus of the amino acid forms a covalcnt bond to PLP, allowing a wide variety of displacement reactions to occur at the alpha carbon. These include decarboxylations, transaminations, and transfers of side chains. Pyridoxine, therefore, plays a central role in the production of many neurotransmitters, such as serotonin, norepinephrine, and histamine. Pyridoxine also is important in the production of heme.

The enzyme LAAAD is present in both the gut and in the brain. This presents a problem because L-dopa is readily transported across the blood brain barrier while dopamine is not. Therefore, orally administered L-dopa is rendered ineffective in the gut by LAAAD. Co-administration of carbidopa resolves this dilemma by inactivating LAAAD in the gut. Like dopamine, carbidopa cannot cross the blood brain barrier leaving LAAAD in the brain free to carry out the production of dopamine at the site of action. In addition to hydrazine containing drugs, penicillamine, used in the treatment of Wilson's disease, cystinuria, and rheumatoid arthritis reacts with and inactivates pyridoxal phosphate.15 Patients treated with penicillamine occasionally develop convulsions, which can be prevented by supplementation with vitamin B6.

Pyridoxine is found in foods such as meats, breads, eggs, soybeans, and many vegetables. A deficiency in pyridoxine can cause facial lesions, depression, peripheral neuropathy, and glossitis. The neurological complications can be directly linked to the effects on neurotransmitter production. Mild pyridoxine deficiencies are sometimes observed in young women taking oral contraceptives. Deficiencies have also been observed in patients with gasteroenteritis or Crohn's Disease, presumably due to poor absorption. Mild deficiencies are of concern due to a correlation with an increased incidence of breast cancer and to the risk of coronary heart disease.16,17 Neurotoxicity has been noted with doses of vitamin B6 in excess of 500 mg/day.

Cobalamin (Vitamin B12)

Cobalamin is vital for cell growth and replication. Its major site of action is at the interface between the folic acid cycle and the active methyl cycle, where cobalamin is a co-factor for the enzyme homocysteine methyltransferase. This enzyme catalyzes the transfer of a methyl group from tetrahydrofolate to homocysteine, forming the amino acid methionine. These 2 cycles impact many other pathways due to the large number of memylation reactions in the body, especially nucleic acid biosynthesis and neurotransmitter biosynthesis. Cobalamin is found in meats and dairy products. Deficiencies are not common because the liver can store a 6-year supply of cobalamin. In addition, cobalamin is highly conserved by enterohepatic recirculation. Strict vegetarians (vegans) may take up to 20 to 30 years to develop a deficiency, whereas inadequate absorption from the ileum due to ileitis or loss of a glycoprotein that complexes with cobalamin prior to absorption may take from 2 to 10 years to become symptomatic.

The anesthetic nitrous oxide inactivates cobalamin and can cause patients with marginal serum levels to develop deficiencies within a week.19 While rare, deficiencies in vitamin B12 are severe and manifest in erythrocytes and nervous tissue. In erythrocytes, nucleic acid biosynthesis is slowed. This results in stem cells that are committed to development into erythrocytes possessing twice their normal protein content but lacking the necessary nuclear material for proper cell division. The resulting anemia, termed megaloblastic anemia, can be quite severe.

Cobalamin deficiency can also cause permanent damage to the nervous system. The results are swelling of neurons and demylination of nerve cells, followed by cell death. These progressive manifestations cause a wide range of neurological symptoms including unsteadiness, decreased reflexes, paresthesias of the extremities, and ultimately confusion, memory loss, hallucinations, and psychosis. The neurological symptoms of cobalamin deficiency may be mistaken for multiple sclerosis. In the elderly and in alcoholics, cobalamin deficiency should be considered as a possible cause of dementia even in the absence of anemia. The neurological symptoms associated with cobalamin deficiency are not directly related to irregularities in the formation of erythrocytes.20 Nutrient depletion or shortage may arise from inadequate food intake, impaired digestion, inefficient absorption or malabsorption, inefficient assimilation and other physiologic or functional disorders. Genetic errors may interfere with or prohibit assimilation and utilization. Gastrointestinal motility disorders may have similar effects. Stress may affect both gastrointestinal tract transit times and quantities or ratios of nutrient needs, leading to depleted nutrients. Americans now spend $2.5 billion yearly on vitamins and minerals, up from $328 million 25 years ago. Roughly 55 per cent of women and 45 per cent of men take some kind of vitamin supplement.

Most of the vitamin supplements consumed by Americans, including those used in"enriched" foods, are making them more deficient! The greatest and most enduring of medical discoveries of the twentieth century were the vitamins. Here, at last, was the discovery of factors that caused health. Vitamin A (1912, the first vitamin to be discovered) was named retinol because, without it, a healthy retina in the eye could not be formed. Then it was found to cure night blindness. And that was just the beginning.

Along came the B-vitamins and the cures for beriberi, pellagra, pernicious anemia, nerve degeneration, enlarged heart, energy production, and much more. These came from unprocessed, whole, natural foods. When man Realised his foods had been devitalized from processing and refinement, he decided to replace the lost vitamins. But, in his effort to extricate himself from the refined food trap,"scientific confusion" led him to try to replace them with"chemical, man-made vitamins." So, again, he refined down natural vitamins into pure, crystalline vitamins, thus robbing them of important food values, just as he had refined natural grains and sweets into their pure, crystalline, carbon-white flour and sugar.

Then, to save money, he duplicated these crystalline vitamins synthetically, mostly from COAL-TAR products. Thus, natural vitamins as well as natural foods were robbed of nature's life and valuable elusive nutrients. The latest round in conventional medicine's ongoing attempts to discredit (and ultimately outlaw) nutritional supplements is found in a highly questionable study published February, 2007 in the Journal of the American Medical Association, which claims that vitamins actually increase the risk of death. The study claims to have Analysed a collection of previous studies on Vitamin A, beta carotene, Vitamin C, Vitamin E and selenium, concluding that most of the nutrients are actually dangerous to human health. Of course, this is research from conventional medicine - an industry that promotes patented chemicals as perfectly safe, even though FDA-approved pharmaceuticals are killing at least 100,000 Americans each year. (Imagine the uproar if vitamins killed even a fraction of that number)

To avoid getting hoodwinked by questionable research on"vitamins," you have to strongly consider the financial interests of the source of this research. JAMA accepts millions of dollars in advertising from drug companies each year, and its pages are absolutely packed with drug ads. The American Medical Association, for its part, has long worked to discredit alternative medicine and has even been found guilty by U.S. federal courts of engaging in a conspiracy to destroy chiropractic medicine. The AMA, which is largely considered a joke by anyone familiar with natural health, is hardly a credible source for publishing scientific findings on nutrition. To protect the multi-billion dollar drug industry, it's likely the AMA would say practically anything. This research published in JAMA does remind us of one important point, however: synthetic chemicals are harmful to human health. If you take cheap"vitamins" made of these synthetic chemicals, you are doing yourself more harm than good. These cheap vitamin manufacturers, by the way, are usually owned by pharmaceutical firms.

Conventional medicine researchers try to blur the line between"junk vitamins" and"quality vitamins" by classifying ALL nutritional supplements as"vitamins," regardless of what they're really made from. By discrediting a few synthetic chemicals, they can effectively dissuade the masses from taking ANY vitamins, including the good ones. And that is, of course, their goal: to use F.U.D. (fear, uncertainty and doubt) to scare consumers away from nutritional supplements so that patients will flock to the patented, synthetic chemicals that earn drug companies billions of dollars in profits. Drugs make money for Big Pharma, and vitamins compete with drug sales.

Round up 100 people who are taking multiple pharmaceuticals, and compare their health to 100 people who are taking vitamins and nutritional supplements. Guess who's healthier? The supplement crowd will be healthier every time. It's the obvious question: If vitamins are so dangerous, where are all the dead vitamin takers? And if pharmaceuticals are so safe, where are all the super-healthy prescription drug patients? They are nowhere to be found. The healthiest people, by far, are those who take supplements, who engage in regular exercise, and who avoid taking prescription drugs.

Only fools believe research about nutrition that comes from the American Medical Association or its journal. Conventional medical researchers declaring that vitamins are worthless is about as credible as Bush Administration climatologists claiming there's no such thing as global warming. With the publication of this research, the distortion of health reality is now complete. According to the Medical Association, vitamins will kill you but pharmaceuticals will make you healthy.

VITAMIN-A

Vitamin-A was the first fat-soluble vitamin to be discovered. In 1919, it was recognized that egg yolk contained a fat-soluble substance essential for life. In 1924, it was demonstrated that the eye disease, xerophthalmia, in children could be prevented by feeding them butter or cod liver oil. Vitamin-A was purified--separated from its natural constituents--and its refined chemical structure proposed in 1931. Synthesis, an artificial chemically-pure duplication, was accomplished in 1947 at Hoffman-LaRoche, a pharmaceutical company in Switzerland. Vitamin-A is actually a whole family of compounds, including retinol, retinal, retinoids, carotenes, and carotenoids, and always occurs in nature with synergists such as fatty acids, chlorophyll, other vitamins, enzymes, minerals, and trace elements. Retinol, retinal, and retinoids are found in foods of animal origin such as eggs, liver, fish, butter, and cheese. The scientific term for vitamin-A is retinol, because of its presence in the retina of the eye.

The role of retinol in vision was elucidated by a number of brilliant scientists, beginning in 1877 with a German, W. Kuhne, who discovered that the purple retinas from dark-adapted frogs turned yellow when exposed to light. The purple Colour is restored in a complex biochemical cycle involving vitamin-A, which makes vision possible. Other scientists demonstrated the role of vitamin A in cell differentiation, bone development, reproduction and immune system function. Dr. Weston Price confirmed the value of vitamin-A in traditional diets during his studies of primitive peoples carried out during the 1930s and 1940s. Due to the outstanding scientific work of these and many other researchers, the administration of cod liver oil to growing children--a tradition found among Arctic peoples such as the Scandinavians and Eskimos--became standard practice until after the Second World War.

In recent decades, much vitamin-A research has focused on its role in preventing cancer, and its use in combination with nontoxic therapies in the treatment of cancer. Dr. Max Gerson treated many cases of terminal cancer with excellent results, using raw liver juice, a rich source of vitamin-A. Modern confinement farming practices effectively prevent vitamin-A from incorporation into animal foods and the processing industry would rather use vegetable oils than animal fats. Some vegetable oils contain carotenes but they do not contain true vitamin-A. Only animal fats contain vitamin-A and vitamin-A is present in large amounts only when the animals have a source of carotenes or vitamin-A in the diet, such as green pasture, insects, and fish meal.

Many popular books on nutrition insist that humans can obtain vitamin-A from fruits and vegetables. Even worse, FDA regulations allow food processors to label carotenes as vitamin-A. Vitamin-A from butter may be two or three times as potent, unit for unit, as vitamin-A from cod liver oil in some people. The food industry, and the low-fat school of nutrition that the industry has spawned, benefit greatly from the fact that the public has only vague notions about vitamin-A. In fact, most of the foods that provide large amounts of vitamin-A--butter, egg yolks, liver, organ meats and shellfish--have been demonized.

Under optimal conditions, humans can indeed convert carotenes to vitamin-A. This occurs in the upper intestinal tract by the action of bile salts and fat-splitting enzymes. Of the entire family of carotenes, beta-carotene is most easily converted to vitamin-A. The accepted ratio of how many units of beta-carotene is needed to produce one unit of vitamin-A is 6:1 and some research suggests an even higher ratio. This means that you have to eat an awful lot of vegetables and fruits to obtain even the daily minimal conversion. But the transformation of carotene to retinol is rarely optimal. Strenuous physical exercise, periods of physical growth, pregnancy, lactation and infection are stresses that quickly deplete vitamin-A stores.

Diabetics and those with poor thyroid function, which could include at least half the adult U.S. population, cannot make the conversion. In fact, the thyroid gland requires more vitamin-A than the other glands, and cannot function without it (mercury toxicity diminishes the thyroid and pituitary glands' functions). And a diet rich in vitamin-A will help protect the diabetic from the degenerative conditions associated with the disease, such as problems with the retina and with healing (mercury binds with the sulfur in insulin).

Children make the conversion very poorly and infants not at all--they must obtain their precious stores of vitamin-A from animal fats--yet the low-fat diet is often recommended for children. Vitamin-A from animal sources (retinol) is primarily absorbed within three to five hours after ingestion in the upper intestinal tract; it is here that the fat-splitting enzymes and bile salts convert carotenes into usable nutrients, though conversion and assimilation takes up to six or seven hours. Carotenes are converted by the action of bile salts (mercury inhibits production of taurocholic acid, a bile salt) and very little bile reaches the intestine when a meal is low in fat. The conversion of carotenes is stimulated by thyroxine, a hormone secreted by the thyroid gland. Zinc, iron, and vitamin-E status also influences carotene conversion. Once converted, the carotenes are absorbed in the same way as retinol and allocated by the liver. Unabsorbed carotene is excreted in the feces.

Butterfat stimulates the secretion of bile needed to convert carotenes from vegetables into vitamin-A, and at the same time supplies very easily absorbed true vitamin-A. Polyunsaturated oils also stimulate the secretion of bile salts but can cause rapid destruction of carotene unless antioxidants are present. It is very unwise, therefore, to depend on plant sources for vitamin-A. This vital nutrient is needed for the growth and repair of body tissues; it helps protect mucous membranes of the mouth, nose, throat and lungs; it prompts the secretion of gastric juices necessary for proper digestion of protein; it helps to build strong bones and teeth and rich blood; it is essential for good eyesight; it aids in the production of RNA; and contributes to the health of the immune system.

Strenuous physical exercise, excessive consumption of alcohol, excessive consumption of iron (especially from"fortified" white flour and breakfast cereal), use of a number of popular drugs, excessive consumption of polyunsaturated fatty acids, zinc deficiency and even cold weather can hinder the conversion of carotenes to vitamin-A, as does the low-fat diet.

Natural vitamin-A provided by liver, eggs, butter, cream and cod liver oil is well recognized as providing excellent protection against birth defects. Vitamin-A deficiency in pregnant mothers results in offspring with eye defects, hydrocephalus, displaced kidneys, harelip, cleft palate and major malformations of the heart and large blood vessels. Vitamin-A stores are rapidly depleted during exercise, fever and periods of stress. Vitamin-A deficiencies are widespread and contribute to high infant mortality, blindness, stunting, bone deformities and susceptibility to infection.

These occur even in communities that have access to plentiful carotenes in vegetables and fruits. It is required for cellular differentiation (determines the function that cells will have): this assures that the cells which are lost from natural turnover, stress, insult, injury, disease, etc. are reproduced in the exact same form as the ones being replaced. Growing children actually benefit from a diet that contains considerably more calories as fat than as protein. Generous amounts of vitamin-A insure healthy reproduction and offspring with attractive wide faces, straight teeth and strong sturdy bodies.

This vitamin assists in normal pregnancy, embryonic development, successful reproduction, fertility, lactation (nursing), and reproductive organ function, spermatogenesis, to adrenal, thyroid, and other gland functions. Vitamin-A is important for normal eyesight; necessary for night vision. A high-fat diet that is rich in vitamin-A will result in steady, even growth, a sturdy physique and high immunity to illness. Vitamin-A-rich foods like liver, egg yolk, cream and shellfish confer resistance to infectious diseases in children and prevent cancer in adults.

Children with measles rapidly use up vitamin-A, which can result in irreversible blindness. An interval of three years between pregnancies, allows mothers to rebuild vitamin-A stores so that subsequent children will not suffer diminished vitality. Kwashiorkor is as much a disease of vitamin-A deficiency, leading to impaired protein absorption, as it is a result of absence of protein in the diet. High-protein, low-fat diets are especially dangerous because protein consumption rapidly depletes vitamin-A stores. Children brought up on high-protein, low-fat diets often experience rapid growth.

The results--tall, myopic, lanky individuals with crowded teeth, and poor bone structure, are commonplace in America. High-protein, low-fat diets can even cause blindness. Scarcity of good quality raw dairy products, a rejection of organ meats as old fashioned or unhealthful, and a substitution of vegetable oil for animal fat in cooking all contribute to the physical degeneration and suffering of Third World peoples. Supplies of vitamin-A are so vital to the human organism that mankind is able to store large quantities of it in the liver and other organs. Even people who can efficiently convert carotenes to vitamin-A cannot quickly and adequately replenish vitamin-A stores from plant foods. Many factors interfere with its absorption and utilization. Inadequate fat in the diet, poor production of bile salts, low enzyme status, and compromised liver function can all interfere with the uptake and usage of vitamin-A, especially when given as a supplement in the form of retinol, rather than as a component of whole foods. One tablespoon of cod liver oil contains at least 15,000 IU and one serving of liver (should be from organically raised animals) can contain up to 40,000 IU vitamin-A.

Foods high in vitamin-A are especially important for diabetics and those suffering from thyroid conditions. Weston Price considered the fat-soluble vitamins, especially vitamin-A, to be the catalysts on which all other biological processes depend. Efficient mineral uptake and utilization of water-soluble vitamins require sufficient vitamin-A in the diet.

Synthetic Vitamin-A

Synthetic vitamin-A interferes with the proper utilization of natural vitamin-A from foods. Pure retinol is added to many fabricated foods like margarine, white flour, extruded breakfast cereals and pizza. A synthetic retinoic acid derivative,isotretinoin (Accutane) is used to treat acne as well as psoriasis, cancer, and oral leukoplakia. Children whose mothers took Accutane during pregnancy were born with nervous system defects, brain damage, congenital heart disease, and more. Depression, crying spells, malaise, and forgetfulness are reported human symptoms.

Inflammation of the lips (cheilitis) develops in 90 per cent of the users; conjunctivitis develops in 40 per cent. Muscle damage has also been added to the list. The synovial membranes in joints are supported by vitamin-A. The vitamin-A complex is supportive to the adrenal glands which produce cortisone and other needed hormones; and vitamin-A plays an important role in any inflammation and repair process. Vitamin-A is important to the external and internal mucous membrane tissues as well as to the immune system. However, synthetic retinoic acid metabolites do not reveal any specific supportive mechanism and results are not seen with synthetic, altered fractions.

The whole vitamin-A complex is known to be very functional, but when distilled or synthetic fractions are used, roles and results are questionable at best. Megadoses of synthetic vitamin-A can result in hypertrophy (excessive growth) of the liver and spleen, with elevation of AST (aspartate aminotransferase), a liver enzyme. Liver fibrosis (abnormal formation of fibrous tissue), fat deposition, obstruction of portal blood flow with portal hypertension (high blood pressure) and sclerosis(hardening) of the blood vessels have been found at liver biopsies. Excess synthetic vitamin-A is also associated with calcium deposits in soft tissues, contributes to attacks of gouty arthritis, hair loss, irregular menses, emotional lability, insomnia, restlessness, night sweats, and more. Vitamin-A is essential to spermatogenesis, the production of sperm in men, and it requires alcohol dehydrogenase, an enzyme to convert retinol to retinal. This enzyme may require the trace element, zinc, as well. Excessive doses of synthetic fractions of vitamin-A can convert to a compound that can be more damaging than a deficiency in its effect on the germinal epithelium of the testes. Vitamin-A is present in almost twice the amount in seminal fluid of highly fertile men compared with that of men deficient in sperm. Zinc increases vitamin-A uptake by the seminiferous tubules of the testes. Vitamin-A is bound to a protein, called retinol-binding protein. The necessity of enzymes, trace mineral activators, amino acids, and other constituents stresses the point that the whole food complex is important and effective whereas an isolated portion cannot produce the same results. Excessive fractionated"vitamins" A and E may produce testicular hemorrhaging. Zinc is required to maintain normal concentrations of vitamin-A in the blood plasma. Diabetic persons tend to have decreased levels of zinc. Since a zinc deficiency may lead to difficulties mobilizing vitamin-A from the liver for use elsewhere in the body, vitamin-A deficiency will worsen the condition. Another essential constituent is fat, as vitamin-A occurs in foods with fatty acids and is a fat-soluble vitamin. It is absorbed much better when natural, unaltered fats are also present. Other fat-soluble vitamins occur with and work with whole vitamin-A complex such as vitamin E complex and essential fatty acids.

When vitamin E is deficient, one develops low blood levels of vitamin-A; when vitamin E is restored to the diet, vitamin-A levels become normal. Yet, using chemically-pure fractions of"vitamins" A and E in large doses creates more problems, more imbalances. Fractionated alpha-tocopherol (so-called vitamin-E) can supplant vitamin A in the body, reducing its effectiveness. Administration of the whole food complex brings excellent responses. Stubborn cases of gastrointestinal irritation and even ulceration in the digestive tract that failed to respond to other treatments have totally resolved after use of real vitamin-A complex. Persons with cystitis (inflammation of the bladder) experience relief after a relatively short time taking food concentrates of Vitamin-A and other supportive nutritional complexes. Edema caused by liver pathology is another instance where vitamin-A complex is of particular value; the liver is the chief location of vitamin-A stores. Colds, chronic rhinitis, sinusitis, psoriasis, and many other conditions--especially those involving epithelial tissues--are responsive to the whole vitamin-A complex. A new strategy that has been the development of and promotion of genetically engineering rice to produce carotenes. Golden rice containing carotenes can't provide true vitamin-A to the world's children but it will further the trend of pushing their parents off the farm and into ghastly slums. The most effective response that consumers have is to boycott processed foods!

VITAMIN B-COMPLEX

The B-COMPLEX factors naturally divide themselves into two groups. One group consists of the factors that are alcohol and water soluble. This includes B1, B12, Pantothenic Acid and Adenine, but the primary factor in this group is called Vitamin B4, an anti-paralysis vitamin. It is one that has not yet been recognized by the FDA as being necessary in human nutrition.

They say it helps pigeons, but not people. So vitamin B4 is a nerve-promoting factor, it promotes the transmission of nerve impulses. This would be needed with heart conditions involving irregular heartbeat, fibrillation, or heart blocks. The B4 factor also causes vessels to constrict. It's very useful for people with low blood pressure--the person whose blood vessels are just hanging open, with no tone. The B4 puts strength in the vessels and snaps them into action by improving the transmission of nerve impulses to the vessels. For this reason, vitamin B would be contraindicated in cases of organic heart trouble, such as angina. Here you would need the G factor. The G factors are not soluble in absolute alcohol. Complex G contains the lipotropic factors that metabolize fat. Complex-G contains B2, B3, B6, Choline, Betaine, Inositol, Folic Acid, Biotin and PABA, and is a vasodilator--it opens up the blood vessels by relaxing the nerves in the vessels. So complex G is for a person with high blood pressurethe tense, coronary type person. It is important to have these factors separated and useable either by themselves or together. This benefit is not achieved with megadoses of synthetic so-called Vitamin-B. Today, the average practitioner of the medical arts, if sufficiently brain washed by the medical-drug-food-dairy-chemical combine of disinformation, cannot conceive that any person could suffer from B-complex deficiency.

Even if they know that refining and processing of food, chemical fertilizers and food additives, all diminish vitamin content and nutrient quality. Because these"foods" are"enriched" with synthetic vitamins. University biochemists and leading health authorities parrot the phrase:"There is no difference between natural and synthetic vitamins." Below are diagrams of the structural formulae of NATURAL vitamin B1 (left) and synthetic B1 (right). Show any high-school chemistry student with a passing grade the diagrams and tell them they are the same, and that the body does not know the difference. They will laugh in your face. Now, leading nutritionists from the university fountainheads of knowledge will tell you:"there is no difference."

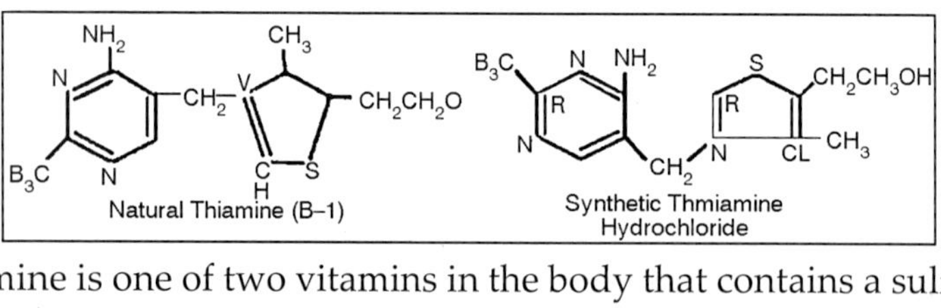

Natural Thiamine (B–1)

Synthetic Thmiamine Hydrochloride

Thiamine is one of two vitamins in the body that contains a sulfur entity; the other is biotin. The high binding affinity that mercury has for the sulfur-containing molecules (thiols and disulfides) gives it the biochemical potential to affect the functions of this critical vitamin and affect critical energy processes in the body. Not only does thiamine have a sulfur atom in its structure to start with, but its involvement as a coenzyme in critical biochemical processes is connected through transfers involving sulfur molecules. Out of about two-dozen substances produced by the metabolism of thiamine in the body and excreted in the urine, science has only identified the function of six of them thus far. There is a lot going on metabolically with vitamin B1, that mercury can affect, that haven't even been considered as potential problems because the complete biochemical function of the vitamin has not been identified. Thiamine is involved in many enzyme reactions in the body. As a coenzyme, B1 is involved in the intermediary metabolism of carbohydrates in all cells of the body. For natural vitamin B1 to perform as a coenzyme, it must be phosphorylated or combined with two molecules of phosphoric acid. As such, it then functions as thiamine pyrophosphate (TPP).

In the biochemically active form, it is also called diphosphothiamine (DPT). Better textbooks of biochemistry use the TPP designation as preferable. Early symptoms of thiamine deficiency, are intriguingly similar to mercury toxicity and are non-specific and hard to pinpoint: fatigue, anorexia, weight loss, gradual loss of muscle strength, peripheral neuritis (numness or increased sensitivity, tingling in the extremities, loss of reflexes), irritability, confusion, depression, lack of initiative, poor memory, gastrointestinal problems (including constipation), and low body temperature. Cardiovascular symptoms can include edema of the ankles, feet and legs, palpitation, tachycardia, abnormal electrocardiograms, decreased blood pressure, and difficulty in breathing after exertion.

Thiamine participates in the metabolism of tryptophan to form nicotinic acid. Because of this, a deficiency of thiamine can cause a deficiency of vitamin B6, which is also required in the metabolic pathway of tryptophan. Thiamine is a cofactor in steroid hydroxylation, fatty acid synthesis, and glucose metabolism in the pentose phosphate cycle, which is another energy system. Mercury inhibits glucose-6-phosphatase, a key enzyme in the glucose cycle. The proportion of glucose metabolized by this route is high in the lactating mammary gland, the adrenal cortex, leukocytes and erythrocytes. Each one of the mentioned also are affected by mercury in ways other than glucose metabolism. Not only mercury, but arsenic and lead impact on the biochemical pathways of these vitamins and minerals. If a person is deficient in thiamine or B1, most likely they are also lacking in all members of the B-complex.

VITAMIN E

The chemical name for vitamin E is tocopherol, which is derived from the Greek tokos (childbirth) and pherin (to bear). The ending ol is the chemical suffix to denote an alcohol. The name tocopherol was bestowed on this vitamin in 1938. Vitamin E is a fat-soluble vitamin, as are vitamins A, D, and K. Fat-soluble vitamins can be stored in the body whereas water-solubles cannot. Fat-soluble vitamins are excreted chiefly by the fecal pathway versus the urinary pathway for the water-soluble vitamins. They are absorbed along with dietary fats, and conditions of extremely low fat intake or impaired uptake of fats will also interfere with their absorption. Antibiotics and certain other drugs, as well as certain disease states such as malabsorption syndromes, decrease the absorption of fat-soluble vitamins from the intestinal tract. One of the primary biological functions of vitamin E relates to its role as an antioxidant. Vitamin E's antioxidant ability is enhanced by selenium. This synergistic effect preserves membranes from destruction by oxidation products and especially retards breakdown (hemolysis) of red blood cells. Vitamin E can reduce the toxic effects of mercury. Vitamin E is able to reduce the chromosomal breakage caused by mercury. Vitamin E also has sulfhydryl-protective activity. When vitamin E is involved in reducing the toxic effects of mercury, there is less available for normal metabolic functions.

Vitamin E protects the enzymes G6P, G6PD, GSH-Px and glutathione reductase from auto-oxidative damage and at the same time protects the biological supply of glutathione. The presence of mercury in combination with a vitamin E deficiency has the potential of producing toxic reactions in the body. The primary lymphoid organs are major targets of selenium and vitamin E dietary deficiencies, contributing to impaired immune function. Vitamin E is not a cofactor, such as the trace elements manganese, copper, zinc, and selenium, it is an independent antioxidant that has its main effect in the lipid phase of the cell, protecting the membrane from auto-oxidative damage.

A distinguishing feature of cystic fibrosis is fat malabsorption. Cystic fibrosis patients usually have a deficiency of vitamin E. Patients with G6PD deficiency have less capacity to generate reduced glutathione, and their red blood cells have a greatly reduced half-life. Vitamin E supplementaton can decrease the level of neonatalhyperbilirubinemia due to G6PD deficiency. Mercury inhibits G6PD activity and also reduces available vitamin E. Alpha and delta tocopherols or mixed tocopherols are not vitamin E. They are merely antioxidant or protective agents (banana peel) for the more complex parts of the vitamin E assemblage (xanthine, phospholipids, lipositols, sex hormone precursors)--see functional architecture of vitamin E. There are seven tocopherols in all, alpha being one of them.

The government has decided to rate any vitamin E product according to the amount of alpha tocopherol it contains. The tocopherols are antioxidants, just like ascorbic acid. They are a part of the E complex, but they are not the active ingredient. Their function is to preserve the active factor. Tocopherols are measured on the basis of cell growth. The entire E complex contains the polyunsaturated fatty acids. (Vitamin F) Vitamin F is part of the E complex. Also, vitamins A and K. There are some forms of vitamin D and manganese in the E complex, as well. The traditional source for vitamin E is vegetable oils. But the best quality vitamin E comes from vegetables. Lettuce is one of the best sources and the pea plant is equally as good. If you take any Labouratory-chelated products, your body has to do some combining to make them work. But if you take your minerals the way nature provides them, combined with protein and vitamins through the growing process of plants, they are bound to work. For example, we could take vitamin E and separate the tocopherols from the manganese. Putting them back together in the Labouratory--chelating them artificially--we would find they wouldn't work like they did before you took them apart. If you take a watch apart piece by piece and Analyse it and then throw all the parts back together, it isn't going to work! Although it is the same watch, it has now lost its function. A nutritional concentrate is an organized mechanism that has a function. If you break the function by separating the parts, the body can no longer use the mechanism. High potency vitamin E is made this way. Tocopherols are broken down and separated from oils--the natural tocopherols are separated from vegetable oil. Then they are packed into a gelatin perle, and you have a high potency vitamin E product--a 220 or 500 International Unit (IU) Vitamin E product. That means 200 or 500 International Units) of alpha tocohpherol--the potency of the product is measured by the amount of alpha tocopherol it contains. Although the label of such a product will say"all natural" or"organic""--it was taken from oil, a natural source--it's still not food.

It's an incomplete product. You're getting lots of tocopherols, but none of the active components of the E complex; the F, A, K, D, manganese, selenium, etc. The Vitamin E2fraction (prostaglandin) is found in beef chromatin material, which is glandular substance. It's a food factor that will stop a heart cramp or a heart pain, usually within about ten minutes. Most cramps are due to calcium deficiency. So E2 gets calcium into the area to relax the muscle. It can be used to relieve any muscle cramp, but it's especially effective to relax cramps in the muscles of the heart. Natural forms of vitamin E complex lose up to 99 per cent of their potency when separated from their natural synergists (Ann. Review Biochemistry, 1943, page 381). These synergists include tannins, fatty acids (vitamin F complex), phospholipids, and other synergistic factors. Chemically purified vitamin E (tocopherols), in high unit doses, reverses its effect and produces the same symptoms as a deficiency (including bone decalcification). Without selenium (the trace mineral activator of vitamin E), vitamin E cannot perform many of its functions.

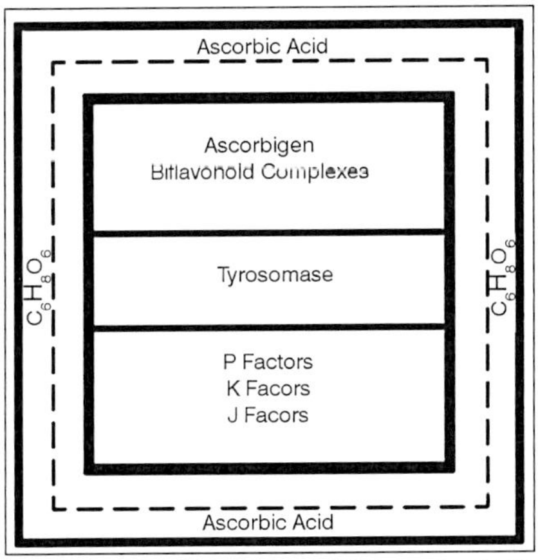

Fig. Function Architecture of Vitamin C-Complex

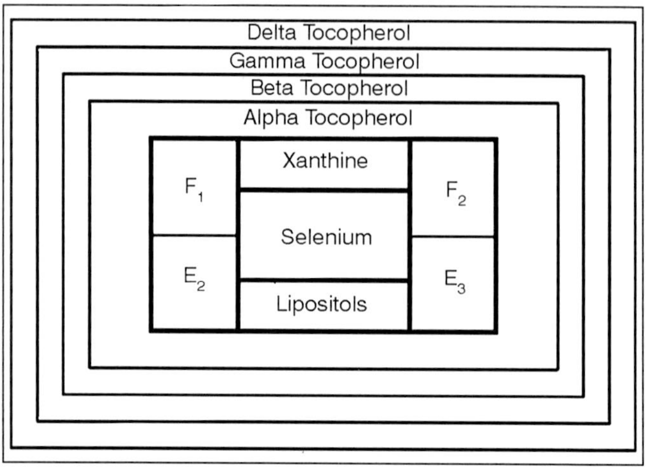

Fig. Functional Architecture of Vitamin E-Complex

Xanthine increases the antioxidative function of the tocopherols as much as 50 per cent. Xanthine plays an important role in the metabolism of purines (nucleoproteins) within the cell. In one instance, hypoxanthine is transformed biochemically to xanthine with the metabolic fate proceeding to uric acid in two forms; enol and keto. Intracellular utilization of purines is dependent upon the said contribution.

Lipositols are essential to cholesterol metabolism in order that cholesterol be altered to a biochemical acceptable to the synthesis of adrenal and sex hormones. F1 is the active form of unsaturated fatty acids described in the Annual Review of Biochemistry, 1949. Vitamin F restores and maintains Calcium in tissue fluids and is anticarcinogenic. F_2 is part of the phospholipid complex in nerve structures; particularly the myelin sheath. It also promotes tissue repair and enhances the benefits of the antiarthritic Wulzen factor. F_2 could very well be the missing link in pernicious anemia according to Dr. Royal Lee. E_2 is a part of E complex that has been compared to nitroglycerine in its function of relieving angina-type pains--the hypoxic spasm of the heart muscle. E_2 reduces pain because it increases the oxygen conserving factor of the blood. The E_3 component of Vitamin E complex is a precursor of sex hormones.

Drugs in five major groups have been shown either to function as vitamin B_6 antagonists, or to increase the turnover of vitamin B_6 in the body. Drugs in ten major groups have been shown to affect the absorption of folic acid; to act as folate antagonists; or to increase the turnover or loss of folate from the body. Drugs in four groups have been shown to affect the absorption of vitamin B_{12}. Depletion of one vitamin can affect the requirement for another vitamin. Mercury can cause similar deficiencies or metabolic changes in the way the body handles these same three vitamins. While man can isolate from foods known factors comprising about 99 per cent plus of total weight, we cannot maintain normal nutrition or"life" by consuming isolated factors. Nature still withholds that"something" that sustains"life." One of the deceptive practices of manufacturers of pseudo-nutrients is to label crystalline-pure and synthetic vitamin fractions as COMPLEXES. The meaning of complex is simply: the sum or combination of various things. Another, but more precise, definition is: Consisting of various connected or interwoven parts; composite. As a result, the consumer buys a"vitamin" product that says on the label:"Vitamins From Natural Sources (usually synthetic, from corn sugar and coal tar) with Bioflavonoids or Rutin." Legally, this can be called a"Vitamin Complex" but it is really no more than a mixture of two or more dead, refined chemicals--not a required for life,"biochemical." Remember, the human anatomy is composed of 100 trillion cells that need 75+ nutrients every day and, in turn, it can require six to eight months or more to overcome a nutritional deficiency. If you try and fool mother nature, she'll make a fool out of you.

4

Carbohydrates

INTRODUCTION

The activities on which the life of the body depends involve a continuous expenditure of energy and a constant exchange of material. Ultimately the body is dependent upon food for the fuel materials which supply energy and for both the substances which are transformed in, and eliminated from, the body, and those whose presence regulates and controls these transformations. The materials leaving the body are to be regarded not merely as wastes but as end products of an orderly and coordinated series of chemical reactions which occur in the body and by virtue of which its functions are performed.

Thus the chief functions of food are:

- To yield energy.
- To build tissue.
- To regulate body processes.

These functions involve reactions which are dependent upon the chemical composition and constitution of the food. Any food constituent which takes part in any of these functions may be regarded as having nutritive value. Most of the nutrient material contained in food requires more or less change to bring it into the exact forms most useful in nutrition. These changes as a rule take place in the digestive tract and together constitute the process of digestion.

The changes which take place in the foodstuffs, after they have been absorbed from the digestive tract, are included under the general term"metabolism." Although the chemical changes and the energy transformations are of course inseparable, it has become customary to speak of the metabolism of matter and the metabolism of energy, and to regard the extent of the metabolism of any material substance as measured by the amount of its end products eliminated, and the extent of the energy metabolism as measured by the amount of heat, or of heat and external muscular work, which the body gives off. The metabolism of matter and the metabolism of energy are normally supported by the food; but if no food is taken, they continue at the expense of the body substance. The expenditure of energy can never cease in the living body because it includes the work involved in carrying on the internal processes which are essential to life itself; and the expenditure of matter cannot cease because the energy for this necessary work is obtained by the breaking down of the organic compounds of the food or of the body substance into simpler compounds, many of which are of no further use to the body and must be eliminated. When the food supplies sufficient energy, the body substance is protected; when the food is insufficient, body substance is burned as fuel.

In order, then, to consider intelligently the nutritive requirements of the body as regards the substances of which it is composed, it is necessary first to know whether the fuel requirements (the requirements of the energy metabolism) have been fully met. The carbohydrates, fats, and proteins of the food all serve as fuel to yield the energy required for the activities of the body, and the proteins serve also as material for the maintenance or growth of body tissue. But of the fifteen chemical elements which are essential to the structure and functions of the body, simple proteins furnish only five. The remaining ten elements are largely constituents of the ash of the food and are known as ash constituents, inorganic foodstuffs, mineral matter, or salts.

Recent investigations have developed the fact that food of sufficient energy value and containing ample amounts of each of the chemical elements known to be essential to the body is not necessarily adequate to meet all the requirements of nutrition. Thus it appears that certain substances occurring in natural foods but not yet chemically identified are also to be included among the nutritive requirements of the body and therefore among the factors which determine the nutritive values of foods. At present these unidentified substances are referred to as"vitamines" or as"fat soluble A" and"water soluble B." *The essentials of a chemically adequate food supply may therefore be summarized as follows*:

- Sufficient of the organic nutrients in digestible forms to yield the needed energy.
- Protein, sufficient in amount and appropriate in kind.
- Adequate amounts and proper proportions of the various ash constituents or inorganic foodstuffs.
- Sufficient of each of the two unidentified"vitamine" factors, the"fat soluble A" and the"water soluble B."

In attempting to give in the following pages a general view of the chemistry of food and nutrition it has seemed best to discuss first the chemical nature and nutritive functions of the substances which serve as sources of energy in nutrition, then the nutritive requirements in terms of energy, protein, the more prominent"inorganic" elements, and the"vitamines," and finally the bearing of these various factors offood value upon problems connected with the economic use of food. Of the constituents of the ordinary mixed food of man the carbohydrates are usually the most abundant and the most economical sources of energy. They are also considered to be the first of the three great groups of foodstuffs to be formed by synthesis from simple inorganic substances in plants;"in the long run, all the energy of living matter comes from them." The synthesis of carbohydrates in nature is therefore a logical starting point for the study of the organic foodstuffs.

In the chlorophyll cells of the leaves of green plants the energy of the sun's rays brings about reaction between carbon dioxide and water which results in the liberation of oxygen and the formation of organic compounds. There is still doubt as to the exact mechanism of the process and no certainty that it is the same in all cases. It has, however, been quite generally found that the volume of oxygen liberated is equal to that of carbon dioxide consumed.

The simplest possible representation of the reaction would be $CO_2 + H_20 \rightarrow CH_20 + O_2$ according to which the first product of the synthesis would be formaldehyde. There is considerable (though not conclusive) evidence that formaldehyde is thus formed and that it is rapidly built up into less reactive compounds.

Whatever the steps in the process* there is normally an early production of carbohydrate. Usually the first product which can be demonstrated as accumulating in the plant as the result of the photosynthesis is a sugar (glucose or sucrose) or starch. Assuming glucose as a typical product and neglecting the intermediate stages, the photosynthesis of carbohydrate may be represented thus:

$$6\ CO_2 + 6\ H_{20} \circledR C_6H_{12}O_6 + 6O_2$$

Glucose is the most familiar representative of a group of simple sugars (monosaccharides or monosaccharoses) which are in composition direct polymers of formaldehyde (CH20) and which are classified, according to the number of carbon atoms in the monosaccharide molecule, as trioses, pentoses, hexoses, etc.

CLASSIFICATION

Definitions of the term"simple sugar" vary somewhat, depending chiefly upon the views of different authors as to how simple a compound may properly be called a sugar.

According to Browne, a simple sugar or monosaccharide may be defined as an aldehyde alcohol or ketone alcohol of the aliphatic series, the molecule of which contains one carbonyl and one or more alcohol groups, one of the latter being always adjacent to the carbonyl group. All simple sugars contain, therefore, as a characteristic group upon the presence of which the chief chemical properties of the sugars depend.

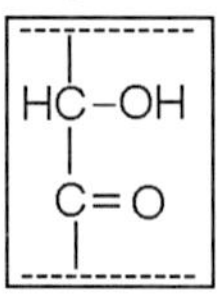

The simplest possible sugar according to this definition is glycolaldehyde, $CH_2OH\text{-}CHO$, which (in analogy with the nomenclature of the familiar sugars) may also be called glyco-lose. The structural formulae of glucose and fructose, the most familiar representatives of the aldehyde-alcohol (aldose) and ketone-alcohol (ketose) sugars, respectively, are as follows:

Glucose

$$\begin{array}{c} CH_2OH \\ | \\ HOCH \\ | \\ HOCH \\ | \\ HCOH \\ | \\ HOCH \\ | \\ HC = 0 \end{array}$$

Fructose

$$\begin{array}{c} CH_2OH \\ | \\ HOCH \\ | \\ HOCH \\ | \\ HCOH \\ | \\ C = 0 \\ | \\ CH_2OH \end{array}$$

Since glucose gives aldehyde reactions but not so readily as the above structural formula would lead one to expect, it is believed that in ordinary solutions of glucose the substance exists partly in the condition indicated by the aldehyde formula and partly in a tautomeric form represented by the lactone or"oxygen bridge" formula. Following are the aldehyde and lactone formulae written without reference to the spatial relationships of the hydrogen and hydroxyl groups:

Aldehyde form

$$CH_2OH\text{-}CHOH\text{-}CHOH\text{-}CHOH\text{-}CHOH\text{-}CHO$$

Lactone form

$$CH_2OH\text{—}CHOH\text{—}\underbrace{CH\text{—}CHOH\text{—}CHOH\text{—}CHOH}_{\text{—O—}}$$

The name monosaccharide ("single sugar") implies that the monosaccharide molecule contains only one sugar radicle - that it cannot be split by hydrolysis into sugars of lower molecular weight. A substance like cane sugar which on hydrolysis splits to two molecules of simple sugar is called adisaccharide or disaccha-rose ("double sugar"). Trisaccharides and tetrasaccharides are also known.

Substances which like starch are of high molecular weight and on complete hydrolysis yield many molecules of simple sugar are called polysaccharides * or polysaccharoses. The term"carbohydrates" covers all the simple sugars and all substances which can be converted into simple sugars by hydrolysis. The term"glucosides" is applied to substances which consist of combinations of carbohydrate radicles with radicles of other kinds and which therefore yield on hydrolysis both a simple sugar and one or more products of other than carbohydrate nature.

CARBOHYDRATES MONOSACCHARIDES

- (Monosaccharoses) Dioses ($C_2H_4O_2$) - Glycolose.
- Moses ($C_3H_6O_3$). Aldoses - Glycerose. Ketose - Dioxyacetone.
- Tetroses ($C_4H_8O_4$).
- Aldoses - Erythrose,3 Threose.3 Ketose - Erythrulose.2
- Pentoses ($C_6H_{10}O_5$).
- Aldoses - Arabinose,2 Xylose,2 Ribose,2 Lyxose.8 Ketoses
- - Araboketose3, Xyloketose (ketoxylose).2 [Methyl pentoses ($C_6H_{10}O_5$) - Rhamnose,2 Fucose2].

Some writers use the term polysaccharides to include all carbohydrates other than monosaccharides. Mathews applies it to all carbohydrates more complex than the disaccharides. Names of a few of the most important carbohydrates are printed in small capitals. Separate mention of the d, l, and dl forms of the various sugars is omitted, since in the study of food and nutrition we are practically concerned only with that one of the three forms which is found in or derived from natural products.

- Hexoses ($C_6H_{12}O_6$).
- Aldoses - Glucose,1 Mannose,2 Galactose,2 Gulose,3 Idose,3Talose,3.
- Allose,2 Altrose.3 Ketoses - Fructose,1 Sorbose,2 Tagatose.3.
- Heptoses ($C_7H_{14}O_7$). Aldose - Mannoheptose.1 Ketose - Sedoheptose.1.

Note: No attempt is here made to summarize the occurrence of any but the tetroses, pentoses, hexoses and heptoses. Glycolose and the trioses if formed in nature are probably too reactive to accumulate sufficiently for identification.

Disaccharides (Disaccharoses)

Dihexoses (Hexobioses) - ($C_{12}H_{22}O_{11}$). Anhydride of glucose + fructose-Sucrose. Anhydrides of glucose + galactose - Lactose, Melibiose. Anhydrides of glucose + glucose- Maltose, Isomaltose, Trehalose, Turanose.

Trisaccharides (Trisaccharoses)

Trihezoses ($C_{18}H_{22}O_{16}$). Anhydride of glucose + galactose +fructose - Raffinose. Anhydride of glucose + glucose + glucose - Melezitose. Anhydride of fructose + fructose + fructose - Secalose.

Tetrasaccharides (Tetrasaccharoses)

Tetrahexoses ($C_{24}H_{42}O_{21}$). Anhydrides of 2 galactose + glucose + fructose - Stachyose, Lupeose.

Polysaccharides (Polysaccharoses)

Pentosans (chief constituents of gums and mucilages). Anhydrides of xylose - Xylans. Anhydrides of arabinose - Arabans.

Hexosans

Anhydrides of glucose-Starch, Cellulose, Glycogen, Dextrin (and other"dextrans"). Anhydrides of mannose - Mannans. Anhydrides of galactose-Galactans (pectins). Anhydrides of fructose - Inulin (and other"levulans").

- Occurs free in nature.
- Not yet found free in nature (or only in small amounts) but obtained by hydrolysis or fermentation of natural product.
- Known only (with certainty) as a Labouratory product.

SIMPLE BIOCHEMICAL TESTS

Materials

- Benedict's solution (not"quantitative")
- Iodine solution (I/KI)
- Waterbaths set at about 100°C
- Boiling tubes Racks
- 1 per cent glucose soln
- 1 per cent fructose soln
- per cent sucrose soln
- other sugars?
- 1 per cent starch sol solution X

HCl, NaOH, litmus/universal indicator paper filter funnels, papers, flasks? measuring cylinders (10, 25, 50, 100 ml) plastic pipettes glass pipettes (1, 5, 10 ml?), pipette pumps

Objectives

Familiarisation with Benedict's test - for reducing sugars modification of Benedict's test for non-reducing sugars iodine test dilution techniques sensitivity of above tests.

Procedure

See also Roberts Students' manual p. 42 You will be expected to write up and hand in a concise account of your techniques and the results. There is no need to use large amounts of test substances or reagents.

SAFETY

Irritant Solutions Wear Safety Glasses And Lab Coats Hot Water And Steam Take Care: Keep Working Area Clear:

- Carry out Benedict's test on the named sugar solutions and starch as provided.
- Use fixed (noted!) quantities of test substance and reagents.
- Do not contaminate stock solutions shared by the group.
- Note results - colour and intensity - separately from conclusions.
  - *Refinement*: Filter the contents of the tube in which you have been testing glucose (using paper previously labelled with pencil). Consider the colour of the filtrate, and accordingly modify the quantities of reagents used. Repeat filtration process. Explain your conclusions in your write-up.
- Dilute the sugar solutions (say ×10, ×100 etc), and carry out Benedict's test on suitable samples (i.e. amounts same as before).
- Explain the dilution technique you used in your write-up.
- What does this tell you about the sensitivity of the test?
- Using sucrose, carry out the test for non reducing sugars. Having obtained a negative result from ordinary benedict's test, a second test is performed after hydrolysing it to reducing sugars (acid hydrolysis, followed by neutralisation), then performing Benedict's test:- Add... drops/ml HCl - leave for... minutes at room temperature (... °C) or... mins at 100 °C - add... drops/ml NaOH (same strength) - check neutrality with litmus (colour......) or universal indicator paper (colour.....).
- Test solution X using a combination of the above tests.
- Test starch solution using iodine solution (I/KI). You could use drops in a recessed tile, but for sensitivity purposes it is probably best to use boiling tubes and drops of iodine.
  - Dilute starch × 10, × 100, etc. and repeat (standardised) test.
  - Warm the last tubes - i.e. the least blue coloured (100 °C waterbath)), and observe colour.
  - Cool (the lower end of) the tube under the tap.
  - Conclusion?

METABOLISM OF CARBOHYDRATES

Carbohydrate metabolism begins with digestion in the small intestine where monosaccharides are absorbed into the blood stream. Blood sugar concentrations are controlled by three hormones: insulin, glucagon, and epinephrine. If the concentration of glucose in the blood is too high, insulin is secreted by the pancreas. Insulin stimulates the transfer of glucose into the cells, especially in the liver and muscles, although other organs are also able to metabolize glucose. In the liver and muscles, most of the glucose is changed into glycogen by the process of glycogenesis (anabolism). Glycogen is stored in the liver and muscles until needed at some later time when glucose levels are low. If blood glucose levels are low, then eqinephrine and glucogon hormones are secreted to stimulate the conversion of glycogen to glucose. This process is called glycogeno-lysis(catabolism).

If glucose is needed immediately upon entering the cells to supply energy, it begins the metabolic process calledglycoysis (catabolism). The end products of glycolysis arepyruvic acid and ATP. Since glycolysis releases relatively little ATP, further reactions continue to convert pyruvic acid to acetyl CoA and then citric acid in the citric acid cycle. The majority of the ATP is made from oxidations in the citric acid cycle in connection with the electron transport chain.During strenuous muscular activity, pyruvic acid is converted into lactic acid rather thatn acetyl CoA. Durlng the resting period, the lactic acid is converted back to pyruvic acid. The pyruvic acid in turn is converted back to glucose by the process called gluconeogenesis(anabolism). If the glucose is not needed at that moment, it is converted into glycogen by glycogenesis. You can remember those terms if you think of"genesis" as the formation-beginning. These processes are summarized in the Metaboism Summary in the graphic on the left. Each of these processes will be developed in greater detail various pages of this module.

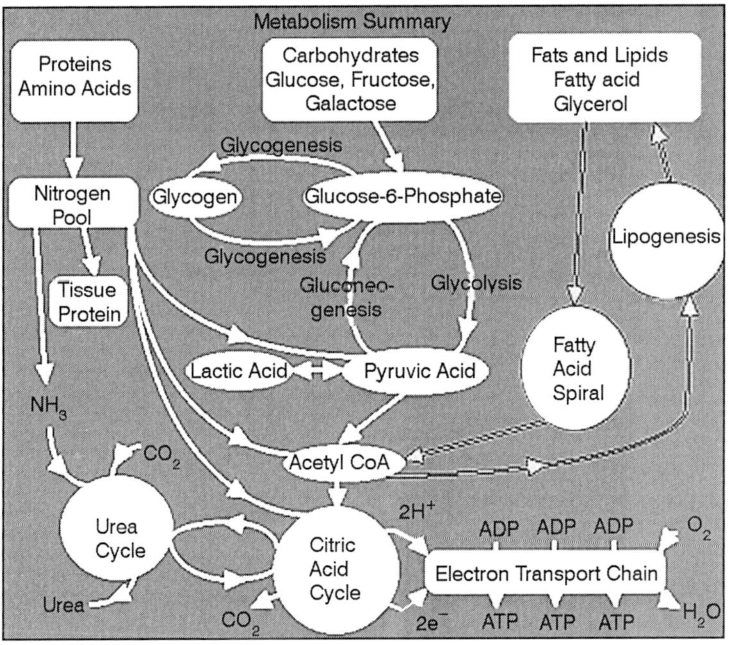

GLYCOLYSIS

All organisms produce ATP by releasing energy stored in glucose and other sugars.

- Plants make ATP during photosynthesis.
- All other organisms, including plants, must produce ATP by breaking down molecules such as glucose.

AEROBIC RESPIRATION

The process by which a cell uses O_2 to"burn" molecules and release energy.

- *The reaction*: $C_6H_{12}O_6 + 6O_2 >> 6CO_2 + 6H_2O$.
- This reaction is the opposite of photosynthesis

This reaction takes place over the course of three major reaction pathways.

- Glycolysis.
- The Krebs Cycle.
- Electron Transport Phosphorylation (chemiosmosis) Glycolysis (glyco = sugar; lysis = breaking).
- *Goal*: Break glucose down to form two pyruvates.
- *Who*: All life on earth performs glyclolysis
- *Where*: The cytoplasm.
- Glycolysis produces 4 ATP's and 2 NADH, but uses 2 ATP's in the process for a net of 2 ATP and 2 NADH.

This process does not require O_2 and does not yield much energy.

FIRST STAGE OF GLYCOLYSIS

- Glucose (6C) is broken down into 2 PGAL's (Phosphoglyceraldehyde - 3Carbon molecules).
- This requires two ATP's.

SECOND STAGE OF GLYCOLYSIS

- 2 PGAL's (3C) are converted to 2 pyruvates.
- This creates 4 ATP's and 2 NADH's.
- The net ATP production of Glycolysis is 2 ATP's.

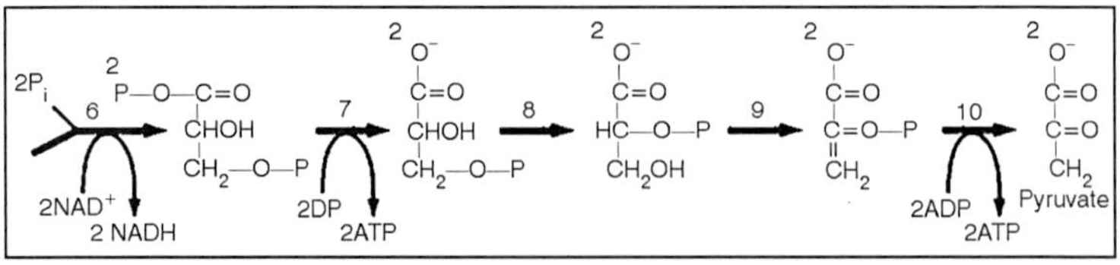

- Goal: take pyruvate and put it into the Krebs cycle, producing NADH and $FADH_2$.
- *Where*: The mitochondria.

- There are two steps.
  - The Conversion of Pyruvate to Acetyl CoA.
  - The Krebs Cycle proper.
- The Krebs cycle and the conversion of pyruvate to Acetyl CoA produce 2 ATP's, 8 NADH's, and 2FADH2's per glucose molecule.

Oxidation of Pyruvate to form Acetyl CoA for Entry Into the Krebs Cycle.

- 2 NADH's are generated (1 per pyruvate).
- 2 CO_2 are released (1 per pyruvate).

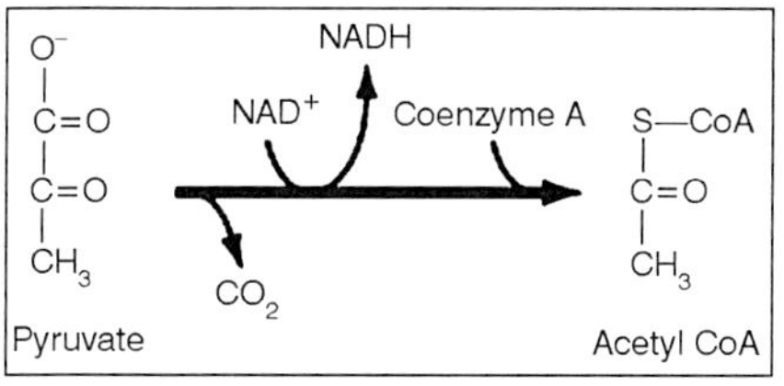

Krebs Cycle

- 6 NADH's are generated (3 per Acetyl CoA that enters).
- 2 $FADH_2$ is generated (1 per Acetyl CoA that enters).
- 2 ATP are generated (1 per Acetyl CoA that enters).
- 4 CO_2's are released (2 per Acetyl CoA that enters).

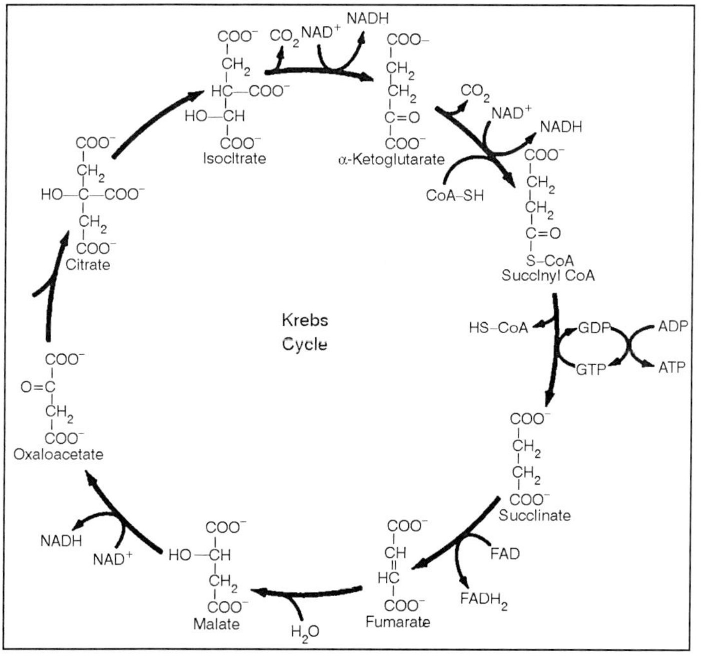

- Therefore, the total numbers of molecules generated in the oxidation of pyruvate and the Krebs Cycle is:
  - 8 NADH
  - 2 $FADH_2$
  - 2 ATP
  - 6 CO_2

Electron Transport Phosphorylation (Chemiosmosis)

- *Goal*: to break down NADH and $FADH_2$, pumping H+ into the outer compartment of the mitochondria
- *Where*: the mitochondria.
- In this reaction, the ETS creates a gradient which is used to produce ATP, quite like in the chloroplast
- Electron Transport Phosphorylation typically produces 32 ATP's.
- ATP is generated as H+ moves down its concentration gradient through a special enzyme called ATP synthase.

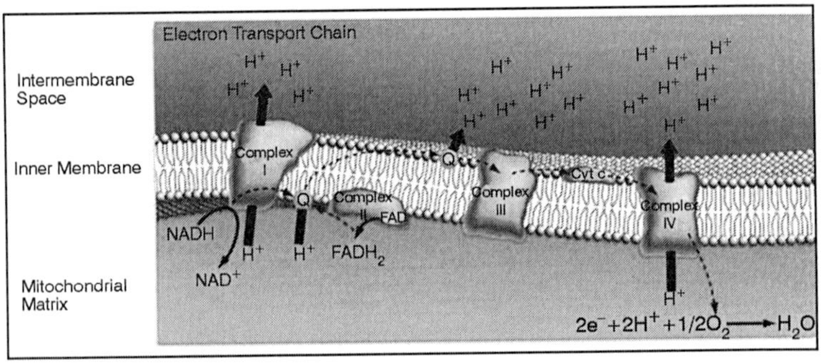

Net Engergy Production from Aerobic Respiration

- *Glycolysis*: 2 ATP.
- *Krebs Cycle*: 2 ATP.
- Electron Transport Phosphorylation: 32 ATP.
  - Each NADH produced in Glycolysis is worth 2 ATP ($2 \times 2 = 4$) – the NADH is worth 3 ATP, but it costs an ATP to transport the NADH into the mitochondria, so there is a net gain of 2 ATP for each NADH produced in gylcolysis.
  - Each NADH produced in the conversion of pyruvate to acetyl COA and Krebs Cycle is worth 3 ATP ($8 \times 3 = 24$).
  - Each FADH2 is worth 2 ATP ($2 \times 2 = 4$).
  - $4 + 24 + 4 = 32$.
- Net Energy Production: 36 ATP!

Anaerobic Respiration

- *Goal*: to reduce pyruvate, thus generating NAD+.
- *Where*: the cytoplasm.
- *Why*: in the absence of oxygen, it is the only way to generate NAD+.
- Alcohol Fermentation - occurs in yeasts in many bacteria.
  - The product of fermentation, alcohol, is toxic to the organism.

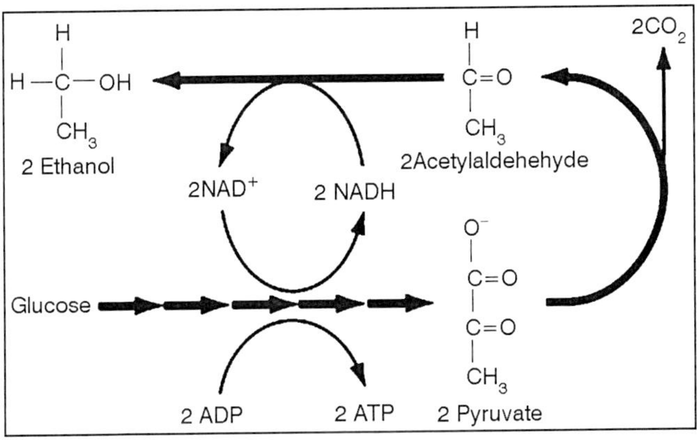

- *Lactic Acid Fermentation*: Occurs in humans and other mammals.
  - The product of Lactic Acid fermentation, lactic acid, is toxic to mammals.
  - This is the"burn" felt when undergoing strenuous activity.

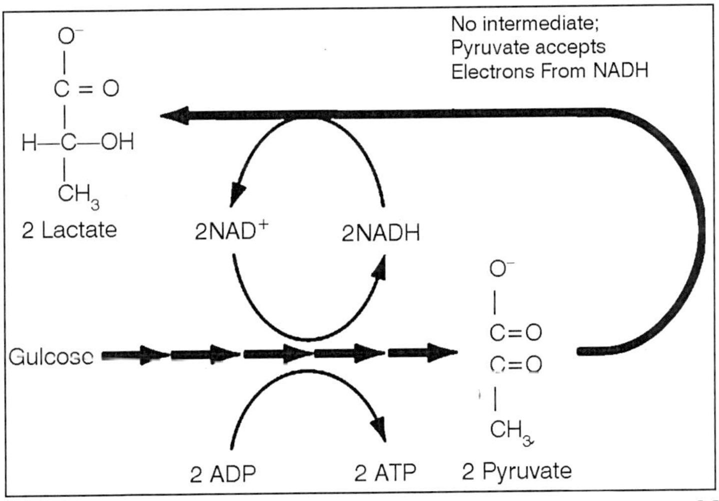

- The only goal of fermentation reactions is to convert NADH to NAD+ (to use in glycolysis).
- No energy is gained.
- Note differences - Anaerobic respiration - 2 ATP's produced (from glycolysis), aerobic respiration - 36 ATP's produced (from glycolysis, Krebs cycle, and Oxidative Phosphorylation).

- Thus, the evolution of an oxygen-rich atmosphere, which facilitated the evolution of aerobic respiration, was crucial in the diversification of life.

Photosynthesis: 6 CO_2 + 6 H_2O >> $C_6H_{12}O_6$ + 6 O_2.

Respiration: $C_6H_{12}O_6$ + 6 O_2 >> 6 CO_2 + 6 H_2O.

We see that these reactions are opposites: This is important since the earth is a closed system. All life has a set amount of natural materials to work with, so it is important that they all be cycled through effectively and evenly.

Energy Yields

- *Glucose*: 686 kcal/mol.
- *ATP*: 7.5 kcal/mol.
- 7.5 × 36 = 270 kcal/mol for all ATP's produced.
- 270/ 686 = 39 per cent energy recovered from aerobic respiration

Related Catabolic Processes-Beta Oxidation

- Fats consist of a glycerol backbone with two or three fatty acids connected to it.
- The body absorbs fats and then breaks off the fatty acids from the glycerol.
- Glycerol is converted to glyceraldehyde phosphate, an intermediate of glycolysis.
- The fatty acids are broken down into two-carbon units which are then converted to acetyl CoA.
  - An eight-carbon fatty acid can produce 4 acetyl CoA's.
  - Each acetyl CoA is worth 12 ATP's (3 NADP, 1 $FADH_2$, 1 ATP).
  - Therefore, this short fatty acid is worth 48 ATP's, a fat with three chains of this length would be worth 144 ATP's!
  - This is why fats are such a good source of energy, and are hard to lose if you want to lose weight.

Comparison Between Plants and Animals

- Animal cells and Plant cells contain mitochondria!
  - However, animal cells contain many more mitochondria than plant cells.
- Animal cells get most of their ATP from mitochondria.
- Plant cells get most of their ATP from the chloroplast
  - The ATP generated from the mitochondria is only used when the plant cannot generate ATP directly from the light-dependent reactions.

Molecules used in Glycolysis and Krebs Cycle

- Not all of the molecules that enter Glycolysis and the Krebs Cycle are used for energy.

- Some are used to synthesize fats, nucleotides, amino acids, and other biologically important molecules.

Regulation of Glycolysis and the Krebs Cycle

- *Step 3 of Glycolysis*: The conversion of Fructose 6-phosphate to Fructose 1,6-bisphosphate.
  - Enzyme catalyzing this reaction = Phospho-fructokinase.
  - *"Committing Step"*: Fructose 6-phosphate can be used by the cell for lots of things, but fructose 1,6-bisphosphate has limited use except in glycolysis.
  - Phosphofructokinase inhibited by high levels of ATP.
- ATP is also a substrate - odd, eh?
- Enzyme has two ATP binding sites, one in the active site and one in an allosteric site.
- Low to mid levels of ATP cause ATP to bind to the active site.
- High levels of ATP also enable ATP to bind to allosteric site, causing a conformation change and shutting down the enzyme.
- Conversion of Pyruvate to Acetyl CoA.
  - Enzyme involved in catalyzing this reaction = pyruvate dehydrogenase.
  - High levels of ATP slow down this reaction by phosporylating the enzyme, changing its shape and shutting it down.
- High levels of NADH and Acetyl CoA also inhibit this enzyme.
  - NAD+, CoA, or AMP (an indicator of low ATP) can speed up the reaction.

DISORDERS OF CARBOHYDRATE

Disorders of carbohydrate metabolism occur in many forms. The most common disorders are acquired. Acquired or secondary derangements in carbohydrate metabolism, such as diabetic ketoacidosis, hyperosmolar coma, and hypoglycemia, all affect the central nervous system. Many forms and variants of peripheral nerve disease also are seen in diabetes. The remaining disorders of carbohydrate metabolism are the rare inborn errors of metabolism (ie, genetic defects).

The inherited defects affecting carbohydrate metabolism that have been discovered so far are inherited as autosomal recessive traits. Although multiple affected siblings may be identified in a kindred, most affected individuals are the first identified in a family. Most of the known defects of carbohydrate metabolism appear to be due to point mutations. Whether this will still hold true after the discoveries that the Human Genome Project will make possible is not known.

Most of the inherited disorders of carbohydrate metabolism fall into a few broad clinical syndromes, which are classified by age of onset as follows:

- Infants and early childhood.

- Episodic lactic acidosis from early infancy, failure to thrive, and hypotonia with or without features that may suggest specific defects.
- Infantile or early childhood mental retardation, hypotonia, failure to thrive, and other features.
- Mental retardation/developmental delay with features that suggest storage disorders.
- Episodic vomiting in infants or young children.

• Childhood or adolescence.
  - Episodic acidosis in childhood or early adulthood, often with features of a specific disorder.
  - Intermittent or episodic ataxia in childhood and adolescence

• In adults and, more rarely, in adolescents or older children.
  - Cramps and weakness, often with some episodes of muscle breakdown and myoglobinuria, during or after heavy anaerobic exercise.
  - Symmetrical neuromuscular disease with weakness and wasting of proximal muscles.

PATHOPHYSIOLOGICAL DISORDERS

The brain, heart, skeletal muscle, and liver depend on ketone bodies or on glucose for energy and for carbon chains to synthesize cellular proteins.

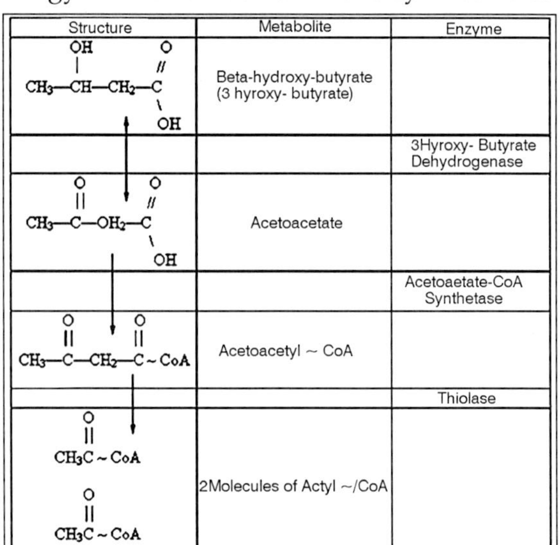

Fig. Ketone Metabolism

For infants as well as individuals with decreased sugar intake because of starvation or dietary choices, ketones are the preferred fuel and source of carbon chains. Nervous and muscle tissues preferentially oxidize ketones over fatty acids, glucose, or amino acids under normal conditions.

KETONE METABOLISM

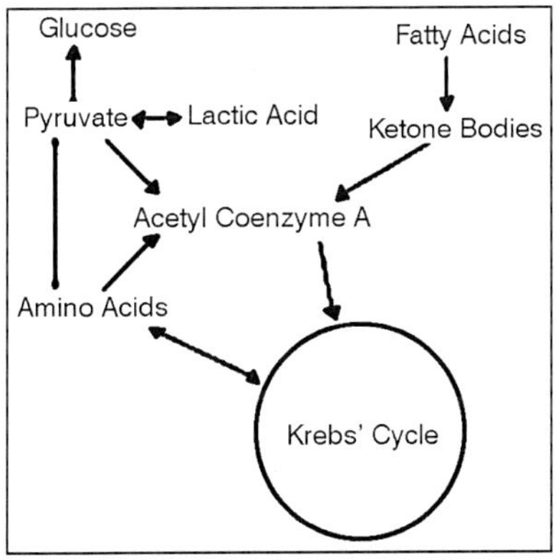

Krebs Cycle

Ketone bodies are largely beta-hydroxybutyrate and acetoacetate. These are formed in the liver and kidneys from the breakdown of fatty acids and a few amino acids. The brains of adults who live on a high-carbohydrate diet lose the ability to synthesize the key enzymes needed to metabolize ketones. These include beta-hydroxybutyrate dehydrogenase and acetoacetyl coenzyme A (CoA) synthetase. For the brain to resynthesize these enzymes to effective levels, starvation for 3-7 days or a high-fat diet is needed. Neural tissue has far more metabolic activity than other tissues of the body. The brain comprises only a small Percentage of total body weight-approximately 1.4 kg in an average 70-kg person-but uses 20 per cent of the body's oxygen at rest. In addition, neurons (and glia) are constantly synthesizing a variety of neurotransmitters, proteins for axonal flow, and proteins and lipids for regeneration of synaptic vesicles and other components of membranes. Many of these chemicals are synthesized in the brain, in part or wholly from glucose or ketone bodies.

Nonglucose carbohydrates (eg, galactose, mannose, inositol) are clinically important in fetal and neonatal nutrition; however, little is known about their metabolism in the neonate. Glucose and galactose increase postprandially and several other carbohydrates contained in milk do not. Brown *et al*, in their attempt to determine whether postprandial changes in plasma carbohydrate and sugar alcohol concentrations are affected by clinical variables such as postnatal age, milk type, feeding volume, or feeding duration in term newborns, found that galactose is almost cleared completely by the neonatal liver.

This provides novel evidence of a marked difference in plasma galactose concentrations [Gal] between relatively younger and older neonates. Acute hypoglycemia, as in an alcoholic or a diabetic who has taken too much insulin, causes the brain to be deprived rapidly and abruptly of blood glucose after it has depended for a long time on glucose rather than ketones as its fuel. Beta-hydroxybutyrate dehydrogenase and acetoacetyl CoA synthetase activity in the brain is insufficient to enable the blood ketones to sustain the brain's metabolism. Without fuel to burn, the neural tissue cannot carry out the high level of metabolism needed to pump ions, fire messages, or transmit intercellular chemical messages, let alone repair itself. The groggy state of sudden memory loss (ie, disorientation) and other intellectual deficits that internists tend to call confusion sets in rapidly with acute hypoglycemia, followed just as rapidly by a downhill course into obtundation, coma, and death. If the average adult brain can withstand only 4-6 minutes of oxygen deprivation, it can withstand only about 40-45 minutes without fuel.

Most physicians, as medical students or interns, have had experience treating diabetic ketoacidosis. Although the experience tends to cast ketone bodies as toxins, they are not. The acidosis produced by ketone bodies is toxic and is the direct cause of the Kussmaul breathing typical of patients with diabetic ketoacidosis. Patients with diabetes in this state usually have had normal to elevated blood glucose for the previous days or weeks. Enough glucose enters the brain cells during this time to inhibit synthesis of beta-hydroxybutyrate dehydrogenase and acetoacetyl CoA synthetase.

Again, the ketone-utilizing enzymes in the brain are not working at effective levels. By the time frank ketoacidosis sets in, too little insulin is left for the blood glucose to be able to sustain the high metabolic activity of the brain. The ketone bodies accumulate but cannot provide energy or carbon for the brain. The hydrogen ion associated with the ketones also accumulates, thereby stimulating (or overstimulating) acid receptors in the floor of the fourth ventricle and interfering with reuptake of potassium in the renal tubules.

The coma of hyperosmolar diabetic disorder seems to be related directly to the addition of large concentrations of glucose in the serum, the cerebrospinal fluid (CSF), the extracellular fluid of the brain and, perhaps to some degree, the intracellular fluid as well. The cause of high concentrations of glucose in these patients but not of ketone bodies remains unknown. The lack of ketones probably explains the lack of acidosis. CSF concentrations of glucose mirror those of the serum with a delay of approximately 2 hours. Patients do not regain consciousness immediately when the blood glucose is lowered.

PATHOPHYSIOLOGY OF INHERITED DISORDERS

Many of the clinical features of the inherited disorders of carbohydrate metabolism are caused by the following:

- Lack of glucose for the metabolism of brain, muscle, liver, or kidney (in circumstances in which ketone bodies cannot be used)

- Inability to break down glucose to pyruvate
- Inability to oxidize pyruvate fully in the Krebs cycle

Some features of the inherited diseases are due to excessive storage of abnormal substrates or to normal substrates that cells cannot degrade normally because they lack a specific enzymatic activity. In some diseases, abnormal metabolites (intermediates of pathways that use carbohydrates) block the normal function of the pathway or of related pathways. Defects of the enzymes of the pentose shunt interfere with the normal production of nucleic acids, which are needed by cells as second messengers and as coenzymes of intermediary metabolism, as well as components of RNA and DNA.

Galactose accumulates in the lens of the eye in galactose kinase deficiency.

Metabolism of galactose. The galactose ring differs from glucose in that the hydroxyl group (OH) at carbon-3 is up and not down. To convert galactose to glucose, the ring is first bound to uridine diphosphate (UDP). In this form, the sugar moiety can be epimerized, putting the OH down and forming glucose. The UDP-glucose then is converted to glucose-1-phosphate, then to glucose-6-phosphate, and subsequently metabolized further.

The osmotic pressure generated causes water to enter the lens; this disruption leads to cataract formation. A similar mechanism may play a role in the pathophysiology of diabetic peripheral neuropathy, which was at one time thought to be due, in part, to the accumulation of sorbitol, a relatively inert but osmotically active metabolite of glucose, and hence to excess water in the cells of peripheral nerves.

Media files 1-8 for the major biochemical pathways involved in carbohydrate metabolism. The expression of a disease often depends on other factors in addition to the causative agent. In inherited disease, variations in the other 30,000 genes, varying diets, infectious disease, access to medical care, and even upbringing and cultural effects can influence the clinical expression of a disease. Intermittent ataxia of childhood from pyruvate decarboxylase deficiency tends to appear in conjunction with certain infections or emotional stress, just as is the case with the neurological manifestations of phytanic acid storage disease

PATTERNS OF OCCURRENCE

Frequency

The acquired disorders of carbohydrate metabolism are fairly common, both in the United States and internationally. Hypoglycemia is a common cause of neurological disease, especially acute mental deterioration, memory loss, disorientation, obtundation, and coma, among both alcoholics and patients with diabetes who are treated with insulin. Hyperinsulinemia from other causes is rare, but pancreatic tumors could be the cause. Diabetes, with its various neurological complications, is among the most common disorders treated in adult patients. Diabetic ketoacidosis still occurs, though education and close medical follow-up make it less common than it was several decades ago. Hyperosmolar coma is also less a problem than when it was first brought to the attention of internists by Plum and Posner's classic monograph Diagnosis of Stupor and Coma.2 Hyperosmolar coma still occurs and needs to be kept in mind while evaluating an obtunded patient.

The inherited disorders of carbohydrate metabolism are rare. Severe defects of the pyruvate dehydrogenase (PDH) complex and the benign chemical anomaly called pentosuria have been reported in very few (2-6) patients. Patients with inherited disorders of carbohydrate metabolism tend to be seen at tertiary Centres, where specialist teams are available. The team may comprise pediatricians, pediatric neurologists, biochemical geneticists, psychiatrists, neurologists with a special interest in mental retardation, and neurologists interested in neuromuscular diseases and the more unusual ataxias.

Neurologists in general practice see few, if any, cases of inherited disorders of carbohydrate metabolism in their careers. This depends partly on whether the practice is in an area in which the carrier state for the inherited disorders is more common and partly on whether the neurologist is on the alert for these disorders. As in every area of medicine, inherited carbohydrate disorders that are looked for tend to be found and those that are not looked for tend to be overlooked. Although each of these disorders was described between 1930 and 1990, certainly they were present in the populations of medically sophisticated countries decades, perhaps centuries, earlier.

Glass *et al* performed a retrospective chart review of new pediatric patients seen during 1998 by specialists of the Division of Clinical and Metabolic Genetics of the Hospital for Sick Children in Toronto, the largest such referral Centre in the country. They found that 81 per cent of specific genetic metabolic diagnoses were made within 1 month of being seen in consultation. Four years after initial consultation, a specific diagnosis was not reached in 5 per cent of patients. The authors concluded that the specific diagnosis of inborn errors of metabolism at a major medical genetic referral Centre tended to be made quickly, or never. Some of the causes of delays in diagnosis include the lack of ready access to existing diagnostic Labouratory testing, technical barriers to the identification of specific metabolic or genetic defects, and incomplete knowledge of genetic defects causing inherited metabolic diseases.

Mortality

Hypoglycemia, diabetic ketoacidosis, and hyperosmolar coma are all potentially fatal but potentially curable conditions. Mortality depends on the time taken by the patient or patient's family to present medical attention, the time taken by the doctor to consider the diagnosis and begin correct treatment, the thoroughness of the treatment, and the time taken to diagnose and thoroughly treat other associated medical problems. The patient with acidosis benefits very little if the ketosis is corrected but an underlying cause is missed, for example, by a delayed spinal tap if meningitis or encephalitis happens to be the illness that has tipped the metabolic balance. Some of the inherited disorders of carbohydrate metabolism are fatal. For some of these conditions, the more profoundly the mutation interferes with enzyme activity, the more likely the disorder will be fatal. Other inherited disorders usually are not fatal if treated early and thoughtfully. Pentosuria is not even a disease in the strict sense; it is more a metabolic curiosity without significant clinical consequences.

In general, patients with inherited carbohydrate disorders that primarily affect the brain or heart tend to die early; those whose disorders affect the liver and kidney die later on; and those whose disorders affect only muscle have lower mortality rates, unless things go very wrong. Some concern exists among physicians who study heat illness in military recruits that subclinical inherited predispositions to disorders of muscle and of RBCs might be among the factors that predispose to heat illness. As heat illness is the major cause of death of military recruits in boot camp, this is an issue that deserves careful study.

Race

No racial predilection has been reported in most inherited disorders of carbohydrate metabolism. One well-known exception is pentosuria, a very rare genetic occurrence of no clinical significance, which is so far confined to Ashkenazi Jews. The pentose excreted in this condition is xylulose.

Sex

No sex predilection has been reported; the known inherited disorders of carbohydrate metabolism are all autosomal recessive.

Age

A general rule exists in inherited diseases. If the degree of deficiency of an enzyme varies from family to family, symptoms are likely to appear earliest in families in which the mutation virtually inactivates the enzyme activity and latest in families in which the activity is only just below a certain minimum symptomatic threshold. Profound defects of activity of the PDH complex are associated with lactic acidosis, hypotonia, profound mental retardation, and often death in the first months or first year of life. Somewhat milder defects and vigourous treatment can be compatible with survival into early to mid childhood.

Moderate defects of the PDH complex are associated with ataxia and other symptoms and signs presenting in childhood and compatible with graduation from college and survival at least into the adult years. Data from studies in the 1970s and 1980s raised the question whether mild defects of this complex might be associated with brain diseases whose onset is in adolescence or early adult life and that are compatible with long survival. These studies have not yet been refuted or disproved.

DIAGNOSIS AND TREATMENT OF ACQUIRED DISORDERS

Diabetic ketoacidosis, hyperosmolar coma, and hypoglycemia present as obtundation or coma. The differential diagnosis is as extensive as it is for obtundation and coma in general. The first clinical task is to make sure the patient survives. Checking the airway, breathing, and circulation (ABCs) and correcting any problems with these is always the first step. The second step is to get baseline blood levels for later analysis before metabolically altering the system during treatment. The third step is to give intramuscular (IM) thiamine followed by 50 mL of 50 per cent intravenous (IV) glucose.

Hypoglycemia can be fatal if not reversed in approximately 45 minutes. A glucose load depletes thiamine stores and precipitates Wernicke syndrome; therefore, thiamine should be given first. Then perform a rapid general examination, looking for clues of serious trouble (eg, abnormal patterns of breathing, unusual body or breath odour, jaundice, pink skin, pallor, petechiae, meningismus, abnormalities of major organs). Next, perform a neurological examination focused on variable conditions associated with coma. Fisher and Plum and Posner have written extensively about the neurological examination of comatose patients.

The neurological examination helps clarify whether the problem is focal or systemic (metabolic). If all signs point to a lesion on one side or in one location, the lesion is focal; imaging studies and IV steroids need to be considered and a neurosurgeon consulted. If the signs are diffuse or do not lateralize, the problem most likely is metabolic. Again, Fisher and Plum and Posner have published schemata that can be followed to go through the myriad possibilities in a logical manner. Subtle clues from the history or the examination can be crucial and therefore must be given serious consideration and addressed immediately.

Bhattacharya *et al*, in their pilot study compared uncooked cornstarch with uncooked physically modified cornstarch (WMHM20) in patients with glycogen storage disease types Ia, Ib, and III. The authors reported longer duration of euglycemia and better short-term metabolic control in most of the patients with WMHM20 when compared with cornstarch.4

Diabetes

The acute severe disorders in patients with diabetes are diabetic ketoacidosis (see Diabetic Ketoacidosis) and diabetic hyperosmolar coma (Hyperosmolar Coma). Hypoglycemia from lack of liver glycogen stores (as in alcoholics) or from excess insulin, whether accidental or intentional, also is an acute emergency. The chronic disorders are more common. These include mononeuritis from diabetes, multiple neuropathy (still called mononeuritis multiplex by some) in patients with diabetes, polyneuropathy in patients with diabetes, and so-called diabetic amyotrophy.

Diabetic Ketoacidosis

Even though the diagnosis may be established clinically, and even though someone experienced in the care of this problem may sometimes treat the ketoacidosis successfully without supporting Labouratory data, confirming the diagnosis by lab studies is still important. Whether the patient will respond easily or with difficulty or what hazards and problems may lie ahead cannot be known at the outset. The extent of Kussmaul breathing is a guide to the severity of acidosis; sudden relief of this breathing, as the patient is being treated, usually means the pH of the blood is back to normal. This can be checked with serial measurements of arterial blood pH (ie, blood gases) while treating the patient.

Ketones and sugars can be demonstrated in the urine and blood by dipstick, but with each repeated step of treatment, sending samples of blood for an accurate analysis by the Labouratory is advisable. Changes on the ECG may suggest hypokalemia, but, again, getting the Labouratory's ongoing help with serial assays of electrolytes, BUN, and blood pH is advisable. The workup and treatment must include any likely coincidental problem (ie, a new illness other than diabetes) that has tipped the metabolic balance.

Diabetic Hyperosmolar Coma and Hypoglycemia

These are the 2 major diagnoses to consider in any patient presenting with obtundation or coma of metabolic rather than structural etiology. Once the comatose patient's ABCs have been secured, several tubes of blood must be drawn for the host of cellular and biochemical analyses a comatose patient may need. Administer 100 mg thiamine IM and then inject an IV bolus of 50 mL of a 50 per cent solution of glucose through a needle of largest possible diametre (the author uses a 15-gauge needle). This amount of glucose is enough to reverse hypoglycemia but represents a mere"drop in the bucket" for a patient with hyperosmolar coma or even diabetic ketoacidosis.

Rapid clinical improvement after a bolus of glucose confirms the diagnosis of hypoglycemia. Patients with hyperosmolar coma do not respond, but the initial blood glucose, as measured in the Labouratory, is very high (ie, usually well over 400 mg/dL), as is the serum osmolality. The CSF glucose corresponds to the blood glucose but with a 2-hour delay; therefore, the CSF glucose remains high even after the blood glucose and electrolytes are brought slowly back to normal. Remember the dangers of central pontine myelinolysis when correcting electrolyte imbalances due to any cause. The chronic neurological complications of diabetes are due to damage to peripheral nerves. Mononeuritis and multiple neuropathy are both due to infarction of the tiny artery supplying a particular nerve, the vasa nervorum. Diabetes is a common cause of these problems, and it often presents with infarction of a single nerve, especially of cranial nerves III and VI and of the long mixed nerves of the limbs. When the distal motor nerves of limbs are involved, the diagnosis may be fairly clear. When the proximal motor nerves are primarily involved, the diagnosis may be difficult. The pattern can be that of weakness of the muscles of one thigh with deep pain (ie, diabetic amyotrophy).

Amyotrophy can be a presentation of diabetes. Sometimes, the opposite limb is affected much later. Amyotrophy tends to be self-limiting and improves without any treatment after a few months. The cause of the waxing and waning of symptoms of amyotrophy is unclear. In some patients, the pattern is of damage to the lumbar plexus, resulting in diabetic plexitis. Again, the reason for the differences in the extent of involvement is unknown. The differential diagnosis of mononeuritis includes local pressure along the course of the nerve or its roots from trauma, hematoma, entrapment, abscess, or growths, and other causes of infarction of the arteria nervorum, of which a current leading cause is collagen-vascular disease, especially the peculiar form of periarteritis nodosa that is confined largely to the peripheral nervous system (ie, not clinically affecting other organ systems).

These may be suggested by repeated abnormalities of the erythrocyte sedimentation rate or the C-reactive protein or by consistent elevations of antibodies specific for certain collagen-vascular diseases. Other causes of multiple neuropathy include Lyme disease, sarcoidosis, acquired and inherited amyloidosis, and, very rarely, first attacks of tumacular neuropathy. Most cases of multiple neuropathies are treatable; therefore, this syndrome is one of the few clinical indications for a nerve biopsy should it be needed to establish the diagnosis.

The sural nerve is almost always the nerve of choice. Consult an experienced surgeon to obtain the specimen. Too small a piece, or one that looked liked nerve but was collagen, does no good, and the patient usually will not consent to a second attempt at biopsy. Many cases of plexitis, whether brachial or lumbar, unilateral or bilateral, are due to infectious or postinfectious problems. These need to be excluded carefully before diabetes is accepted as the cause of plexitis. The problem is complicated by treatment.

For example, the nondiabetic causes of plexitis often respond to high-dose steroids. Most of the author's patients have responded well to pulse-dose steroids for 5 days as currently used for other immune-mediated neurological disorders. Steroids elevate blood sugars. This means that diabetes needs to be sought and excluded before a therapeutic trial and that a patient with diabetes who is to be given even pulse-dose steroids needs careful monitoring, often with testing of blood glucose 4 times a day for monitoring excessive elevations of glucose during treatment. In the era before CT scans of the abdomen and pelvis were available, the worsening of what appeared to be a lumbar plexitis with high-dose steroids pointed to the correct problem, pressure palsy from a pelvic abscess.

Polyneuropathy

Diffuse disorders of peripheral nerves exist, both sensory and motor, but often the sensory disorders are the most profound, clinically affecting the longest Fibres first. This is a common problem in neurology in general, and it is also a common problem in people with diabetes. The patients have numbness and tingling in the toes and feet and, later, up the legs. By the time the sensory loss reaches the knees, the fingertips begin to be involved. Often, small-Fibre sensation (eg, superficial touch, pinprick, appreciation of hot and cold) is involved early and most predominantly. These patients are likely to scald themselves by getting into too hot a bathtub. They test the water with their feet, but their feet can no longer warn them of the high temperature.

To some extent, the severity of the neuropathy correlates with the severity of the diabetes and with episodes of lack of metabolic control. To some extent, the neuropathy may improve if control is restored. However, the correlation is not at all tight. Some patients presenting with neuropathy have mild diabetes; in others, the neuropathy occurs and does not get better readily, regardless of the degree of control of blood glucose levels.

Some patients present with a marked dysesthetic pain, others do not. In many patients, later rather than sooner, a marked autonomic component is present, which is not present in others. When the autonomic system is involved, the distal skin often becomes mottled, cyanotic or pink, smooth, and loses hair follicles to become shiny. Capillary refill may become poor. Postural hypotension occurs in advanced stages of autonomic involvement. Treatment of complications of the peripheral nervous system in a patient with diabetes is still poor. Good control of blood glucose levels helps prevent or delay onset of these complications; therefore, young patients today often are put on insulin pumps, and blood glucose is monitored much more closely.

The neuropathic pain of diabetes often responds to modern antiepilepsy drugs, as does neuropathic pain of other etiologies. The authors' patients have responded to gabapentin (Neurontin), usually started at very low doses but often increasing to the large doses used in epilepsy. Pain can be well controlled by maintaining trough levels of 6-12 mg/dL.

5

Amino Acids

INTRODUCTION

One of a class of organic compounds containing the amino (NH_2) and the carboxyl (COOH) group, occurring naturally in plant and animal tissues and forming the chief constituents of protein; many of them are necessary for human and animal growth and nutrition and hence are called Essential Amino Acids. Amino Acids are the building blocks of protein.

PRIMARY FUNCTIONS OF AMINO ACIDS

Depending upon the source of dated reference material studied, some reports state"20" Amino Acids have been identified. More modern reference materials state"22" Amino Acids have been identified. In my personal research, identifying all possibilities available to me at this time, I have listed"28" Amino Acids/ or Amino Acid Derivatives. Of my 28 listings, 22 have been identified as"Commonly Occurring" in Proteins. In living things, Amino Acids serve as the primary unit of synthesis both of Tissue Proteins and other Non-Protein Nitrogenous Compounds. Amino acids may undergo complex inner conversions in the body into macromolecules such as carbohydrates, nucleic acids, lipids, etc. Mineral co-factors also play an important role in that they interact with enzymes to assist with their catalytic activities. They are also capable of complexing with the individual free amino acids. For example, magnesium (MG2+) is important in the phosphorylation process or the transfer of high-energy phosphate groups into the body.

Upon entering the body any free amino acids undergo complex inner-conversions into those substances necessary to maintain the metabolic pathways in the body. For example, Phenylalanine goes to Tyrosine and Arginine goes to Citrulline and Ornithine. (The Arginine pathway is a very important cycle.) Amino acid inner-conversions are not limited to other amino acids. Many free form amino acids play important roles as precursors for metabolic intermediaries. An example of this, Tyrosine's and therefore Phenylalanine's ability to give rise to two hormones Thyroxsine and Epinephrine. The Sulfur containing amino acids often precurse Taurine, which is an important bio-acid component. As W.C. Rose observed in 1938,"Perhaps undo emphasis was placed upon the quantity of protein ingestion and that scant consideration was given to possible difference in nutrient quality." Upon ingestion of foodstuff all nutrients and amino acids do not immediately diffuse into the surrounding tissue of the blood stream, but must first undergo a series of steps of biochemical reactions in the digestive tract. These reactions reduce the protein into its' individual amino acids. Protein cleaving enzymes are activated in the digestive process, but in order for this cleavage to occur energy must be supplied on the part of the individuals metabolism to break the peptide bonds. By starting with a free amino acid mixture in which all amino acids are in their L conformation little digestion is necessary. The result is an energy saving and rapid absorption into the blood stream and surrounding tissues."Amino acid supplementation of foods and feeds bring about many beneficial results. Of practical importance is the attainment of a better balance of protein and therefore a better balanced diet, an extension or savings of the available protein supply, and an improvement in the efficiency of the protein and food utilization.

ACID–BASE CHEMISTRY OF AMINO ACIDS

AMINO ACIDS ARE WEAK POLYPROTIC ACIDS

From a chemical point of view, the common amino acids are all weak polyprotic acids. The ionizable groups are not strongly dissociating ones, and the degree of dissociation thus depends on the pH of the medium. All the amino acids contain at least two dissociable hydrogens. Consider the acid–base behaviour of glycine, the simplest amino acid. At low pH, both the amino and carboxyl groups are protonated and the molecule has a net positive charge. If the counterion in solution is a chloride ion, this form is referred to as glycine hydrochloride. If the pH is increased, the carboxyl group is the first to dissociate, yielding the neutral zwitterionic species Gly^0 (Figure).

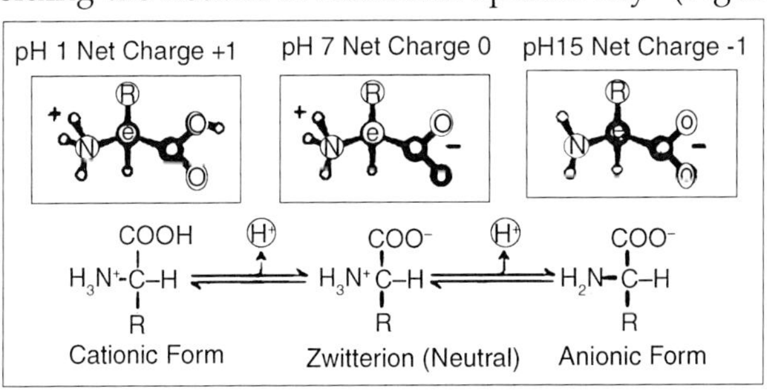

Fig. The Ionic Forms of the Amino Acids, Shown without Consideration of any Ionizations on the Side Chain.

The Cationic Form is the Low pH Form, and the Ttitration of the Cationic Species with base Yields the Zwitterion and finally the Anionic Form. *(Irving Geis)* Further increase in pH eventually results in dissociation of the amino group to yield the negatively charged glycinate. If we denote these three forms as Gly^+, Gly^0, and Gly^-, we can write the first dissociation of Gly^+ as

$$Gly^{+}+H_2O \leftrightharpoons Gly^{0}+H_3O^{+}$$

and the dissociation constant K_1 as

Values for K_1 for the common amino acids are typically 0.4 to 1.0×10^{-2} *M*, so that typical values of pK_1 center on values of 2.0 to 2.4. In a similar manner, we can write the second dissociation reaction as

$$Gly^{0} + H_2O \rightleftarrows Gly^{-} + H_3O^{+}$$

and the dissociation constant K_2 as:

$$K_2 = \frac{\left[Gly^{-}\right]\left[H_3O^{+}\right]}{Gly^{0}}$$

Table. pK_a Values of Common Amino Acids

Amino Acid	**α–COOH pK_a**	**α–NH31 pK_a**	**R group pK_a**
Alanine	2.4	9.7	
Arginine	2.2	9.0	12.5
Asparagine	2.0	8.8	
Aspartic acid	2.1	9.8	3.9
Cysteine	1.7	10.8	8.3
Glutamic acid	2.2	9.7	4.3
Glutamine	2.2	9.1	
Glycine	2.3	9.6	
Histidine	1.8	9.2	6.0
Isoleucine	2.4	9.7	
Leucine	2.4	9.6	
Lysine	2.2	9.0	10.5
Methionine	2.3	9.2	
Phenylalanine	1.8	9.1	
Proline	2.1	10.6	
Serine	2.2	9.2	~ 13.00
Threonine	2.6	10.4	~ 13.00
Tryptophan	2.4	9.4	
Tyrosine	2.2	9.1	10.1
Valine	2.3	9.6	

Typical values for pK_2 are in the range of 9.0 to 9.8. At physiological pH, the α–carboxyl group of a simple amino acid (with no ionizable side chains) is completely dissociated, whereas the α–amino group has not really begun its dissociation. The titration curve for such an amino acid is shown in Figure.

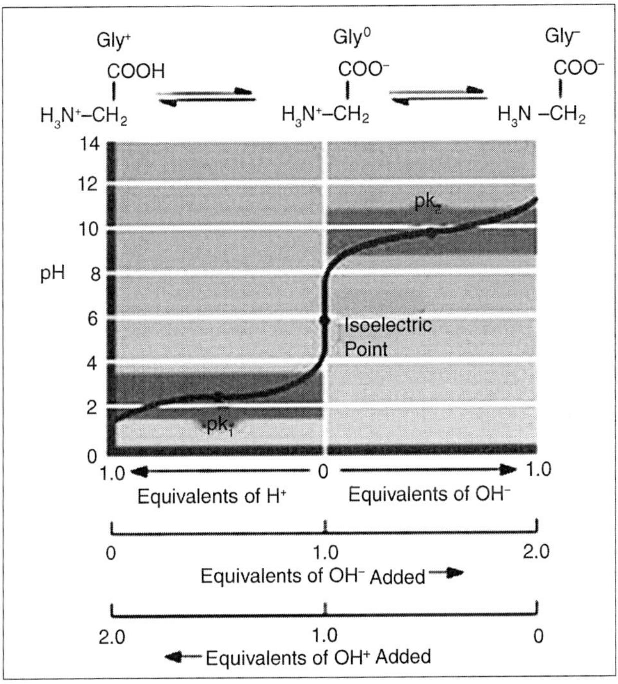

Fig. Titration of Glycine, a Simple Amino Acid. The Isoelectric Point, pI, the pH where the Molecule has a Net Charge of 0, is defined as $(pK_1 + pK_2)/2$

Example: What is the pH of a glycine solution in which the $\alpha\text{-}NH_3^+$ group is one–third dissociated?

Solution: The appropriate Henderson–Hasselbalch equation is

$$pH = pK_a + \log_{10}\frac{[Gly^-]}{[Gly^0]}$$

If the α–amino group is one–third dissociated, there is one part Gly^- for every two parts Gly^0. The important pK_a is the pK_a for the amino group. The glycine α–amino group has a pK_a of 9.6. The result is

$$pH = 9.6 + \log_{10}(1/2)$$

$$pH = 9.3$$

Note that the dissociation constants of both the α-carboxyl and α–amino groups are affected by the presence of the other group. The adjacent α-amino group makes the α-COOH group more acidic (that is, it lowers the pK_a) so that it gives up a proton more readily than simple alkyl carboxylic acids. Thus, the pK_1 of 2.0 to 2.1 for α-carboxyl groups of amino acids is substantially lower than that of acetic acid (pK_a = 4.76), for example. What is the chemical basis for the low pK_a of the α-COOH group of amino acids? The α-NH_3^+ (ammonium) group is strongly electron–withdrawing, and the positive charge of the amino group exerts a strong field effect and stabilizes the carboxylate anion.

Ionization of Side Chains

As we have seen, the side chains of several of the amino acids also contain dissociable groups. Thus, aspartic and glutamic acids contain an additional carboxyl function, and lysine possesses an aliphatic amino function. Histidine contains an ionizable imidazolium proton, and arginine carries a guanidinium function. The β–carboxyl group of aspartic acid and the γ–carboxyl side chain of glutamic acid exhibit pK_a values intermediate to the α–COOH on the one hand and typical aliphatic carboxyl groups on the other hand.

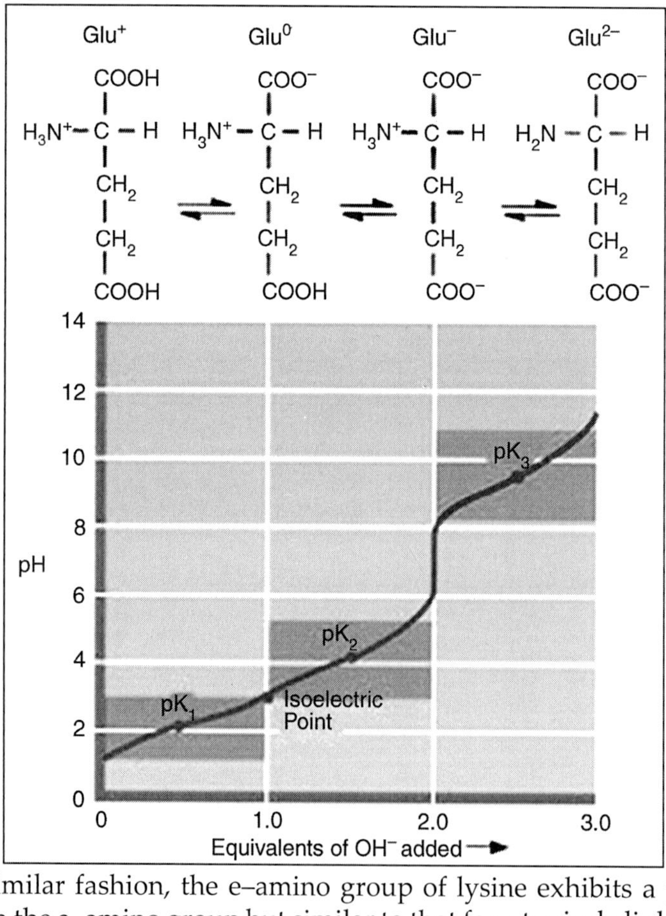

In a similar fashion, the e–amino group of lysine exhibits a pK_a that is higher than the a–amino group but similar to that for a typical aliphatic amino group. These intermediate values for side–chain pK_a values reflect the slightly diminished effect of the a–carbon dissociable groups that lie several carbons removed from the side–chain functional groups. Figure shows typical titration curves for glutamic acid and lysine, along with the ionic species that predominate at various points in the titration. The only other side–chain groups that exhibit any significant degree of dissociation are the *para*–OH group of tyrosine and the –SH group of cysteine. The pK_a of the cysteine sulfhydryl is 8.32, so that it is about 12 per cent dissociated at pH 7. The tyrosine *para*–OH group is a very weakly acidic group, with a pK_a of about 10.1. This group is essentially fully protonated and uncharged at pH 7.

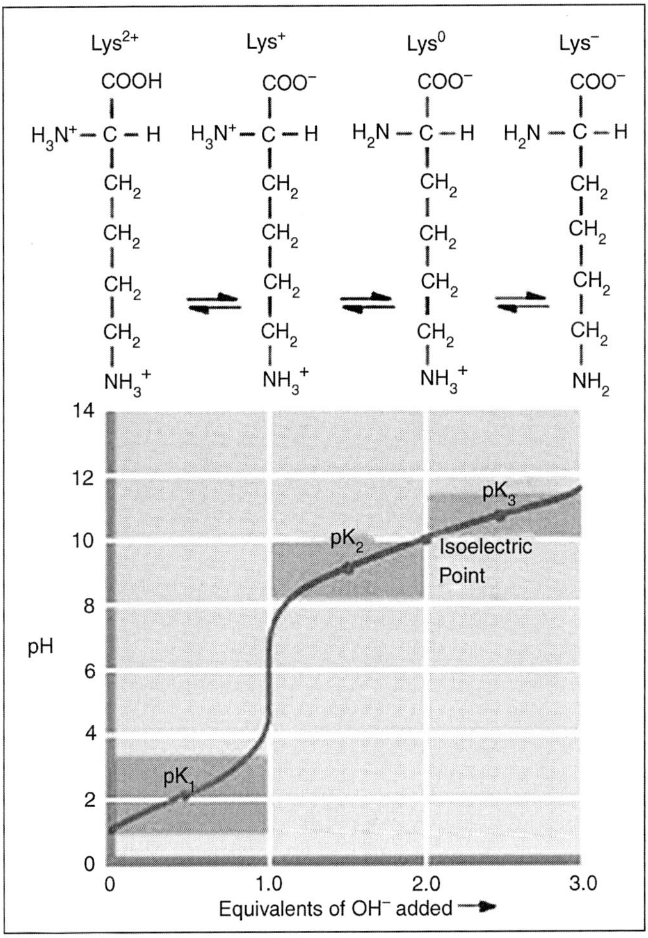

Fig. Titrations of Glutamic Acid and Lysine

CRITICAL DEVELOPMENTS IN BIOCHEMISTRY

Aquorea victoria, a species of jellyfish found in the northwest Pacific Ocean, contains a green fluorescent protein (GFP) that works together with another protein, aequorin, to provide a defence mechanism for the jellyfish. When the jellyfish is attacked or shaken, aequorin produces a blue light. This light energy is captured by GFP, which then emits a bright green flash that presumably blinds or startles the attacker. Remarkably, the fluorescence of GFP occurs without the assistance of a prosthetic group—a "helper molecule" that would mediate GFP's fluorescence. Instead, the light–transducing capability of GFP is the result of a reaction between three amino acids in the protein itself. As shown below, adjacentserine, tyrosine, and glycine in the sequence of the protein react to form the pigment complex—termed a chromophore. No enzymes are required; the reaction is autocatalytic.

Because the light– transducing talents of GFP depend only on the protein itself (upper photo, chromophore highlighted), GFP has quickly become a darling of genetic engineering laboratories. The promoter of any gene whose cellular expression is of interest can be fused to the DNA sequence coding for GFP. Telltale green fluorescence tells the researcher when this fused gene has been expressed.

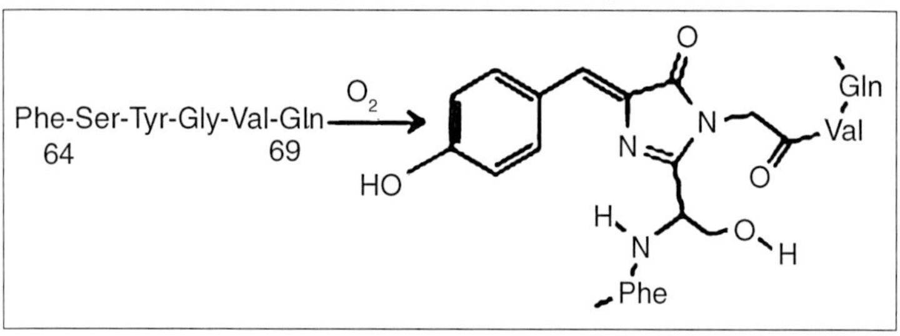

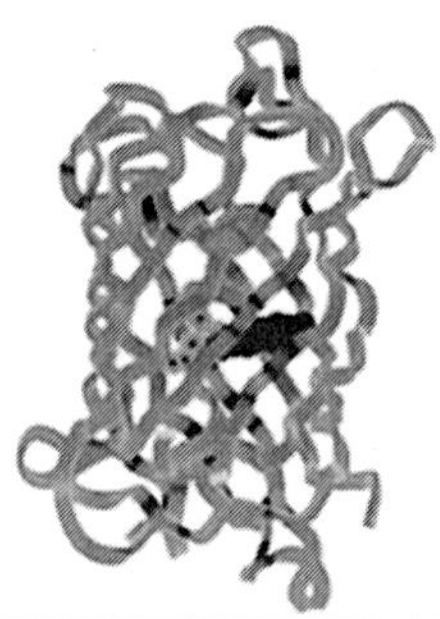

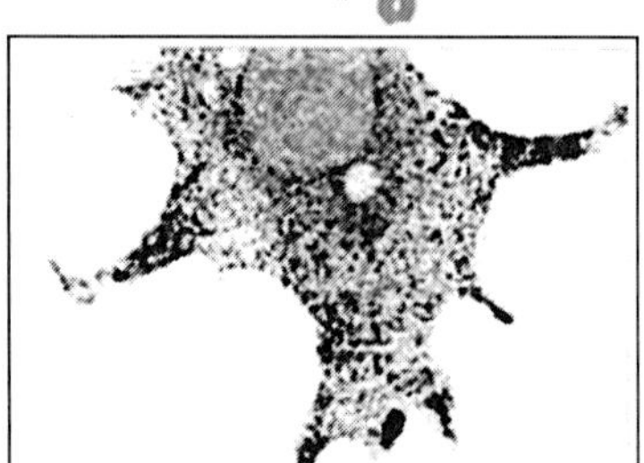

Autocatalytic oxidation of GFP amino acids leads to the chromophore shown on the left. The green fluorescence requires further interactions of the chromophore with other parts of the protein.

FUNCTIONAL IDENTITY OF AMINO ACIDS

Of all the fundamental amino acids, 10 (ten) are classified as"ESSENTIAL" and are NOT biosynthesized by the body. All commonly occurring amino acids can be subdivided into 5 fundamental groups each of which has functional significance.

- *Group* #1: This group includes all the amino acids possessing aliphatic properties (those whose carbon atoms are joined in open chains), Glycine, Alanine, Leucine, w, Serine, Threonine, and Isoleucine.
- *Group* #2: These are the Sulfur containing amino acids, Methionine, Cystine, Cysteine.
- *Group* #3: These amino acids are the acidic amino acids, (those that can donate a hydrogen ion), Aspartic Acid, Glutamic Acid, and their associated amids.
- *Group* #4: These are the diamino acids (those with additional NH_2 groups), Arginine and Lysine.
- *Group* #5: These are the cyclic and aromatic amino acids, Phenylalanine, Tyrosine, Tryptophan, Histidine, Proline, and Hydroxyproline that is synthesized from Prolamine.

ESSENTIAL AMINO ACIDS

Table. Estimated Minimum Requirements of Adults for Essential Amino Acids

Amino Acid	Women/gms/day	Men/gms/day
Cystine	0.055	1.010
Isoleucine	0.450	0.700
Leucine	0.620	1.100
Lysine	0.500	0.800
Methionine	questionable	1.100
Phenylalanine	0.220	1.010
Threonine	0.310	0.500
Tryptophan	0.160	0.250
Tyrosine	0.900	questionable
Valine	0.650	0.800

Cystine

An essential, Sulfur-containing amino acid, produced by digestion or acid hydrolysis of proteins, sometimes found in the urine and kidneys, and readily reduced to two molecules of Cysteine.

- Reported helpful in dermatological conditions.
- Promotes faster recovery of tissue after surgery.
- Part of the Insulin molecule.
- Found high in hair (sulfur bonds).

Isoleucine

An essential amino acid produced by hydrolysis of fibrin and other proteins; ESSESNTIAL for optimal INFANT growth and for nitrogen equilibrium in adults.

- Needed along with other branch chain amino acids for all rebuilding of muscle tissue.
- Role in the release of energy during muscular work. (Metabolizes along the same pathway as fat.)

Leucine

An essential amino acid, ESSENTIAL for optimal growth in INFANTS, and for nitrogen equilibrium in adults.

- Metabolized along the same pathways as fat.
- Precursors of cholesterol.
- Involved in the role of energy release during any work of the muscles.

Lysine

An essential, naturally occurring amino acid, ESSESNTIAL for optimal growth in human INFANTS, and for maintenance of nitrogen equilibrium in adults.

- Reported to inhibit growth and replication of herpes simplex and Epstein Barr viruses (EBV).
- Promotes bone growth in infants.
- Stimulates secretion of gastric juices.
- Found in abundance in muscle tissue, connective tissue and collagen.
- Low in vegetarian diets.

Methionine

An essential, naturally occurring amino acid, which is an essential component of the diet, furnishing both methyl groups and sulfur necessary for normal metabolism.

- Prevents deposits and cohesion of fats in the liver.
- Gives rise to Taurine (an important inhibitory neuro modulator in the brain).
- Involved in synthesis of Choline (Must be given with Vitamin B6 (paradoxial-5-phosphate) to inhibit synthesis of homocysteine which promotes plaque deposition in the arteries.)

Phenylalanine

An essential, naturally occurring amino acid, ESSESNTIAL for optimal growth in INFANTS, and for nitrogen equilibrium in human adults.

- May be useful in appetite control by stimulating CCK secretion.

- Shown to be useful in management of certain types of depression.
- Increases blood pressure in hypotension.
- Gives rise to Tyrosine.
- It is one of the amino acids in the dypeptide sweetener Aspartain. (Aspartic acid and Phenolalanine).

Threonine

An essential, naturally occurring amino acid, essential for human metabolism.

- Rises to three times it's normal value at pregnancy.
- Acts as a lipotropic factor.
- Recently found to increase brain Glycine content, greatly reducing ALS symptoms.

Tryptophan

An essential, naturally occurring amino acid, existing in proteins and essential for human metabolism.

- Reported useful in the management of depression and schizophrenia.
- Produces Serotonin, which induces sleep. Has a Serotonengenic effect.
- Precursor of the vitamin Niacin.
- Vasoconstrictor, which appears to aid in blood clotting mechanism, aids in elevating the threshold of pain.

Tyrosine

An essential, naturally occurring amino acid present in most proteins. It is a product of Phenylalanine metabolism and a precursor of thyroid hormones, catecholamines, and melanin.

- Along with Phenylalanine, it is a useful anti depressant due to increased production of Tacolamine. (Recommended to be taken with Vitamin B6 (paradoxial-5-phosphate).)
- Reported to stabilize blood pressure by lowering energy in some cases and elevating it in others.
- Involved in tissue pigmentation
- Is important in the formation of thyroid hormone

Valine

An essential, naturally occurring amino acid, essential for human metabolism.

- Needed for all muscle building.
- Required in the precursors of cholesterol.

Histidine

An amino acid obtainable from many proteins by the action of sulfuric acid and water; it is ESSENTIAL for optimal growth in INFANTS. Its decarboxylation results in formation of histamine.

- Release of histamines from body stores are required for sexual arousal.
- Reported useful in alleviating pain associated with rheumatoid arthritis. (Should be taken with vitamin C; Should be used with caution with Manic Depressives with elevated histamines; Should be used with caution in woman with severe depression or suicidal tendencies due to Premenstrual Syndrome; Is considered a mild neurotransmitter).

TYPES OF AMINO ACIDS

AMINOACETIC ACID

A nonessential amino acid, derived from many proteins, used as a gastric antacid and dietary supplement, and in the treatment of various myopathies and peripheral vascular insufficiencies. A solution is used as an irrigation fluid. Its hydrochloride salt is use as a source of hydrochloric acid in the treatment of achlorhydria.

Alanine

A natural amino acid occurring in two forms, alpha-Alanine, and beta-Alanine.

- Hypoglycemia (Prolongs the stabilization of blood glucose over long periods of time.)
- Alanine may be used for the production of glucose in gluconeogenesis.

Arginine

An amino acid occurring in proteins; it is involved in the urea cycle, which converts ammonia to urea.

- Induces growth hormone release from the pituitary gland.
- Major component of seminal fluid.
- Helpful in burn treatment, elevated ammonia levels, and cirrhosis of the liver.
- Stimulates immune response by enhancing the production of T-Cells.
- Protective effect of toxicity of hydrocarbons and intravenous diuretics.

(Should be used with caution in schizophrenic cases.)

Intake should be kept low in persons with Herpes simplex and Epstein Barr Virus (EBV. Mechanism of interaction is believed because of possible replications of the Herpes and EB viruses.)

Aspartic Acid

A nonessential, natural dibasic amino acid, involved in transamination reactions, the ornithine cycle, and the formation of carnosine, anserine, purines, and pyrimidines.

- Has a protective function over the liver.
- Helps in detoxification of ammonia.
- Promotes mineral uptake in the intestinal tract.
- It is part of the sweetener Aspartain that is a dypeptide of the amino acids; Aspartic Acid and Phenylalanine.

Carnitine

A betaine derivative involved in the transport of fatty acids into mitochondria, where they are metabolized.

- Suggested as useful in mobilization of surface fats such as cellulite.
- Helpful in treatment of fatigue and when muscle weakness is present. Useful in the oxidation of long chain fatty acids. A major source of energy for tissues.
- Reported as useful in myocardial ischemia (low energy metabolism by the heart).
- Useful in clearing triglycerides from the blood.
- Useful in metabolic abnormalities.

Citrulline

An alpha-amino acid involved in urea production; formed from ornithine and is itself converted into Arginine in the urea cycle.

- Produces the amino acids, Arginine and Ornithine.
- Detoxifies nitrogen containing waste products such as ammonia (a cellular toxin) and is therefore part of ammonia detoxification (urea cycle).
- Stimulates growth hormone production.

Cysteine

A sulfur-containing amino acid produced by enzymatic or acid hydrolysis of proteins, readily oxidized to Cystine, sometimes found in urine.

- Assists tissues damaged by alcohol abuse, cigarette smoke and air pollution through the detoxification of Acetalaldahide.
- Helps maintain skin flexibility and texture by slowing abnormal cross linking of collagen (connective tissue protein which holds the skin together).
- Promotes red and white blood cell reproduction and tissue restoration in lung affecting diseases.
- Useful in iron deficient anemia and promotion of iron absorption.
- Helpful in prevention of peroxidized fats and free radicals.
- Converts to Cystinine in the absence of vitamin C.

- One of the amino acids found in tripeptide Glutathione and GTF.
- Hair contains between 10 to 16 per cent Cysteine.
- Heavy metal chelator. (Should be used with caution in diabetics due to possible three dimensional structure changes in the insulin cycle)

Cystine

An essential, Sulfur-containing amino acid, produced by digestion or acid hydrolysis of proteins, sometimes found in the urine and kidneys, and readily reduced to two molecules of Cysteine.

- Reported helpful in dermatological conditions.
- Promotes faster recovery of tissue after surgery.
- Part of the Insulin molecule.
- Found high in hair (sulfur bonds).

GABA - (Gamma Amino Butyric Acid)

An amino acid that is one of the principle inhibitory neurotransmitters in the central nervous system.

- Useful in schizophrenia and epilepsy, depression, high blood pressure, high stress levels, manic Behaviour, and acute agitation due to increased regulation of nerve firings and enhancement of Niacinamide binding receptors.
- May induce calmness and tranquility by inhibiting neurotransmitters, which decrease the activity in neurons.
- May be useful in reducing an enlarged prostate due to the suppression of Prolactia released by the pituitary gland.

Glutamic Acid

Glutamate is a salt of Glutamic Acid; in biochemistry, the term is often used interchangeably with Glutamic Acid. Glutamic Acid is a crystalline dibasic nonessential amino acid, widely distributed in proteins, which is thought to be a neurotransmitter, inhibiting neural excitation in the central nervous system; its hydrochloride salt is used as a gastric acidifier. The monosodium salt of L-Glutamic acid (sodium glutamate) is used in treating encephalopathies associated with hepatic diseases, and to enhance the flavour of foods and tobacco.

- Reported useful in Muscular Dystrophy.
- Does not cross blood brain barriers.
- It is an excitatory neurotransmitter in the CNS; (Precursor of GABA.)

Glutamine

The monoamide of Glutamic Acid, an amino acid occurring in proteins; it is an important carrier of urinary ammonia and is broken down in the kidney by the enzyme glutaminase.

- Is useful in treatment of Alcoholism by reducing the desire to drink.
- May be helpful in the improvement of Autism and mental retardation in children.
- Reported improvement in mental functions such as memory and dexterity.
- Aids peptic ulcer healing due to antacid quality.
- Crosses Blood brain barriers.
- Participates in nucleonic acid synthesis.
- Converts to Glutanic acid.

Glutathione

Reduced Glutathione, a tripeptide of Glutamic Acid, Cysteine, and Glycine, which serves as a reducing agent in many biochemical reactions being converted to oxidized Glutathione (GSSG) in which the Cysteine residues of two Glutathione molecules are connected by a disulfide bridge. Reduced Glutathione is important in protecting erythrocytes (red blood cells) from oxidation and hemolysis; deficiency causes sensitivity to oxidant drugs.

- It is a tripeptide containing amino acid that contains: Cysteine, Glycine, and Glutamate.
- Inhibits peroxide formations, reducing free radical damage.
- Detoxifies aromatic hydrocarbons common in air pollution, such as chlorine. Reported to have protective function against radiation therapy.
- Transports other amino acids into the interior of the cell.

Glycine

A nonessential amino acid, occurring as a constituent of proteins and functioning as an inhibitory neurotransmitter in the central nervous system; used as a gastric antacid and dietary supplement, and in the treatment of various myopathies.

- Is the simplest and sweetest of the amino acids. Can be used as a sweetener in herbal beverages.
- Reported useful in degenerative diseases such as MD (Muscular Dystrophy)
- Detoxifies benzoic acid (a common food additive) and aromatic hydrocarbons in the liver.
- Involved in synthesis of nucleic acids and bio-acids. May be useful in conditions characterized by abnormal nerve firing such as epilepsy, inhibition of tripeptides Glutathione and GTF.
- Is one of the amino acids in the tripeptides Glutathione and GTF; low brain concentrations of Glycine have recently been found in ALS. (Amyotrophic Lateral Sclerosis or"Lue Gehrig's Disease.")

Histidine

Is not listed as essential for adults, but is very essential for infants. An amino acid obtainable from many proteins by the action of sulfuric acid and water; it is ESSENTIAL for optimal growth in INFANTS. Its decarboxylation results in formation of histamine.

- Release of histamines from body stores are required for sexual arousal.
- Reported useful in alleviating pain associated with rheumatoid arthritis. (Should be taken with vitamin C; Should be used with caution with Manic Depressives with elevated histamines; Should be used with caution in woman with severe depression or suicidal tendencies due to Premenstrual Syndrome; Is considered a mild neurotransmitter).

Hydroxyproline

An amino acid produced in the digestion of hydrolytic decomposition of proteins, especially collagens.

Isoleucine

An essential amino acid produced by hydrolysis of fibrin and other proteins; ESSESNTIAL for optimal INFANT growth and for nitrogen equilibrium in adults.

- Needed along with other branch chain amino acids for all rebuilding of muscle tissue.
- Role in the release of energy during muscular work. (Metabolizes along the same pathway as fat.)

Leucine

An essential amino acid, ESSENTIAL for optimal growth in INFANTS, and for nitrogen equilibrium in adults.

- Metabolized along the same pathways as fat.
- Precursors of cholesterol.
- Involved in the role of energy release during any work of the muscles.

Lysine

An essential, naturally occurring amino acid, ESSESNTIAL for optimal growth in human INFANTS, and for maintenance of nitrogen equilibrium in adults.

- Reported to inhibit growth and replication of herpes simplex and Epstein Barr viruses (EBV).
- Promotes bone growth in infants.
- Stimulates secretion of gastric juices.
- Found in abundance in muscle tissue, connective tissue and collagen.
- Low in vegetarian diets.

Methionine

An essential, naturally occurring amino acid, which is an essential component of the diet, furnishing both methyl groups and sulfur necessary for normal metabolism.

- Prevents deposits and cohesion of fats in the liver.
- Gives rise to Taurine (an important inhibitory neuro modulator in the brain).
- Involved in synthesis of Choline (must be given with Vitamin B6 (paradoxial-5-phosphate) to inhibit synthesis of homocysteine which promotes plaque deposition in the arteries)

Autistic Patients

"Taurine is deficient in Autistic patients; Methionine has an altered abnormally normal to high level, with anything distal to the Methionine pathway being low. Since Taurine is so important in the intelligence quotients of most species, this seems to be a significant." Alpha A - Leverton, Protein and Amino Acid Nutrition (1959) B - Rose, Journalism of Biological Chemistry 217:977, 1955.

Ornithine

An amino acid obtained from Arginine by the splitting of urea; it is an intermediate in urea biosynthesis.

- May reduce fat and increase muscle mass by promoting fat metabolism and stimulating growth hormone production.
- Helps in detoxification of ammonia in the urea cycle.
- May be useful in autoimmune disease such as arthritis.

Phenylalanine

An essential, naturally occurring amino acid, ESSESNTIAL for optimal growth in INFANTS, and for nitrogen equilibrium in human adults.

- May be useful in appetite control by stimulating CCK (Chaleceptokinin Enzyme) secretion.
- Shown to be useful in management of certain types of depression.
- Increases blood pressure in hypotension.
- Gives rise to Tyrosine
- It is one of the amino acids in the dypeptide sweetener Aspartain.

Proline

A cyclic amino acid occurring in proteins; it is a major constituent of collagen.

- Is an anti-hypotensive agent in lowering high blood pressure.
- Reported helpful in repairing muscle and tendon damage.
- Reported useful in promoting skin flexibility in relation to aging and sun exposure.

- A major amino acid found in collagen (connective tissue) in the presence of Vitamin C.

Serine

A naturally occurring Amino Acid, present in many proteins.

Taurine

A crystallized acid, ethylamine sulfonic acid, from the bile; found also in small quantities in lung and muscle tissue.

- Low levels seen in newborn infants fed low Taurine diets.
- Associated with retinal degenerations.
- The role of Taurine as a nutrient is to protect the cell membranes by attenuating such toxic compounds as oxidants, secondary bioacids and antibiotics.
- Recommended for children on long-term parenteral nutrition.
- Helpful in balancing calcium and potassium flux in heart muscle.
- Helps patients suffering from congenitive heart failure by alleviating their physical signs and symptoms.
- Increases left ventricular performance without any significant changes in atrial pressure.
- Often times considered a neuro modulator.
- Helpful in treating some types of epilepsy.

Autistic Patients:"Taurine is deficient in autistic patients; Methionine has an altered abnormally normal to high level, with anything distal to the Methionine pathway being low. Since Taurine is so important in the intelligence quotients of most species, this seems to be a significant.

Threonine

An essential, naturally occurring amino acid, essential for human metabolism.

- Rises to three times it's normal value at pregnancy.
- Acts as a lipotropic factor.
- Recently found to increase brain Glycine content, greatly reducing ALS symptoms.

Tryptophan

An essential, naturally occurring amino acid, existing in proteins and essential for human metabolism.

- Reported useful in the management of depression and schizophrenia.
- Produces Serotonin, which induces sleep. Has a Serotonengenic effect.
- Precursor of the vitamin Niacin.

- Vasoconstrictor, which appears to aid in blood clotting mechanism, aids in elevating the threshold of pain.

(Should be taken with Vitamin B6 (paradoxial-5-phosphate) and in the presence of carbohydrates such as fruit or vegetable juice to maximize uptake in the brain.)

Tyrosine

An essential, naturally occurring amino acid present in most proteins. It is a product of Phenylalanine metabolism and a precursor of thyroid hormones, catecholamines, and melanin.

- Along with Phenylalanine, it is a useful anti depressant due to increased production of Tacolamine. (Recommended to be taken with Vitamin B6 (paradoxial-5-phosphate).)
- Reported to stabilize blood pressure by lowering energy in some cases and elevating it in others.
- Involved in tissue pigmentation
- Is important in the formation of thyroid hormone (Should not be used when MAO's are prescribed or when cancer melanoma is present)

Valine

An essential, naturrally occurring amino acid, essential for human metabolism.

- Needed for all muscle building.
- Required in the precursors of cholesterol.

REACTIONS OF AMINO ACIDS

CARBOXYL AND AMINO GROUP REACTIONS

The α–carboxyl and α–amino groups of all amino acids exhibit similar chemical reactivity. The side chains, however, exhibit specific chemical reactivities, depending on the nature of the functional groups. Whereas all of these reactivities are important in the study and analysis of isolated amino acids, it is the characteristic behaviour of the side chain that governs the reactivity of amino acids incorporated into proteins.

There are three reasons to consider these reactivities. Proteins can be chemically modified in very specific ways by taking advantage of the chemical reactivity of certain amino acid side chains. The detection and quantification of amino acids and proteins often depend on reactions that are specific to one or more amino acids and that result in colour, radioactivity, or some other quantity that can be easily measured. Finally and most importantly, the biological functions of proteins depend on the behaviour and reactivity of specific R groups.

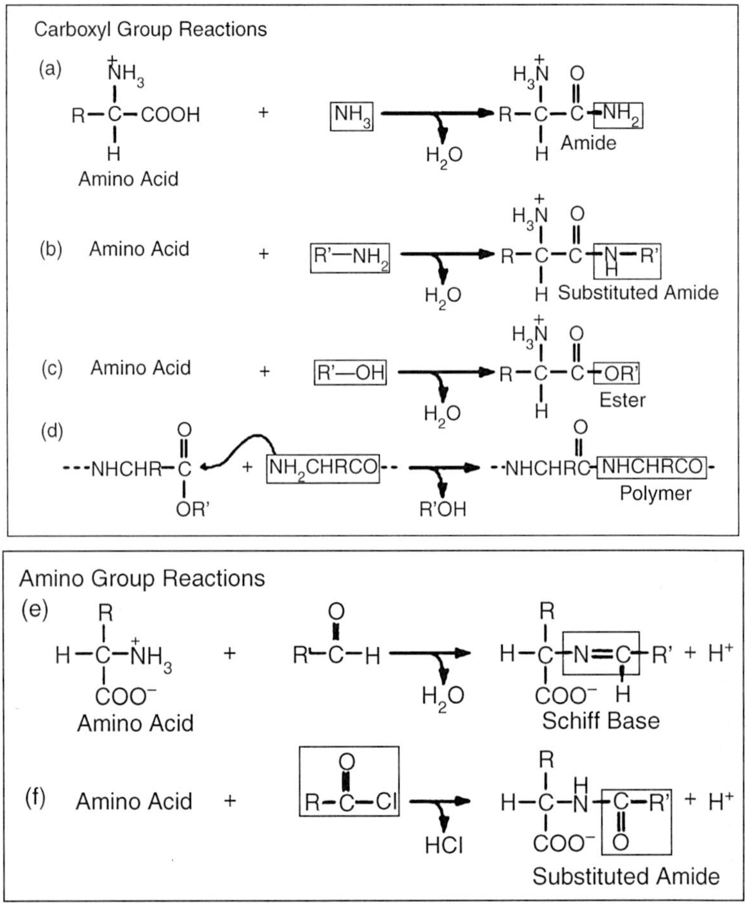

Fig. Typical Reactions of the Common Amino Acids

The carboxyl groups of amino acids undergo all the simple reactions common to this functional group. Reaction with ammonia and primary amines yields unsubstituted and substituted amides, respectively (Figure). Esters and acid chlorides are also readily formed. Esterification proceeds in the presence of the appropriate alcohol and a strong acid. Polymerization can occur by repetition of the reaction shown in Figure. Free amino groups may react with aldehydes to form Schiff bases (Figure) and can be acylated with acid anhydrides and acid halides.

THE NINHYDRIN REACTION

Amino acids can be readily detected and quantified by reaction with ninhydrin. As shown in Figure, *ninhydrin*, or triketohydrindene hydrate, is a strong oxidizing agent and causes the oxidative deamination of the α–amino function. The products of the reaction are the resulting aldehyde, ammonia, carbon dioxide, and hydrindantin, a reduced derivative of ninhydrin. The ammonia produced in this way can react with the hydrindantin and another molecule of ninhydrin to yield a purple product (Ruhemann's Purple) that can be quantified spectrophotometrically at 570 nm.

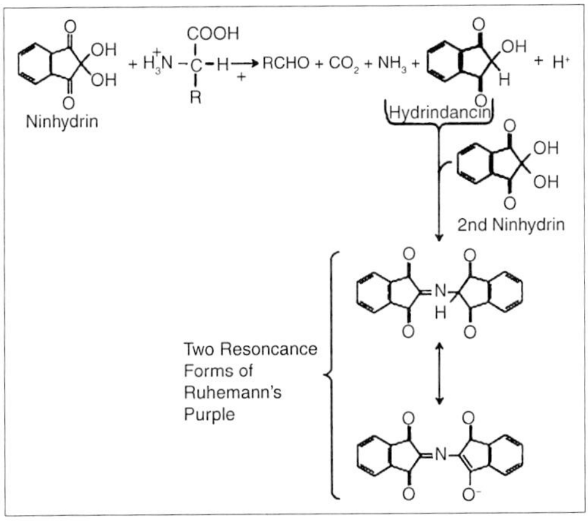

Fig. The Pathway of the Ninhydrin Reaction, which Produces a Coloured Product called "Ruhemann's Purple" that Absorbs Light at 570 nm

The appearance of CO_2 can also be monitored. Indeed, CO_2 evolution is diagnostic of the presence of an α–amino acid. α-amino acids, such as proline and hydroxyproline, give bright yellow ninhydrin products with absorption maxima at 440 nm, allowing these to be distinguished from the α–amino acids. Because amino acids are one of the components of human skin secretions, the ninhydrin reaction was once used extensively by law enforcement and forensic personnel for fingerprint detection. (Fingerprints as old as 15 years can be successfully identified using the ninhydrin reaction.) More sensitive fluorescent reagents are now used routinely for this purpose.

SPECIFIC REACTIONS OF AMINO ACID SIDE CHAINS

A number of reactions of amino acids have become important in recent years because they are essential to the degradation, sequencing, and chemical synthesis of peptides and proteins. In recent years, biochemists have developed an arsenal of reactions that are relatively specific to the side chains of particular amino acids. These reactions can be used to identify functional amino acids at the active sites of enzymes or to label proteins with appropriate reagents for further study. Cysteine residues in proteins, for example, react with one another to form disulfide species and also react with a number of reagents, including maleimides (typically *N*–ethylmaleimide), as shown in Figure.

Crysteines also react effectively with iodoacetic acid to yield *S*–carboxymethyl cysteine derivatives. There are numerous other reactions involving specialized reagents specific for particular side chain functional groups. Figure presents a representative list of these reagents and the products that result. It is important to realise that few if any of these reactions are truly specific for one functional group; consequently, care must be exercised in their use.

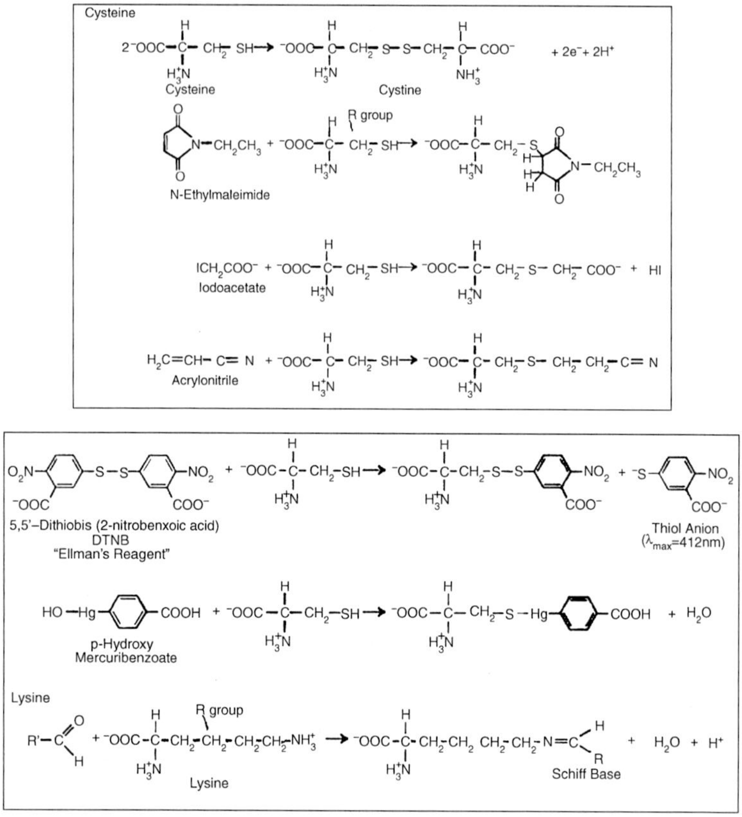

Fig. Reactions of Amino Acid Side-chain Functional Groups

6

Nucleus: The Cell Structure

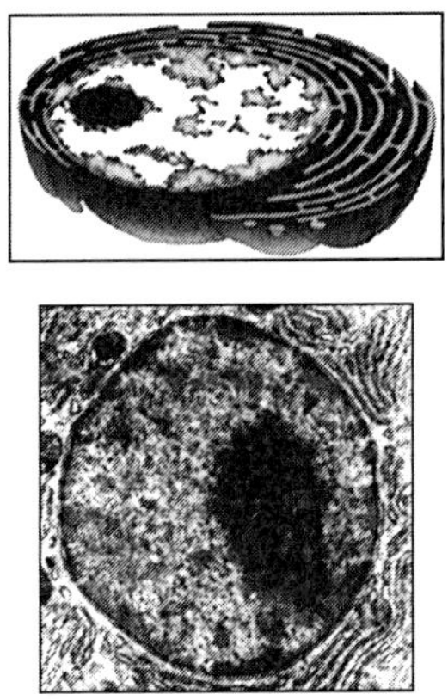

Fig. Nucleus

Look down:

- Notice the slight curve? You're standing on a somewhat spherical structure about 50 feet in diameter. It's the nucleus—basically the cell's brain.

The nucleus is the most prominent organelle and can occupy up to 10 per cent of the space inside a cell. It contains the equivalent of the cell's gray matter—its genetic material, or DNA. In the form of genes, each with a host of helper molecules, DNA determines the cell's identity, masterminds its activities, and is the official cookbook for the body's proteins. Go ahead—jump.

It's a bit springy, isn't it? That's because the nucleus is surrounded by two pliable membranes, together known as the nuclear envelope. Normally, the nuclear envelope is pockmarked with octagonal pits about an inch across (at this scale) and hemmed in by raised sides. These nuclear pores allow chemical messages to exit and enter the nucleus. But we've cleared the nuclear pores off this area of the nucleus so you don't sprain an ankle on one. If you exclude the nucleus, the rest of the cell's innards are known as the cytoplasm.

Eukaryotic Cells	Prokaryotic Cells
The cells of "complex" organisms, including all plants and animals	"Simple" organisms, including bacteria and blue–green algae
Contain a nucleus and many other organelles, each surrounded by a membrane (the nucleus and mitochondrion have two membranes)	Lack a nucleus and other membrane–encased organelles
Can specialize for certain functions, such as absorbing nutrients from food or transmitting nerve impulses; groups cells can form large, multicellular organs and organisms	Usually exist as single, virtually identical cells
Most animal cells are 10–30 micrometers across, and most plant cells are 10–100 micrometers across	Most are 1–10 micrometers across

THE NUCLEOLUS

The presence of nucleoli, which appear as dense granules within the nucleus, was first described by Fontana in 1781. By the end of the 19 th century a relationship between the size of the nucleolus and the synthetic activity of the cell was postulated. It was found that nucleoli were small or absent in cells exhibiting little protein synthesis (sperm cells, blastomeres, muscle cells, and so forth), whereas they were large in oöcytes, neurons, and secretory cells — those in which protein synthesis is a prominent feature. In the living cell, nucleoli are highly refringent bodies.

This is the result of a large concentration of solid material, which may constitute 40 to 85 per cent of the dry mass. The light microscope generally reveals the nucleolus as structurally homogeneous, although small corpuscles or vacuoles are sometimes noted. The nucleolus stains with pyronine and other stains and absorbs ultraviolet light at 260 nm. Treatment with ribonuclease shows that the capacity to absorb this basophilic stain and ultraviolet ratiation depends on the presence of RNA.

The nucleolus may be surrounded by a ring of Feulgen-positive chromatin, which represents the heterochromatic regions of the chromosomes associated with the nucleolus. In large nucleoli some Feulgen-positive granules can be seen in portions of the chromosomes that penetrate the nucleolus. The nucleolus is formed at the nucleolar organizer, which is a chromosomal site that contains tandem repeats of the genes coding for 18S and 28S and rRNA. This DNA becomes uncoiled and penetrates the nucleolus, where it is actively transcribed. In addition, other ribosomal components, such as 5S RNA and the ribosomal proteins, which are synthesized in other parts of the cell, converge on the nucleolus, where the assembly of ribosomal subunits starts. Electron microscopy has confirmed the existence of a definite submicroscopic organization within the nucleolus. In some cells nucleoli have a compact structure; in others the structure may be more or less open and include clear regions that communicate with the nucleoplasm. Two characteristic components may be recognized: the granular zone and the fibrillar zone.

Granular Zone

The *granular zone* consists of electron-dense granules of 15 to 20 nm, *i.e.*, smaller in diameter than the ribosomes. Frequently the granular zone occupies the more peripheral region of the nucleolus, which is surrounded by the associated chromatin. Both the granular and the fibrillar zones are most clearly distinguished.

The *fibrillar zone* consists of fine fibres 5 to 10 nm in diameter. It generally occupies the central region of the nucleolus. Both the granular and fibrillar zones are digested by ribonuclease. In some cells the nucleolus is surrounded by a ring of heterochromatin, called the *nucleolarassociated* chromatin, which sometimes extends into the nucleolus. Autoradiographic studies indicate that ^{3}H-uridine is incorporated first into the fibrillar area and is only later found in the granular part. This suggests the relationship:

$$\text{nucleolar DNA} \rightarrow \text{fibrillar area} \rightarrow \text{granular area}$$

The fibrillar area contains the ribosomal DNA and its initial transcripts (called 45S RNA), and the granular part contains ribosomal precursor particles at different stages of maturation. During mitosis nucleoli undergo cyclic changes. Nucleoli are formed around the DNA loop that extends from the nucleolar organizer. There may be several nucleoli per cell, but frequently they tend to fuse into one or few nucleoli.

During late prophase the DNA loop containing the rRNA genes gradually retracts and coils into the nucleolar organizer of the corresponding chromosome. Since this DNA is very extended as a consequence of intense RNA synthesis, the nucleolar organizer region is one of the last to undergo condensation, thus producing a *secondary constriction* on the chromosome. The fibrillar and granular components are gradually dispersed into the nucleoplasm. After the cell divides, during telophase, the nucleolar organizer DNA uncoils and the nucleolus is reassembled.

A knowledge of the behaviour of the chromosomes during meiosis and especially in what has already been said about the genetic consequences. However, in 1856 when Gregor johann Mendel discoverd the fundamental laws of inheritance, nothing was known about the chromosomes and meiosis. His discovery was based on the precise quantification of experimental crosses and exceptional abstract thinking. He studied crosses between peas (*Pisum sativum*) which had pairs of differential or contrasing characteristics. For example, he used plant that have white and red flowers, smooth and roguh seeds, yellow and green seeds, long and short stems, and so forth. After crossing the parental generation (P_1), he observed the resulting *hybrids* of the first filial generation, F1. Then he crossed the hybrids (F_1) among themselves an studied the result in the second filial generation. F_2. In a cross between parents with yellow and green seeds in the first generation Mendel found that all the hybrids had yellow seeds and thus, the characteristic of only one parent In the second cross (F_2), the characteristics of both parents reappeared in the proportion of 75 percent to 25 per cent, or 3:1. Mendel postulated that the colour of the seeds was controlled by a "factor" that was transmitted to the offspring by means of the gametes. This hereditary factor, which is now called the *gene,* could be transmitted without mixing with other genes. He postulated that the gene could be segregated in the hybrid into different games to be *distributed* in the offspring of the hybrid. For this reason this is called the *law* or *principle of segregation of the genes.*

Mendel found that the plants with yellow colour, had different genetic constutions. One-third of this group always gave yellow seeds, but the other two-thirds of the F_2 generation produced planats with yellow and green seeds in the ratio of 3:1 When the 25 per cent of plants in F_2 with green seeds wrer a pure strain for this character. If we represent the genes in the crossing by letters, designating by *A* the gene with yellow character and by *a* the gene with green character, we have the folloing. In the first generation (F_1) both *A* and *a* genes are present, but only A is revealed because it is *dominant;* gene *a* remains hidden and is called *recessive.* In the hybrid F_1 both genes are segregated and enter different gametes. Half of them will have the gene *A,* and the other half,*a.* Since each individual produces two types of games in each sex, there are four possible combinations in $F_{2.}$ This gives as a result the proportion 1:2:1, corresponding to 25 per cent of plants with pure yellow seeds (*AA*), 50 per cent with hybrid yellow seed (*Aa*), and 25 per cent with pure green seeds (*aa*).

Mendel's results can now be explained in terms of the behavioural of chromosomes and genes. The genes present in the chromosomes are found in pairs called *alleles.* In each homologous chromosome the gene for each trait occurs at a particular point called a *locus* (plural *loci*). The gray mouse will have two GG genes, one in each homologue. Since the two homologues pair and then separate to enter the gametes.

The mechanism is the same in a dominant as in a recessive white mouse. In the hybrid F_1, one chromosome bears gene G and the homologous chromosome bears gene *g*. When hybrids are self-fertilized, the gametes unite in the combinations shown by the checkerboard method illustrated in the individuals having two similar alleles are called *homozygous* (i. e., *GG* or *gg*); those with different alleles, *heterozygous.*

GENOTYPE AND PHENOTYPE

Johanssen proposed the term *genotype* for the genetic constitution and *phenotype* for the visible characteristics shown by the individual in (91). For example, in the case of the peas with green or yellow seeds there are two phenotypes in $F_{2:}$ yellow seeds and green seeds in the ratio of 3:1, respectively. However, according to the genetic constitution, there are three different genotypes: *AA, Aa,* and *aa* in the ratio of 1:2:1. The phenotype includes all characteristics of the individual that are an expression of gene activity. For example, in humans phenotypic characteristics include eye colour, baldness, the different hemoglobins, the blood groups, and a difference in the ability to taste thiourea.

In crossings of certain plants that have white and red flowers, such as *Mirabilis jalapa,* it is possible to find in F_2 three phenotypes (red, pink, and white flowers), which correspond to the three genotypes. This is due to incomplete dominance. The principle of dominance and recessiveness in not always accomplished completely; dominance may be complete in most cases, but incomplete in others. In this case there is a mixture of characteristics, called *intermediary heredity.*

Whereas the law of segregation applies to the behaviour of a single pair of genes, the *principle of independent assortment* describes the simultaneous behaviour of two or more pairs of genes located in different pairs of chromosomes. Genes that lie in separate chromosomes are independently distributed during meiosis. The resulting off-spring is a hybrid (also called a dihybrid) at two loci.

All the above-mentioned examples of genetic crosses illustrated the fact that during meiosis there is a random distribution of the chromosomes which leads to the segregation of the genes in the gametes. However, when this type of study was carried out in the fruit fly *Drosophila melanogaster* by Morgan and collaborators, it became evident that the law of independent assortment was not universally applicable and that in certain crosses of two or more allelic pairs of genes, there was a certain limitation of free segregation. In each case there was a marked tendency for parental combinations to remain linked and for a lesser proportion of new combinations to be produced. In *Drosophila* there are only four pairs of chromosomes, and this increases the chances for genes to occupy loci in the same chromosome.

If two genes A and B, with the corresponding alleles a and b, are in the same chromosome, only two classes of gametes will be obtained. After studying a considerable number of different crosses in *Drosophila,* Morgan reached the conclusion that all genes of this fly were clustered into four linked groups corresponding to the four pairs of chromosomes.

Recombination During Prophase

Further studies along these lines revealed that the linkage is not absolute and that it may be broken with a certain frequency. We have seen in the previous chapter that *genetic crossing over* or *recombination* takes place between the DNA molecules and is not visible microscopically. However, the *chiamata* observed at diplonema morphologically represent in some way the sites of these molecular events. Thus, there should be a correlation between the two phenomena.

It is also known that the frequency of recombination of two linked genes is a function of the distance which separates them along the chromosome. When two genes are close to one another, the probability of crossing over is less than when they are far apart. If the distance between genes is estimated by linkage analysis, it is possible to construct a map indicating the relative position of each gene along the chromosome.

An appraisal of the possible number of new recombinations may be obtained by counting the number of chiasmata during meiosis. The so-called *recombination index* is calculated by adding to the number of bivalents the number of chiasmata detected in the same cell at diplonema. In a species with a higher indes, the possibility of new recombinations is higher, and this implies a greater possibility of variation.

There are several other pieces of evidence to support the relationship between chiasmata and crossing over. The number of chiasmata is related to the length of the chromosomes in the bivalent. The presence of one chiasma reduces the possibility of another occurring in the immediate vicinity. This phenomenon has been called *positive interference.* The chiasma frequency is roughly constant in a given species, but can be modified by genetic or environmental action. Chromosome pairing, chiasma formation, and crossing over are under genetic control. Lines of rye with high and low chiasma frequencies have been obtained by inbreeding. Among environmental factors that may affect the number of chiasmata are temperature, radiation, chemicals, and nutrition.

Neurospora

Among the different organisms studies in genetics, the mold *Neurospora* osccupies a special place.

The advantage of this material is twofold:

- It is possible to identify and to follow the fate of each of the four chromatids present in the bivalent meiotic chromosome and thus to determine whether the crossing over involves two, three, or all four chromatids.
- It is possible to make a close correlation between genetic constitution and biochemical expression of genes.

The four cells resulting from the two meiotic division undergo a mitotic division, which gives rise to eight haploid ascospores. Each of these ascospore can be isolated by dissection and cultured separately, giving rise to haploid individuals having the genetic constitution carried in each of the four original chromatids of the bivalent chromosome.

CELL MEMBRANE

The membrane that surrounds a cell is made up of proteins and lipids. Depending on the membrane's location and role in the body, lipids can make up anywhere from 20 to 80 per cent of the membrane, with the remainder being proteins. Cholesterol, which is not found in plant cells, is a type of lipid that helps stiffen the membrane. You may not remember it, but you crossed a membrane to get in here. Every cell is contained within a membrane punctuated with special gates, channels, and pumps. These gadgets let in—or force out—selected molecules. Their purpose is to carefully protect the cell's internal environment, a thick brew (called the cytosol) of salts, nutrients, and proteins that accounts for about 50 per cent of the cell's volume (organelles make up the rest).

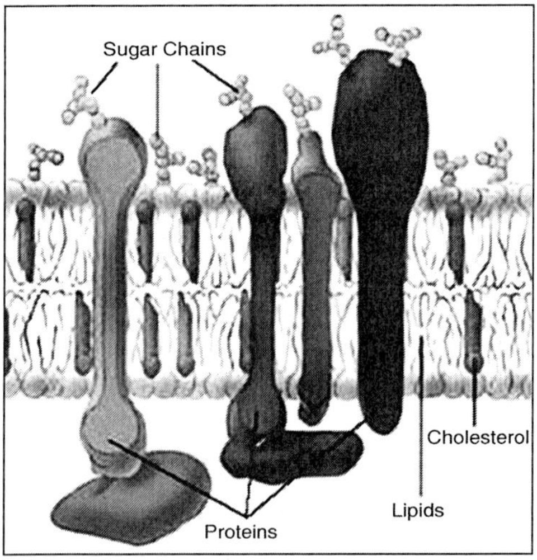

The cell's outer membrane is made up of a mix of proteins and lipids (fats). Lipids give membranes their flexibility. Proteins transmit chemical messages into the cell, and they also monitor and maintain the cell's chemical climate. On the outside of cell membranes, attached to some of the proteins and lipids, are chains of sugar molecules that help each cell type do its job. If you tried to bounce on the cell's outer surface as you did on the nuclear membrane, all these sugar molecules and protruding proteins would make it rather tricky (and sticky).

CHEMICAL COMPOSITION

Regarding chemical composition of the plasma membrane of the human red cell is more than about that of any other cell. Protein represents approximately 52 per cent of its mass, *lipids* 40 per cent, and *carbohydrates* 8 per cent. Oligosaccharides are bound to lipids for example *glycolipids* and, mainly, to proteins for example *glycoproteins).* There is a wide variation in the lipids-protein ratio between different cell membranes. Myelin is an exception, in the sense that the lipid predominates; in the other cell membranes there is higher protein lipid ratio. It may be observed that in myelin the area occupied by the protein is insufficient to cover that of the lipids, whereas in a red cell ghost the opposite situation is found.

The main lipid components of the plasma membrane are phospholipids, cholesterol, and galactolipids; their proportion varies in different cell membranes. The major proportion of membrane phospholipids is represented by phosphatidyl-choline, phosphatidylethanolamine and sphingomyelin, all of which have no net charge at neutral pH (*i.e., neutral phospholipids)* and tend to pack tightly in the bilayer.

This property is also shared by cholesterol.) Five to 20 per cent of the phospholipids are acidic, including: phosphatidylinositol, phosphatidylserine, cardiolipin, phosphatidylglycerol, and sulfolipids. *Acidic phospholipids* are negatively charged and in the membrane are associated principally with proteins by way of lipid-protein interactions. One of the main characteristics of the molecular organization of the plasma membrane is the asymmetry of all of its chemical components. This refers to their non-uniform distribution between the two surfaces, *i.e.*, the inner or *protoplasmic surface* (P_s), in contact with the ground cytoplasm, and the outer or *external surface* (E_s) in contact with the surrounding fluid medium.

Regarding non-permeant reagents and various *phospholipases (i.e.,* enzymes that hydrolyze different parts of the phospholipid molecule), it has been demonstrated that the distribution of the; *phospholipids is* highly asymmetrical. While the *outer layer* consists mainly of lecithin and sphingomyelin, the *inner layer* is composed mainly of phosphatidylethanolamine and phosphatidylserine. Furthermore, the *glycolipids* are mainly in the outer half of the bilayer. It is assumed that this asymmetry is rather stable and that there is no exchange of lipids across the bilayer.

Carbohydrates

The distribution of the oligosaccharides is also highly asymmetrical. In both *glycolipids* and *glycoproteins* they are confined exclusively to the external membrane surface. In erythrocyte membranes, hexose, hexosamine, fucose, and sialic acid are bound mainly to proteins. In fact all The proteins present at the outer surface are glycosylated.

Because of the presence of sialic acid residues, as well as carboxyl and phosphate groups, the outer surface of the membrane is negatively charged; consequently, positively charged proteins may be bound by electrostatic interactions to the plasma membrane. Only a small amount of sialic acid exists in the form of *gangliosides* (*i.e.*, glycolipids) in the plasma membrane of liver. However, gangliosides are important constituents of the neuronal surface and are probably involved in ion transfers.

Membrane Proteins

Proteins represent the main component of most biological membranes. They play an important role, not only in the mechanical structure of the membrane, but also as carriers or channels, serving for transport) they may also be involved in regulatory or ligand-recognition properties. In addition, numerous enzymes, antigens, and various kinds of receptor molecules, are present in plasma membranes. Membrane proteins have been classified as *integral (intrinsic) or peripheral (extrinsic) according* to the degree of their association with the membrane and the methods by which they can be solubilized.

Peripheral proteins are separated by mild treatment, are soluble in aqueous solutions, and are usually free of lipids. Examples include: the above-mentioned *spectrin,* which may be removed from red cell ghosts by chelating agents; cytochrome *c,* found in mitochondria; and acetylcholinesterase, in electroplax membranes, which are easily removed in high salt solutions.

Integral proteins represent more than 70 per cent of the two protein types and require drastic procedures for isolation. Usually they are insoluble in water solutions and need the presence of detergents to be maintained in a non-aggregated form. The study of integral proteins from different membranes has shown that they are rather heterogeneous in relation to molecular weight. These proteins may be attached to oligosaccharides, thus forming glycoproteins.

Polypeptides of the Red Cell Membrane

The usual method used to separate the membrane polypeptides is to dissolve the erythrocyte ghosts in sodium dodecyl sulfate (SDS) (an ionic detergent) and then to use polyacrylamide gel electrophoresis (PAGE). The gel can be stained for proteins (with Coomassie Blue) or for carbohydrates. Densitometric scan of the PAGE which, by the position of the bands, permits estimation of the molecular weight of the polypeptides and, by the surface of the profile, enables determination of its relative mass. For example, *polypeptides 1* and *2* corresponding to *spectrin* (*i.e.,* myosin), represent about 30 per cent of the total protein. *Polypeptide* 3 is the major intrinsic protein and represents about 25 per cent.

Protiens molecular organization is highly asymmetrical, and this can be demonstrated by the use of reagents that are unable to cross the membrane. The reagent is first applied to the *intact erythrocyte* and then to the *white ghost* the difference between the two preparations may give information about the position of a particular protein with respect to the outer or inner surface. Some specific labels can be detected on either side of the membrane by using cytochemical and electron microscopic techniques. For example, using antibodies against spectrin or actin (polypeptide 5) it is possible to show that each is in the cytoplasmic surface of the red cell membrane. On the other hand, acetylcholinesterase can be shown to be in the outer surface by its inactivation with proteolytic enzymes. It can be said with certainty that *every protein constituent is asymmetrically distributed in the red cell membrane.* All readily soluble polypeptides are localized at the cytoplasmic surface, and those that are at the outer surface are tightly bound to the lipid structure of the membrane.

Major Polypeptides of the Red Cell Membrane

On the cytoplasmic side of the red cell membrane there are *polypeptides 1 and 2* (the myosinlike *spectrin*) and *polypeptide 5 (actin).* These proteins are associated into supramolecular structures forming *microfilaments* which can be identified by electron microscopy, especially by scanning electron microscopy. Under the red cell membrane these structures form a filamentous network that gives stability to the membrane by providing a kind of skeletal support to the fluid lipid bilayer, with its intrinsic proteins. This system of microfilaments may also control the characteristic shape of the erythrocyte. Thus, a possible alteration of the microfilaments has been implicated in a hereditary abnormality of the erythrocytes in which they are spheroidal instead of biconcave discs *(hereditary spherocytosis).*

The molecular weight of the major intrinsic protein is 90,000 daltons, spans the thickness of the membrane and has a small amount of carbohydrate on the pole at the outer surface. This polypeptide is present in the membrane as a dimer held together by S-S bonds. There are between 500,000 and 600,000 such dimers per cell, enough to account for the 8 nm particles observed in freeze-fractured membranes. This dimeric protein appears to be involved in the facilitated *diffusion of anions* (*i.e.,* chloride, bicarbonate) across the membrane.

As per the glycophorin has a molecular weight of 55,000, of which 60 per cent is carbohydrate. Near the COOH end of the molecule there is a region that is very hydrophobic and which interacts with the lipids of the membrane. The COOH end is probably exposed to the interior of the red cell. The NH_2 end is more hydrophilic, is exposed to the external environment, and has the attached oligosaccharides that are at the outer surface of the membrane.

At this surface glycoproteins contain protein-bound antigens of the ABO blood groups and other antigens, such as: the MN groups reacting with rabbit antisera, the influenza virus, phytohemagglutinin, and wheat germ agglutinin. It has been calculated that there are some 700,000 copies of glycophorin per human red cell. This protein accounts for 80 per cent of the carbohydrate and 90 per cent of the negatively charged sialic acid present in the cell surface. Proteolipids among the most hydrophobic integral proteins of the membrane, are characterized by their strong association with lipids and the fact that they are, soluble in organic solvents. First isolated by Folch and Lees from myelin, proteolipids are found in practically all cell membranes, and in many of them they represent receptor proteins for synaptic transmitters or form channels across those membranes.

Asymmetrical Distribution of Enzymes

More than 30 enzymes have been detected in isolated plasma membranes. Those most constantly found are 5'-nucleotidase, Mg^{2+} ATPase, Na^+-K^+ activated-Mg^{2+} ATPase, alkaline phosphatase adenyl, cyclase, acid phosphomono-esterase, and RNAse. Some enzymes have a preferential localization; for example, alkaline phosphatase and ATPase are more abundant at the bile capillaries, while disaccharidases are present in microvilli of the intestine. A specific localization with a mosaic arrangement has been postulated for some of these enzymes. Disaccharidase forms 5 to 6 nm globular units coating the membrane of the microvilli. The plasma membrane lacks the respiratory chain and glycolytic activity.

Of all the membrane-associated enzymes, Na^+-K^+ activated-Mg^{2+} ATPase is one of the most important because of its role in ion transfer across the plasma membrane. This enzyme is dependent on the presence of lipids and is inactivated when all lipids are extracted. Enzymes also show an *asymmetrical distribution*. Such as in the outer surface of erythrocytes there are acetylcholinesterase, nicotinamide-adenine dinucleotidase, and the ouabain binding site of the Na^+ K^+ ATPase. In the inner surface there is NADH-diaphorase, G3PD, adenylate cyclase, protein kinase, and ATPase.

MOLECULAR MODELS OF THE CELL MEMBRANE

On the molecular structure of the membrane before the isolation of plasma membranes, theories were generally based on indirect information. Since substances soluble in lipid solvents penetrate the plasma membrane easily, Overton postulated in. 1902 that the plasma membrane is composed of a thin layer of lipid. In 1926 Gorter and Grendell found that the lipid content of hemolyzed erythrocytes was sufficient to form a double layer of lipid molecules over the entire cell surface. This theory was also supported by electrical measurements that indicated a high impedance at the plasma membrane. The high impedance is because of the faex fact that it is difficult for ions to penetrate a lipid layer.

Tension at a water-oil interface is about 10 to 15 dynes per centimeter, whereas surface tension of cells is almost nil. It has been postulated that the low tension is due to the presence of protein layers on the lipid components. In fact, when a very small amount of protein is added to a model lipid-water system, the surface tension is lowered comparably. In 1935 Danielli and Davson explained about all these properties. They proposed that the plasma membrane contained a *lipid bilayer,* with protein adhering to both lipid-aqueous interfaces. Other models containing globular lipid micelles, globular proteins, or a combination of both have been proposed a globular-bilayer transition has also been postulated. These globular models do not account satisfactorily for the high electrical impedance. On the fine structure of the plasma membrane, electron microscopy has thrown some light and has revealed the numerous structural differentiations that this membrane and the underlying cell cytoplasm have in different cell types. To resolve the structure of the plasma membrane, extremely thin sections (~ 20 nm) must be used, otherwise one would observe different orientations of the plasma membrane with respect to the plane of the section.

The membrane appears most thin when it is exactly perpendicular to the plane of the section. Membranes of 6 to 10 mm have been observed at the surface of all cells. The membranes of two cells that are in close contact appear as dense lines separated by a space of 11 to 15 nm, which is strikingly uniform and contains a material of low electron density. This intercellular component can be considered as a kind of cementing substance.

The Unit Membrane–Re evaluation of the EM Image

There are three layer in cell membrane outer dense layers of 2.0 nm and a middle clear one of about 3.5 nm led Robertson to postulate the so-called *unit membrane* model. The electron microscopic image was interpreted along the lines of the Danielli-Davson model, so that the clear layer corresponded to the hydrocarbon chains of the lipids, and the dense layers, to the proteins. This unit membrane model was also extended to all kinds of intracellular membranes. That this concept is an oversimplication which does not account for the many proteins traversing the membrane can be readily seen by observing some high resolution micrographs in which fine bridges cross the unit membrane.

On the basis of some experiments the unit membrane has been re-evaluated recently methods designed to avoid the removal of proteins from membranes during the preparation for conventional electron microscopy or freeze-fracturing. With both methods a granular appearance of the membrane is obtained, which is more in keeping with the fact that a high proportion of the proteins traverse the membrane. According to these findings the unit membrane image is, to a great extent, artifactual.

The Fluid Mosaic Model

The main source of the knowledge of the molecular organization of biological membrane comes from an integration of the data on chemical analysis and from the application of several biophysical techniques.

The important concepts are:

- That the lipid and integral proteins are disposed in a kind of mosaic arrangement;
- That biological membranes are quasi-fluid structures in which both the lipids and the integral proteins are able to perform translational

movements within the overall bilayer. The concept of fluidity implies that the main components of the membrane are held in place only by means of non-covalent interactions.

For better study of the molecular organization of the membrane, it is necessary to remember that, not only the lipids, but also many of the intrinsic proteins and glycoproteins of the membrane, are amphipathic molecules. The term *amphipathy,* coined by Hartley in 1936, refers to the presence, within the same molecule, of hydrophilic and hydrophobic groups these amphipathic molecules constitute liquid crystalline aggregates in which the polar groups are directed towards the water phase and the non-polar groups are situated inside the bilayer.

The integral proteins of the membrane are intercalated to a greater or lesser extent into a rather continuous lipid bilayer. This arrangement is based on the fact that these integral proteins are also amphipathic, with polar regions protruding from the surface and non-polar regions embedded in the hydrophobic interior of the membrane. It is well recognized that a protein of appropriate size or a cluster of protein subunits may pass across the entire membrane (transmembrane proteins). Such traversing proteins could be in contact with the aqueous solvent on both sides of the membrane. A similar model has been postulated for the cholinergic receptor. This model also reflects the generally accepted view that the major proportion of the phospholipids in the membrane are arranged in a bilayer form. This mosaic model of the membrane, with intercalating proteins, comes from the use of freeze-fracturing techniques in erythrocytes and other cell membranes. The red cell ghosts show a large number of particles, about 8 nm in diameter, that have been interpreted as representing proteins embedded within the plane of cleavage which passes through the middle of the lipid bilayer. There are between 500,000 and 600,000 such particles in a cell and more are attached to the inner half of the membrane (PF) than to the outer half (EF). We mentioned earlier that these particles represent dimers of the polypeptide found in *band 3,* and that they may also correspond to anionic permeation channels.

The degree of dispersion or aggregation of these particles within the membrane appears to be influenced by the *spectrin-actin system.* This suggests that some link between this system and the dimers of band 3 could exist in the living membrane. The mosaic model allows for the characteristic *asymmetry* of the membrane, in which specific components predominate in the outer half or the inner half of the membrane. This implies that there are constraints on the transmembrane rotation of those macromolecules that are well oriented for the carrying of information across the cell membrane. In this model the fluidity of the membrane also implies that both the lipids and the proteins have considerable freedom of movement within the bilayer.

The fluidity of the lipids depends:

- On the degree of saturation of the hydrocarbon chains

- On the ambient temperature. A considerable proportion of the lipids in the membrane are unsaturated, so that the melting point of the bilayer is below body temperature.

Membrane Fluidity

For studying purpose, it can be classified as physical and biological.

The physical techniques are of two main classes:

1. Those that involve a minimal perturbation of the membrane, such as *X-ray diffraction* and *nuclear magnetic resonance (NMR) spectroscopy,*
2. Those that use certain added molecules to monitor specific sites of the membrane. Into this second class fall *fluorescence microscopy,* which uses fluorescent probes and *electron spin resonance (ESR)* spectroscopy.

In practice, these physical techniques can be subdivided on the basis of their ability to examine mainly either the lipid or the protein components, or both. ESR spectroscopy uses paramagnetic probes (*e.g.*, nitroxide-containing amphipathic molecules) which are introduced into the lipid bilayer. This can be done with natural and artificial membranes The data obtained provide information on lipid-protein packing, lateral diffusion of lipids, lipid-protein interactions, the fluidity of the membrane, and the rate of transmembrane rotation (so-called flip-flop) of molecules across the bilayer.

The *biological techniques* involve light and fluorescence microscopy and electron microscopy, including freeze-fracturing and radioisotope labeling methods. One of the simplest methods consists of binding gold or carbon particles to the cell surface and observing under the light microscope the movement of those particles on the surface. More information has come from studies in which different ligands, such as antibodies and plant lectins, interact with cell surface receptors. If these ligands are labeled with fluorescent dyes their movement can be followed by fluorescence microscopy.

For example, in a classical experiment by Frye and Edidin mouse L cells and human transformed cells which have different surface antigens, were marked with the corresponding antibodies labeled with two distinct fluorescent dyes. These cells were then induced to fuse under the influence of Sendai viruses. While at the beginning both cell surfaces could be recognized by their differing labels, after 40 minutes considerable intermixing of the antigens had occurred, so that the two labels could no longer be recognized. This intermixing was retarded by temperatures below 20° C.

In surface antigens of cultured muscle cells, a diffusion rate of 1×10^{-9} cm per second was calculated. Also, the so-called clustering or capping effect observed in lymphocytes may result in the displacement of antigenic molecules within the surface membrane. In this case, upon treatment with a labeled antibody, the antigens are first randomly distributed, but after some time they become clustered, forming patches, and agglomerate at one pole of the cell (*i.e.*, capping) where pinocytosis of the antigen-antibody complexes takes place. This process is also inhibited by temperatures that produce solidification of the lipid bilayer.

A number of histochemical techniques using the electron microscope may give information about the fluidity of the membrane. For example, antigens labeled with ferritin or peroxidase can be used. Another extremely valuable tool is freeze-fracturing with which it is possible to demonstrate the movement of the particles in response to different experimental conditions that produce changes in the molecular structure of the membrane.

Membrane Fluidity

The fluid mosaic model has been used to explain the fact that many receptors present in the cell membrane are coupled to enzymes such as adenylate and guanylate cyclase. We can say that they are macromolecules, in most cases of a protein nature, that act at the interface between the cell surface and the environment. A receptor has the dual function of recognizing a chemical signal (*i.e.*, one of the many regulatory agents, such as peptide hormones, neurotransmitters, prostaglandins, antigens plant lectins, and some bacterial toxins) and of initiating a biological response. The function of recognition is determined by the presence of a site that binds specifically to the ligand (*i.e.*, to the binding portion of a particular regulatory agent). This interaction produces a conformational change that may induce the translocation of ions or the activation of membrane enzymes, such as adenylate cyclase, thus producing cyclic AMP (cAMP).

In a cell there may be several kinds of receptors (at least eight in a fat cell) coupled to the same adenylate cyclase. The ligands do not compete for each specific receptor when binding, and the effect of multiple ligand-receptor bindings on adenylate cyclase is not additive. These findings have been interpreted as indicating that each ligand reacts with a unique receptor and that several receptors may be coupled to a certain number of enzyme molecules in the membrane. According to the *mobile hypothesis* both the receptors and the enzymes can diffuse independently within the plane of the membrane The greater degree of freedom that this model includes within the lipid realm of the membrane would allow different receptors to couple to a single enzyme.

Fusing cells in which the adenylate cyclase was inactivated but the receptor remained intact, with other cells which contain only the enzyme, resulted in receptor-enzyme coupling. In other words, the receptor of one cell (*e.g.*, the b-adrenergic receptor of an erythrocyte) was coupled to the adenylate cyclase of the other cell (*e.g.*, a tumor cell lacking the receptor). The coupling is manifested by a great increase in cAMP after receptor interaction with the specific ligand (in this case isoproterenol). The diagram shown below interprets in a highly simplified manner the fusion of two different cells that results in coupling between a receptor from one cell and adenylate cyclase from the other.

The Myelin Sheath and the Photoreceptors

Important information about the molecular structure of the cell membrane is derived from studies of natural multi-layered lipoprotein systems, such as the myelin sheath and the photoreceptors.

The Myelin Sheath: The myelin sheath is lipoprotein membrane that surrounds the axon of the nerve fibre. In peripheral nerves this sheath is formed by the Schwann cells. In central nerves the myelin sheath is produced by the activity of the oligodendroglial cells. It is a very special membrane structure in which the lipids are more abundant than the protein. In central myelin there are two main protein moieties: the proteolipid and the basic protein. The Myelin Sheath has been known for over a century that the myelin sheath has a strong birefringence, which indicates a high degree of organization at a submicroscopic level. Studies with x-ray diffraction have revealed a spacing of 17 nm in amphibian and 18.0 to 18.5 nm in mammalian peripheral nerves. Within this period, the proportion corresponding to the lipid, protein, and water content has been estimated. The existence of lecithin-cholesterol and Sphingomyelin-cholesterol complexes is postulated. The first type can be accommodated within the thickness of the lipid layer (L), but in the second, the longer sphingomyelin molecules must interdigitate in order to fill the same space.

This model also accounts for the localization of protein (HP) and water (HL) and is in accord with the view that each x-ray diffraction period of 18 nm corresponds to two unit membranes, *i.e.,* to two lipid bilayers separated by a watery space (H), that originally corresponds to extracellular space. It has confirmed by electron microscopic studies that myelin has a multi-layered membranous structure. In most cases, however, the x-ray diffraction period is reduced to 10 to 12 nm, thereby introducing a great many artifacts. Recently, with a highly polar embedding medium based on polymerized glutaraldehyde-urea, the periodicity obtained was of the same order as by x-ray diffraction. With this technique the lipids are not extracted; this explains the lacks of collapse of the structure.

In nerve conduction to function as an insulator, preventing the dissipation of energy into the surrounding medium. It might act not only as a dielectrics (insulating) material but also as a kind of resonant conductor in which the energy waves resonating in the lipid layers between the protein membranes could pass with maximum speed and minimum loss of energy.

Photoreceptors

The retinal rods and cones are highly differentiated cells that have at their outermost segment a lipoprotein structure that is specialized for photoreception. Studies with the polarization microscope suggest a submicroscopic organization consisting of transversely oriented protein layers alternating with lipid molecules arranged longitudinally along the axis of the photoreceptor. This type of layered organization has been demonstrated by electron microscopy in fragmented rod outer segments and in thin sections of the retina. These observations indicate that the rod consists of a stack of superimposed disks (several hundred) along the axis. These disks are really flattened sacs made of two membranes, which surround a thin space of 3 nm and become continuous at the edges. The space between the rod sacs is 5 to 12 nm. The cone outer segments, with minor differences, have a similar structure. To osmotic change rod sacs are highly sensitive. In hypotonic solutions they swell considerably, and the inner space between the membranes becomes very large. The rod outer segments can easily be isolated in the frog and be submitted to chemical analysis. The protein composition is simpler than in other membranes. The photopigment protein, *rhodopsin,* represents 40 per cent of the total mass of the rod sacs; another 40 per cent is lipid. Rhodopsin is an integral type of protein requiring detergents for extraction. In the pline of the membrane the rhodopsin molecules have a liquid-like distribution. X-ray diffraction studies have led to the interesting conclusion that this molecule may change its position within the lipid layer. In the dark (non-stimulated condition) the molecule is about one third embedded in the lipid and two thirds exposed to the water surface. Upon activation by light (bleached rhodopsin) the molecule becomes more deeply embedded within the lipid bilayer. This change is presumably caused by changes in surface charges of the molecule.

The process of photoreceptors transform light energy into another type of energy that can be conducted as nerve impulse. It is based on a cycle of chemical reactions which involve the visual pigments present in the protein membranes of the rod and cone sacs. This multilayered structure is a very effective system that facilitates the maximum absorption and utilization of light by the chromophoric groups present in the visual pigments (retinenes). The acute sensitivity of the photoreceptors, which can react to a single photon, can be explained by the fact that the possibility of striking a sensitive molecule is increased by a factor of hundreds or thousands by the molecular organization of the photoreceptor. It is known that the interaction with photons of light produces liberation of Ca^{2+} ions which, in turn, inhibit the sodium current that is characteristic of the photoreceptor in the dark condition. In this way photoreceptors act as energy transducers, transforming light energy into electrical signals.

ENDOPLASMIC RETICULUM

If you peer over the side of the nucleus, you'll notice groups of enormous, interconnected sacs snuggling close by. Each sac is only a few inches across but can extend to lengths of 100 feet or more. This network of sacs, the endoplasmic reticulum (ER), often makes up more than 10 per cent of a cell's total volume.

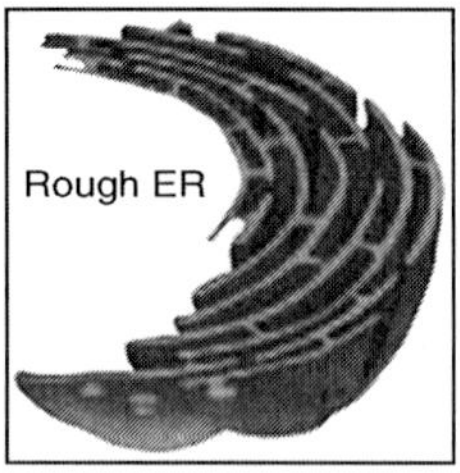

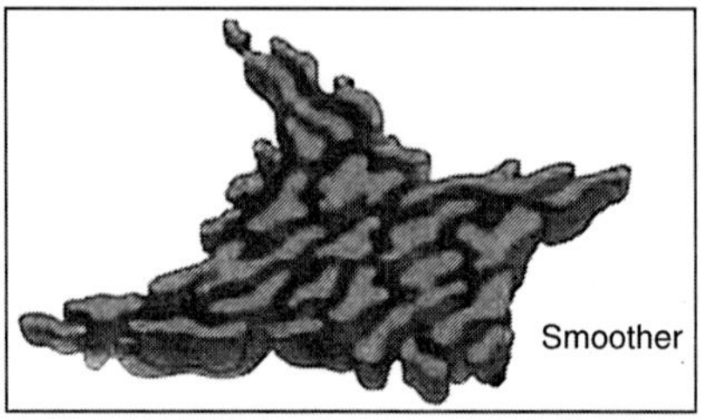

The endoplasmic reticulum comes in two types: Rough ER is covered with ribosomes and prepares newly made proteins; smooth ER specializes in making lipids and breaking down toxic molecules. Take a closer look, and you'll see that the sacs are covered with bumps about 2 inches wide. Those bumps, called ribosome's, are sophisticated molecular machines made up of more than 70 proteins and 4 strands of RNA, a chemical relative of DNA. Ribosomes have a critical job: assembling all the cell's proteins. Without ribosomes, life as we know it would cease to exist. To make a protein, ribosomes weld together chemical building blocks one by one. As naked, infant protein chains begin to curl out of ribosomes, they thread directly into the ER. There, hard-working enzymes clothe them with specialized strands of sugars.

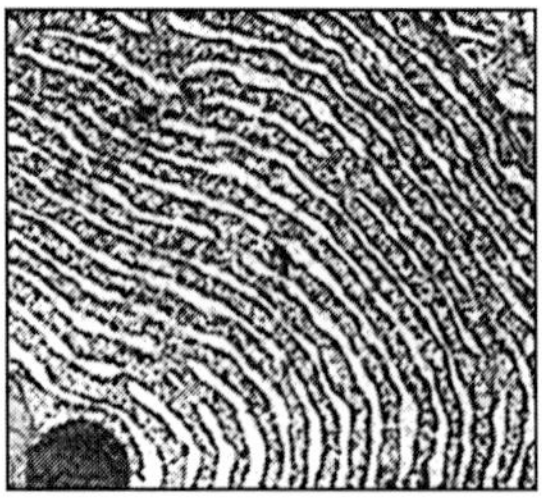

Fig. Rough ER

Now, climb off the nucleus and out onto the ER. As you venture farther from the nucleus, you'll notice the ribosomes start to thin out. Be careful! Those ribosomes serve as nice hand—and footholds now. But as they become scarce or disappear, you could slide into the smooth ER, unable to climb out.

In addition to having few or no ribosomes, the smooth ER has a different shape and function than the ribosome–studded rough ER. A labyrinth of branched tubules, the smooth ER specializes in synthesizing lipids and also contains enzymes that break down harmful substances. Most cell types have very little smooth ER, but some cells—like those in the liver, which are responsible for neutralizing toxins—contain lots of it.

Next, look out into the cytosol. Do you see some free–floating ribosomes? The proteins made on those ribosomes stay in the cytosol. In contrast, proteins made on the rough ER's ribosomes end up in other organelles or are sent out of the cell to function elsewhere in the body. A few examples of proteins that leave the cell (called secreted proteins) are antibodies, insulin, digestive enzymes, and many hormones.

RX: RIBOSOME BLOCKERS

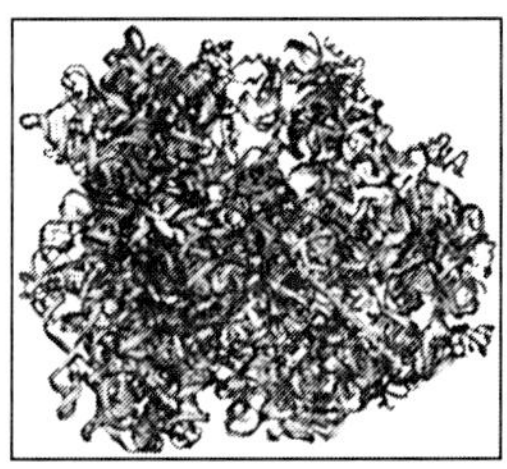

In a dramatic technical feat, scientists obtained the first structural snapshot of an entire ribosome in 1999. This more recent image captures a bacterial ribosome in the act of making a protein (the long, straight spiral in the lightest shade of blue). It also shows that—unlike typical cellular machines, which are clusters of proteins (shown here as purple ribbons)–ribosomes are composed mostly of RNA (the large, light blue and grey loopy ladders). Detailed studies of ribosomal structures could lead to improved antibiotic medicines.

All cellular organisms, including bacteria, have ribosomes. And all ribosomes are composed of proteins and ribosomal RNA. But the precise shapes of these biological machines differ in several very specific ways between humans and bacteria. That's a good thing for researchers trying to develop bacteria–killing medicines called antibiotics because it means that scientists may be able to devise therapies that knock out bacterial ribosomes (and the bacteria along with them) without affecting the human hosts.

Several antibiotic medicines currently on the market work by inhibiting the ribosomes of bacteria that cause infections. Because many microorganisms have developed resistance to these medicines, we urgently need new antibiotics to replace those that are no longer effective in fighting disease. Using sophisticated imaging techniques like X–ray crystallography, researchers have snapped molecular pictures of antibiotics in the act of grabbing onto a bacterial ribosome. Studying these three-dimensional images in detail gives scientists new ideas about how to custom design molecules that grip bacterial ribosomes even more strongly. Such molecules may lead to the development of new and more effective antibiotic drugs.

GOLGI

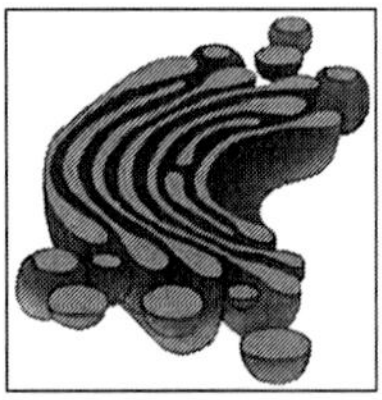

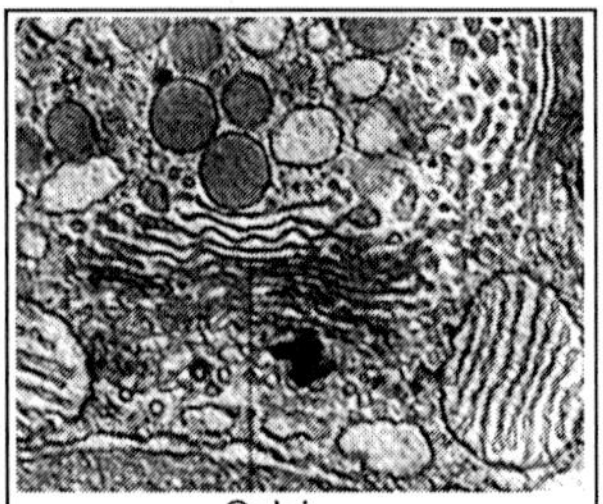

Golgi

TINA CARVALHO

Now, let's slog through the cytosol a bit. Notice that stack of a half dozen flattened balloons, each a few inches across and about 2 feet long?

That's the Golgi complex, also called the Golgi apparatus or, simply, the Golgi. Like an upscale gift shop that monograms, wraps, and mails its merchandise, the Golgi receives newly made proteins and lipids from the ER, puts the finishing touches on them, addresses them, and sends them to their final destinations. One of the places these molecules can end up is in lysosomes.

LYSOSOMES

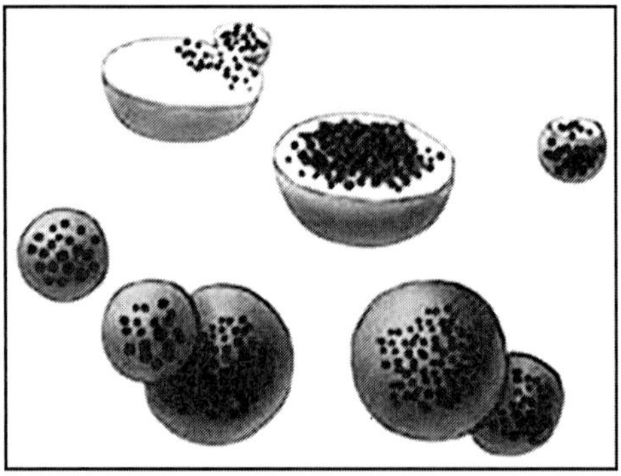

See that bubble about 10 feet across? That's a lysosome. Let's go—I think you'll like this. Perhaps even more than other organelles, lysosomes can vary widely in size—from 5 inches to 30 feet across. Go ahead, put your ear next to it. Hear the sizzling and gurgling? That's the sound of powerful enzymes and acids chewing to bits anything that ends up inside.

But materials aren't just melted into oblivion in the lysosome. Instead, they are precisely chipped into their component parts, almost all of which the cell recycles as nutrients or building blocks. Lysosomes also act as cellular garbage trucks, hauling away unusable waste and dumping it outside the cell. From there, the body has various ways of getting rid of it.

THE ARCHITECTURE OF CELLS

Although generalizations in biology usually lack the theoretical underpinnings found in physics, there are very clear commonalities among living systems that give biology a unity. One is the style of cellular construction. The biological universe consists of two types of cells— *prokaryotic cells,* which lack a defined nucleus and have a simplified internal organization, and *eukaryotic cells,* which have a more complicated internal structure including a defined, membrane-limited nucleus. Detailed analysis of the DNA from a variety of prokaryotic organisms in recent years has revealed two distinct types: bacteria (often called "true" bacteria or eubacteria) and archaea (also called *archaebacteria* or *archaeans*). As we discuss the archaea are in some respects more similar to eukaryotic organisms than to the true bacteria.

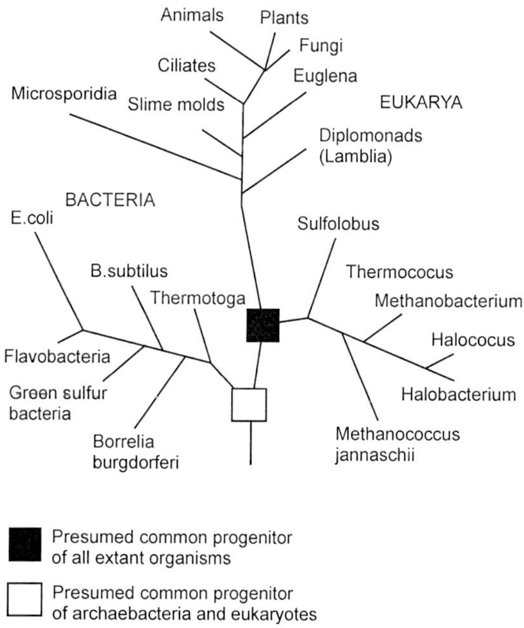

Fig. The three Kingdoms of Organisms are Related through Common Sequences of their Ribosomal RNAs.

Based on the assumption that organisms with more similar genes evolved from a common progenitor more recently than those with more dissimilar genes, researchers have developed the lineage tree. According to this tree, the archaea and eukarya (eukaryotes) are thought to have diverged from the bacteria before they diverged from each other. Despite the differences in the organization of prokaryotic and eukaryotic cells, all cells share certain structural features and carry out many complicated processes in basically the same way.

CELLS ARE SURROUNDED BY WATER- IMPERMEABLE MEMBRANES

A cell, because it is a limited space, must have an outer border. The construction of that border represents one of the most fundamental considerations in biological organization. The outer shell of cells, like any shell, is built to keep the interior contents from leaking out into the surrounding environment. The chemical processes of cellular life generally take place in a watery solution, and the intracellular constituents of cells are largely molecules that are easily dissolved in water. Similarly, the environment around cells is a watery one, the blood and other bodily fluids being solutions in water. Cells then, in order to maintain their integrity, need to be surrounded by an environment through which water cannot flow. A membrane composed of fatty molecules serves this purpose.

We all know from common experience that "oil and water don't mix." That maxim is all one needs to appreciate how a cell is constructed. When oil is poured on water, the oil spreads into a thin film; that film is analogous to the film of fat that surrounds cells, called the plasma membrane. Biological membranes differ from a pure oil film in that the molecules that make the membrane have both oily and watery portions; they have long fatty chains, but they also have a head group that is water-soluble by virtue of being electrically charged.

Thus membranes are formed because these bipartite molecules, called phospholipids, spontaneously orient themselves to form a double layer, or bilayer, having a fatty interior with external surfaces bonded to the surrounding water by the charged head groups. The membrane is given rigidity by interspersion of cholesterol, a molecule we have come to hate because of its association with heart disease, but one that is required to build the outer membrane of all our cells. Hence from an understanding of the contrasting properties of watery solutions and oily layers, an understanding of cellular construction emerges.

In spite of the rigidity provided by cholesterol, membranes composed of fat are not very strong, so numerous mechanisms for strengthening the borders of cells have evolved. In plants the plasma membrane is surrounded by a rigid cell wall. Although most animal cells lack a cell wall, proteins attached to their exterior surfaces provide some stability; the linking of cells together through these proteins helps maintain the integrity of tissues. Tissues and organs are often covered by strong networks of proteins and other molecules that strengthen and protect them, and also wall off the various compartments of the body. Single-celled organisms, like bacteria, have special outer coats to protect them.

MEMBRANES SERVE FUNCTIONS

Although membranes are valuable as a way to segregate the watery interior of the cell from its environment, or to segregate intracellular events from one another, they have other important functions, including energy storage. Because membranes separate watery compartments from one another, if an ion or a molecule dissolved in water is moved through a membrane into a new cellular compartment, it will not be able to diffuse freely out of the compartment into which it was moved. It takes energy to move the molecule, but once moved, the molecule stores that energy by virtue of its entrapment. Formally, this storage of energy is just like the storage of energy in a battery. Therefore, membranes not only delineate compartments, but also serve as active participants in the cell's dynamism.

The functions of many proteins depend on their mode of association with membranes. For instance, the passage of water-soluble molecules through membranes is carried out by protein transporters that are embedded in the membrane. Also, cells send information to one another by releasing signaling molecules. The outer membranes of cells have proteins, known appropriately as *receptors,* that bind the circulating signaling molecules. These signaling molecules allow the individual activities of the many cells in the body to be coordinated. The receipt of a signaling molecule by a receptor causes the transient organization of particular types of intracellular proteins, called *signal-transduction proteins,* into an activated complex at the interior face of the cell's outer membrane, from which it directs alterations of events in the cell's cytoplasm or nucleus.

PROKARYOTES

All prokaryotes are single-celled organisms, or protists. The bacterial lineage includes *Escherichia coli,* found in animal intestines and a favourite experimental organism, and the photosynthetic organisms formerly known as *blue-green algae* but better known today as *cyanobacteria.* (Because most prokaryotes studied in laboratories are bacteria, discussions of prokaryotic structure or metabolism throughout this book refer to these organisms, not archaeans, unless noted otherwise.) Many members of the archaeal lineage grow in unusual, often extreme, environments. For instance, the halophiles require high concentrations of salt to survive, and the thermoacidophiles grow in hot (80°C) sulfur springs, where a pH of less than 2 is common. Other archaeans, called *methanogens,* live in oxygen-free milieus and generate methane (CH_4) by the reduction of carbon dioxide. Archaeal cells have a similar structure. In general, prokaryotes consist of a single closed compartment containing the cytosol and bounded by the plasma membrane. Although bacterial cells do not have a defined nucleus, the genetic material, DNA, is condensed into the central region of the cell. In addition, most ribosomes — the cell's protein-synthesizing particles — are found in the DNA-free region of the cell. Some bacteria also have an invagination of the cell membrane, called a *mesosome,* which is associated with synthesis of DNA and secretion of proteins. Thus bacterial cells are not completely devoid of internal organization.

Bacterial cells possess a cell wall, which lies adjacent to the external side of the plasma membrane. The cell wall is composed of layers of peptidoglycan, a complex of proteins and oligosaccharides; it helps protect the cell and maintain its shape. Some bacteria (*e.g., E. coli*) have a thin cell wall and an unusual outer membrane separated from the cell wall by the periplasmic space. Such bacteria are not stained by the Gram technique and thus are classified as gram-negative. Other bacteria (*e.g., Bacillus polymyxa*) that have a thicker cell wall and no outer membrane take the Gram stain and thus are classified as gram-positive.

EUKARYOTIC CELLS

Eukaryotes comprise all members of the plant and animal kingdoms, including the unicellular fungi (*e.g.,* yeasts, mushrooms, molds) and protozoans. Eukaryotic cells, like prokaryotic cells, are surrounded by a plasma membrane. However, unlike prokaryotic cells, most eukaryotic cells also contain extensive internal membranes that enclose specific compartments, the organelles, and separate them from the rest of the cytoplasm, the region of the cell lying outside the nucleus.

Most organelles are surrounded by a single phospholipid membrane, but several, including the nucleus, are enclosed by two membranes. Each type of organelle plays a unique role in the growth and metabolism of the cell, and each contains a collection of specific enzymes that catalyze requisite chemical reactions. The membranes defining these subcellular compartments control their internal ionic composition so that it commonly differs from that of the cytosol (the portion of the cytoplasm outside the organelles) and among the various organelles.

The largest organelle in a eukaryotic cell is generally the nucleus, which houses most of the cellular DNA. In addition to the nucleus, several other organelles are present in nearly all eukaryotic cells: the mitochondria, in which much of the cell's energy metabolism is carried out; the rough and smooth endoplasmic reticula, a network of membranes in which glycoproteins and lipids are synthesized; Golgi vesicles, which direct membrane constituents to appropriate places in the cell; and peroxisomes, in which fatty acids and amino acids are degraded. Animal cells, but not plant cells, contain lysosomes, which degrade worn-out cell constituents and foreign materials taken in by the cell. Chloroplasts, where photosynthesis occurs, are found only in certain leaf cells of plants and some single-celled organisms. Both plant cells and some single-celled eukaryotes contain one or more vacuoles, large, fluid-filled organelles in which nutrients and waste compounds are stored and some degradative reactions occur.

The cytosol of eukaryotic cells contains an array of fibrous proteins collectively called the cytoskeleton. Three classes of fibres compose the cytoskeleton: microtubules (20 nm in diameter), built of polymers of the protein tubulin; microfilaments (7 nm in diameter), built of the protein actin; and intermediate filaments (10 nm in diameter), built of one or more rod-shaped protein subunits.

The cytoskeleton gives the cell strength and rigidity, thereby helping to maintain cell shape. Cytoskeletal fibres also control movement of structures within the cell; for example, some cytoskeletal fibres connect to organelles or provide tracks along which organelles move. The rigid cell wall, composed of cellulose and other polymers, that surrounds plant cells contributes to their strength and rigidity. Fungi are also surrounded by a cell wall, but its composition differs from that of bacterial or plant cell walls.

CELLULAR DNA

The DNA in the nuclei of eukaryotic cells is distributed among 1 to more than 50 long linear structures called chromosomes. The number and size of the chromosomes are the same in all cells of an organism, but vary among different types of organisms. Each chromosome comprises a single DNA molecule associated with numerous proteins, and the total DNA in the chromosomes of an organism is referred to as its genome.

Chromosomes, which stain intensely with basic dyes, are visible in the light microscope only during cell division when the DNA becomes tightly compacted. In all prokaryotic cells, most of or all the genetic information resides in a single circular DNA molecule, about a millimeter in length; this molecule lies, folded back on itself many times, in the central region of the cell. Although the large genomic DNA molecule in prokaryotes is associated with proteins and often is referred to as a chromosome, the arrangement of DNA within a bacterial chromosome differs greatly from that within the chromosomes of eukaryotic cells.

MITOCHONDRIA: CELLULAR POWER PLANTS

Blink. Breathe. Wiggle your toes. These subtle movements—as well as the many chemical reactions that take place inside organelles—require vast amounts of cellular energy. The main energy source in your body is a small molecule called ATP, for adenosine triphosphate.

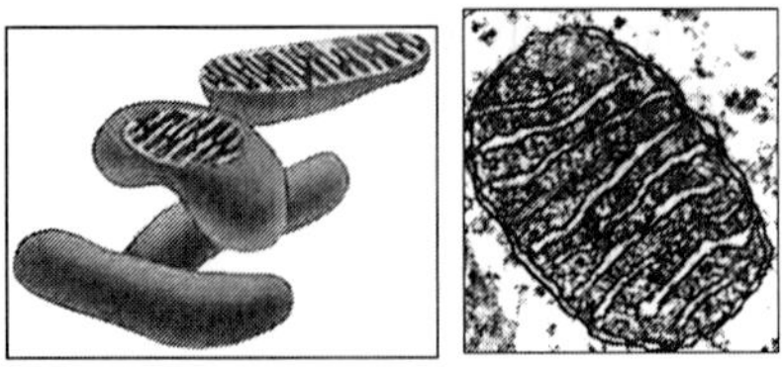

Fig. Mitochondrion

ATP is made in organelles called mitochondria. Let's see if we can find some. They look like blimps about as long as pickup trucks but somewhat narrower. Oh, a few of them are over there. As we get nearer, you may hear a low whirring or humming sound, similar to that made by a power station. It's no coincidence. Just as power plants convert energy from fossil fuels or hydroelectric dams into electricity, mitochondria convert energy from your food into ATP. Like all other organelles, mitochondria are encased in an outer membrane. But they also have an inner membrane. Remarkably, this inner membrane is four or five times larger than the outer membrane. So, to fit inside the organelle, it doubles over in many places, extending long, fingerlike folds into the center of the organelle. These folds serve an important function:

They dramatically increase the surface area available to the cell machinery that makes ATP. In other words, they vastly increase the ATP–production capacity of mitochondria. The mazelike space inside mitochondria is filled with a strong brew of hundreds of enzymes, DNA (mitochondria are the only organelles to have their own genetic material), special mitochondrial ribosomes, and other molecules necessary to turn on mitochondrial genes.

	Actual Size (Average)	**Perceived Size when Magnified 3 Million Times**
Cell diameter	30 micrometers	300 feet
Nucleus diameter	5 micrometers	50 feet
Mitochondrion length	Typically 1–2 micrometers, but can be up to 7 micrometers long	18 feet
Lysosome diameter	50–3,000 nanometers	5 inches to 30 feet
Ribosome diameter	20–30 nanometers	2–3 inches
Microtubule width	25 nanometers	3 inches
Intermediate filament width	10 nanometers	1.2 inches
Actin filament width	5–9 nanometers	0.5–1 inch

EXAMINATION OF MITOCHONDRIA

Although the examination of mitochondria in living cells is somewhat difficult because of their low refractive index, they can be observed easily in cells cultured in vitro, particularly under dark-field illumination and phase contrast. This examination has been greatly facilitated by colouration with a dilute solution of *Janus green*. The resultant greenish blue stain is due to me action of the cytochrome oxidase system present in mitochondria, which maintains the vital dye in its oxidized (coloured) form. In the surrounding cytoplasm the dye is reduced to a colourless leukobase.

Fluorescent dyes, which are more sensitive, have been used in isolated mitochondria and recently in intact cultured cells. Such dyes are suitable for metabolic studies of mitochondria in situ. This has been demoustrated by the micromanipulation that mitochondria is relatively stable, and without alternation it can be displaced by microneedles. Micromanipulation has demonstrated that mitochondria are relatively stable and can be displaced by the microneedle without alterations. The *specific gravity* is greater than that of the cytoplasm. By ultracentrifugation of living cells at 200,000 to 400,000 g, mitochondria are deposited intact at the centrifugal pole.

Observation in vivo is of particular interest when supplemented by time-lapse cinematography. In cultured fibroblasts, continuous and sometimes rhythmic changes in volume, shape and distribution of mitochondria can be observed. The two main types of mitochondrial motion are *displacement* from one part of the cell to another, mainly by cytoplasmic streaming, and *bending* and *stretching cycles*. It has been found that the zones of bending are constant and correspond to regions in which the mitochondrial crests are vesicular. Active changes in volume and shape of mitochondria may be caused by chemical, osmotic and mechanochemical changes. In living cells, low amplitude contraction cycles associated with oxidative phosphorylation have been observed.

Shape of Mitochondria

The *shape* of mitochondria is variable, but in general these organelles are *filamentous or granular*. During certain functional stages, other derived forms may be seen For example, a long mitochondrion may swell at one end to assume the form of a *club* or be hollowed out to take the form of a *tennis racket*. At other times mitochondria may become *vesicular* by the appearance of a central clear zone. The morphology of mitochondria varies from one cell to another, but it is more or less constant in cells of a similar type or in those performing the same function.

Size of Mitochondria

The size of mitochondria is also variable; however, in most cells the width is relatively constant (about 0.5 mm), and the length is variable, reaching a maximum of 7 *p.m*. Depending on the functional stage of the cell, however, it possible to find rods that are either thinner or thicker. The size and shape of the fixed mitochondria depend also on the osmotic pressure and the pH of the fixative. Using buffered fixatives at physiologic pH has been shown to be important.

Distribution of Mitochondria

Mitochondria are, in general, uniformly distributed throughout the cytoplasm, but there are many exceptions to this rule. In some cases, they accumulate preferentially around the nucleus or in the peripheral cytoplasm. Overloading with inclusions, such as glycogen and fat, displaces these organelles. During mitosis, mitochondria are concentrated near the spindle, and upon division of the cell they are distributed in approximately equal number between the daughter cells.

The distribution of mitochondria within the cytoplasm should be considered in relation to their function as energy suppliers. In some cells they can move freely, carrying ATP where needed, but in others they are located permanently near the region of the cell where presumably more energy is needed. For example, in certain muscle cells (*e.g.*, diaphragm), mitochondria are grouped like rings or braces around the I-band of the myofibril. In the rod and cone cells of the retina all mitochondria are located in a portion of the inner segment. The basal mitochondria of the kidney tubule are intimately related to the infoldings of the plasma membrane in this region of the cell. It is assumed that this close relationship with the membrane is related to the supply of energy for the active transport of water and solutes.

Orientation of Mitochondria

Mitochondria may have a more or less definite orientation. For example, in cylindrical cells they are generally oriented in the basal-apical direction, parallel to the main axis. In leukocytes, mitochondria are arranged radially with respect to the centrioles. It has been suggested that these orientations depend upon the direction of the diffusion currents within cells and are related to the organization of the cytoplasmic matrix.

Number of Mitochondria

Mitochondria are found in the cytoplasm of all eukaryotic cells but not in prokaryotes. In bacteria the respiratory enzymes are located in the plasma membrane. The mitochondrial content of a cell is difficult to determine, but, in general, it varies with the cell type and functional stage. It is estimated that in liver, mitochondria constitute 30 to 35 per cent of the total protein content of the cell, and in kidney, 20 per cent. In yeast, ultrastructural evidence, based on serial sections, points to the existence of a single branched organelle per cell.

A normal liver cell contains between 1000 and 1600 mitochondria, but this number diminishes during regeneration and also in cancerous tissue. This last observation may be related to decreased oxidation that accompanies the increase to anaerobic glycolysis in cancer. Another interesting finding is that there is an increase in the number of mitochondria in the muscle after repeated administration of the thyroid hormone, thyroxin. An increased number of mitochondria has also been found in human hyperthyroidism. D oöcytes contain as many as 300,000 mitochondria—the largest number recorded for a cell. There are fewer mitochondria in green plant cells than in animal cells, because some of their functions are taken over by chloroplasts.

STRUCTURE OF MITOCHONDRIA

A mitochondrion consists of two membranes and two compartments, the larger of which contains the *mitochondrial matrix.* An outer limiting membrane, about 6 nm thick, surrounds the mitochondrion. Within this membrane, and separated from it by a space of about 6 to 8 nm, is an inner membrane that projects into the mitochondrial cavity complex infoldings called *mitochondrial crests.*

This inner membrane, also about 6 nm thick, divides the mitochondrion into two chambers or compartments:

1. The outer chamber contained between the two membranes and in the core of the crests
2. The inner chamber, bound by the inner membrane.

This inner chamber is filled with a relatively dense proteinaceous material usually called the *mitochondrial matrix.* This is generally homogeneous, but in some cases it may contain a finely filamentous material, or small, highly dense granules. These granules are now considered to be sites for binding divalent cations, particularly Mg^{2+} and Ca^{2+}. The mitochondrial crests that project from the inner membrane are, in general, incomplete septa or ridges that do not interrupt the continuity of the inner chamber; thus, the matrix is continuous within the mitochondrion.

The Mitochondrial Matrix Contains Ribosomes and a Circular DNA

Within the mitochondrial matrix are e-mail ribosomes and a circular DNA. The outer and inner membranes and the crests can be considered to be fluid molecular films with a compact molecular structure, the matrix is gel-like and contains a high concentration of soluble proteins and smaller molecules. This double (solid-liquid) structure is important in providing an explanation for some of the mechanical properties of mitochondria.

F_1 Particles

The use of negative staining has enabled recognition of other details of mitochondrial structure. If a mitochondrion is allowed to swell and break in a hypotonic solution and is then immersed in phosphotungstate, the inner membrane in the crest appears covered by particles of 8.5 nm that have a stem, linking each with the membrane. These so-called "elementary" or "F_1" particles, are regularly spaced at intervals of 10 nm on the inner surface of these membranes. According to some estimates, there are between 10^4 and 10^5 elementary particles per mitochondrion. These particles correspond to a special ATPase involved in the coupling of oxidation arid phosphorylation. The presence of such F_1 particles on the matrix side (M side) confers to the inner mitochondrial membrane a characteristic asymmetry which is of basic importance to its function. It is interesting to recall here that bacterial plasma membranes contain F_1 particles with a similar asymmetric distribution (*i.e.,* towards the bacterial matrix). On the other hand, the opposite orientation is found in chloroplasts where there are F_1 particles on the outer or stromal side of the thylakoid membrane.

Different Cell and Mitochondrial Sturctural Variation

Some investigators assume that a common pattern of mitochondrial structure developed at an early stage of evolution and was subsequently transmitted, without considerable modifications, from protozoa to mammals and from algae to flowering plants. Detailed structural variations can be observed, however. The crests may be arranged longitudinally or they may be simple or branched, forming complex networks. In protozoa, insects, and adrenal cells of the glomerular zone, the infoldings may be tubular instead of lamellar and the tubules may be packed in a regular way. The number of crests per unit volume of a mitochondrion is also variable. Mitochondria in liver and germinal cells have few crests and an abundant matrix, whereas those in certain muscle cells have numerous crests and little matrix. In some cases the crests are so numerous that they may have a quasi-crystalline disposition. The greatest concentration of crests is found in the flight muscle of insects. In general, there seems to be a correlation between the number of crests and the oxidative activity of the mitochondrion. On majore feature of the mitochondria is the enlargement of the space within the crests or tubules which, corresponds to the outer compartment of the mitochondrion. This is very conspicuous and appears as tubular or vesicular openings, which are much less opaque than the inner matrix. These structures seem to be related to the specific activity of the gland. In this case, mitochondria, in addition to functioning in cell oxidations, are actively engaged, in the synthesis of steroid hormones.

CYTOSKELETON: THE CELL'S SKELETON

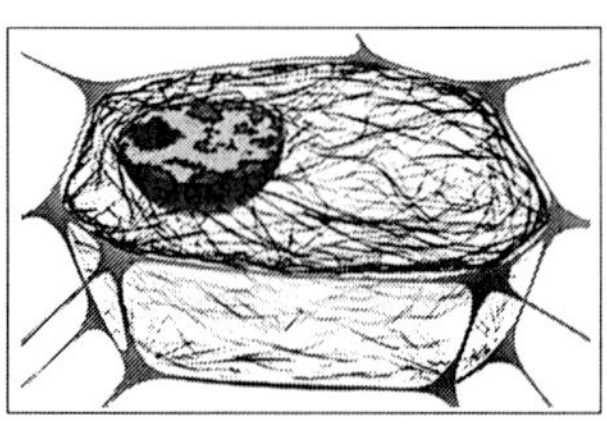

Now, about all those pipes, ropes, and rods you've been bumping into. Together, they are called the cytoskeleton—the cell's skeleton. Like the bony skeletons that give us stability, the cytoskeleton gives our cells shape, strength, and the ability to move, but it does much more than that. Think about your own cells for a moment. Right now, some of your cells are splitting in half, moving, or changing shape. If you are a man, your sperm use long tails called flagella to swim. If you are a woman, hairlike fibres called cilia sweep newly released eggs from your ovaries into your uterus. And all that is thanks to the cytoskeleton. As you can see, the cytoskeleton is incredibly versatile. It is made up of three types of fibres that constantly shrink and grow to meet the needs of the cell: microtubules, intermediate filaments, and actin filaments. Each type of fibre looks, feels, and functions differently.

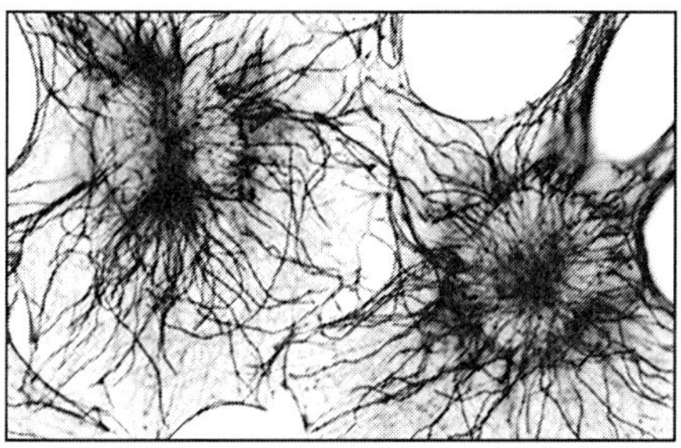

The 3–inch–wide flexible pipes you just banged your head on are called microtubules. Made of the strong protein tubulin, microtubules are the heavy lifters of the cytoskeleton. They do the tough physical labour of separating duplicate chromosomes when cells copy themselves and serve as sturdy railway tracks on which countless molecules and materials shuttle to and fro. They also hold the ER and Golgi neatly in stacks and form the main component of flagella and cilia. Grab one of those inch–thick ropes. Yeah, you can swing on it—it won't snap. These strands, called intermediate filaments, are unusual because they vary greatly according to their location and function in the body. For example, some intermediate filaments form tough coverings, such as in nails, hair, and the outer layer of skin (not to mention animal claws and scales). Others are found in nerve cells, muscle cells, the heart, and internal organs.

In each of these tissues, the filaments are made of different proteins. So if doctors analyse intermediate filaments in tumors, they can determine the origin of—and possible treatments for—some kinds of cancer. See that bundle of long rods near the edge of the cell? You can touch it, but don't try to bend the rods. They shatter easily. These rods, slightly thinner than intermediate filaments, are actin filaments. They are made up of two chains of the protein actin twisted together. Although actin filaments are the most brittle of the cytoskeletal fibres, they are also the most versatile in terms of the shapes they can take. They can gather together into bundles, weblike networks, or even three–dimensional gels. They shorten or lengthen to allow cells to move and change shape. Together with a protein partner called myosin, actin filaments make possible the muscle contractions necessary for everything from your action on a sports field to the automatic beating of your heart.

GOLGI SPELUNKING

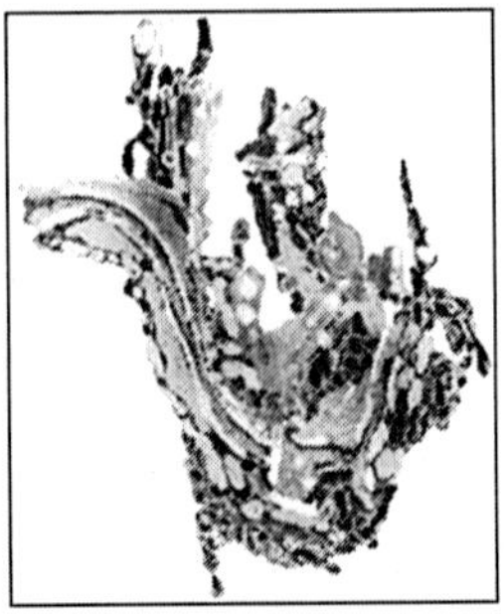

Scientists use a variety of techniques to study organelles like the endoplasmic reticulum and Golgi, gaining ever more detailed understanding of these minute but very complicated structures. For example, Kathryn Howell of the University of Colorado School of Medicine in Denver uses a specialized high–voltage electron microscope, rapid freezing methods, and a computer modeling programme to obtain a vivid three–dimensional view of the Golgi and the pathways that proteins use to exit it. Howell begins by quick–freezing living cells, embedding them in plastic, and slicing the plastic–coated sample into thin sections. As she tilts the microscope stage, she can capture many images of the same region of the sample. A computer assembles these images to form a three–dimensional view, called a tomogram, of the Golgi and other organelles. Based on the tomogram, Howell's research team can produce a movie of a virtual journey through the cell. Howell's research shows that there are several pathways for proteins and other molecules to exit the Golgi. The findings are revealing, as earlier studies using different methods had suggested that there was only one road out of this organelle. No doubt new chapters to this story will be written as biologists and computer scientists create even more sophisticated tools for imaging cells.

MORPHING MITOCHONDRIA

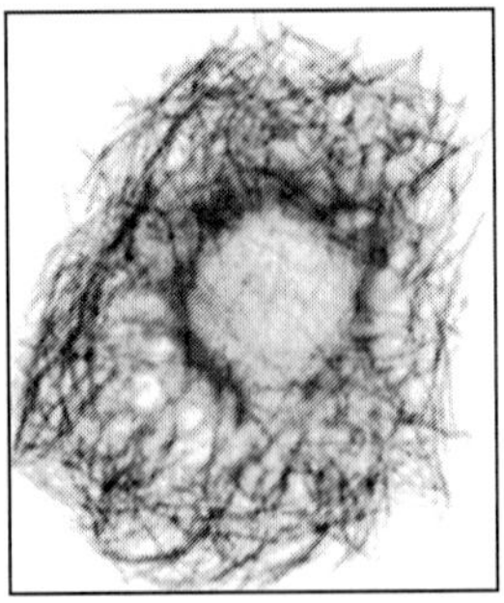

Scientists such as Michael P. Yaffe of the University of California, San Diego, study what mitochondria look like and how they change throughout a cell's life. To approach this research problem, Yaffe uses simple organisms—such as yeast or fruit fly cells—which, like your own cells, have membranes, a nucleus, and other organelles. This similarity makes these organisms important models for understanding human biology. Yaffe's work helped change the textbook depiction of mitochondria as kidney bean–shaped organelles.

Using advanced microscopy, Yaffe and others have unveiled many different shapes for mitochondria, ranging from the classic beans to long snakes and weblike structures, all of which are thought to change on a constant basis. Researchers are discovering that the different mitochondrial shapes accompany changes in cellular needs, such as when growing cells mature into specific types or when a cell responds to disease. Many scientists believe that mitochondria—which divide on their own, have their own genome and protein–making machinery, and resemble prokaryotes in many ways—are descendents of oxygen–loving microorganisms that were taken in by primitive cells. This historical event set the stage for advanced life forms like plants and animals.

SUGAR MONOMERS

Single sugar molecules are typically 3 to 7 carbons long and are termed monosaccharides. Figure shows that each carbon on the sugar molecule is decorated with a hydroxyl group (OH) except for one carbon that forms a carbonyl group. If the carbonyl group is at the end of the molecule, it forms an aldehyde; if the carbonyl group is in the middle of the molecule, it forms a ketone as indicated in Figure. All sugars can exist as linear molecules in solution and those greater than 5 carbons long can also circularize with the carbonyl group attacking a hydroxyl on one of the other carbons.

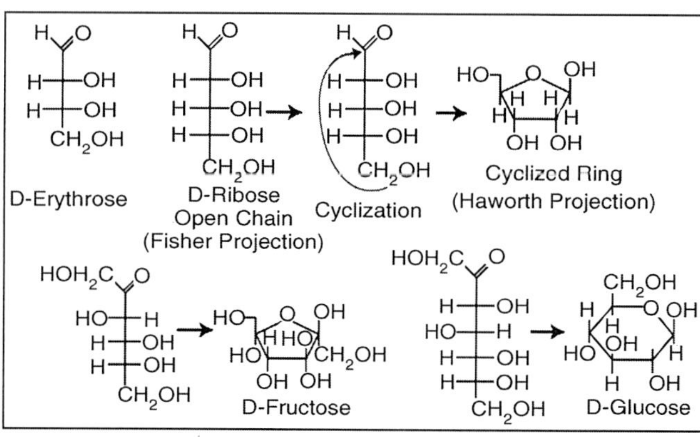

Fig. Common Monosaccharides

The circular sugar contains 5 to 7 members in the ring with one of the members being oxygen. Glucose, fructose and ribose are some of the more common sugars found in the cell. The structures of some important monosaccharides are shown. These often serve as building blocks for major structures in the cell, such as the cell wall and the genome. Both the linear and circular forms are shown.

SUGAR POLYMERS

Sugars readily polymerize: the combination of two sugars is called a disaccharide, while the combination of three is called a trisaccharide, and polymers of greater than three sugars are referred to as polysaccharides. Sugars are connected by α or β linkages. If the hydrogen is pointing up, it is a α linkage, if the hydrogen is pointing down, it is a β linkage as shown in Figure.

This distinction may seem trivial, but it greatly influences the properties of the molecule. (The name of the linkage further depends upon the orientation of the hydrogen on the lowest numbered carbon that forms the bond. The lowest numbered carbon is decided by the rules of organic chemistry, but this is really not important for our discussion. Just realise there is a set pattern of numbering for each organic molecule and this helps determines the α or β linkage.)

For example, starch is made up of glucose linked by α-1,4 bonds Starch is also water–soluble and can serve as a food source for many organisms. In contrast, cellulose, containing glucose linked by β–1,4 bonds, is insoluble in water and is not as readily degraded by most microbes. Polysaccharides might contain only one type of sugar monomer or a variety of different ones, sometimes in repeating units. Sugars can be linked together to form more complex polymers.

Lactose is a common sugar in milk, while maltose is found in many grains. Starch found in potatoes and other vegetables is a long polymer of glucose units. A common form of starch contains 20 per cent amylose and 80 per cent amylopectin. Polymers of sugars can serve as storage products for the cell. The breakdown of sugar yields a huge amount of energy, which means that they are terrific molecules for effectively storing energy for later utilization. Starch and glycogen are two examples of sugar polymers that are used in this way. Polysaccharides also serve as structural components of many different molecules in the cell including nucleic acids and the cell wall.

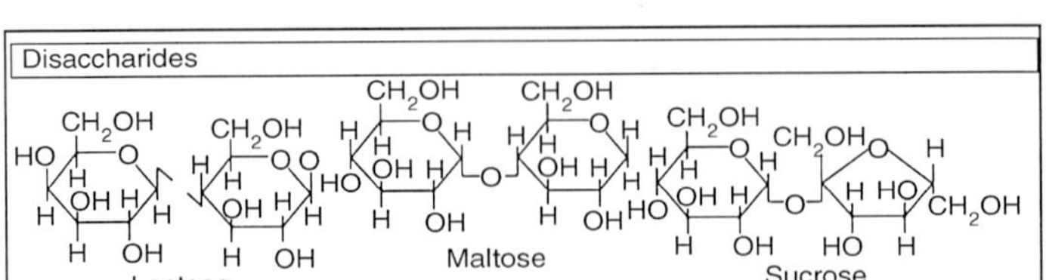

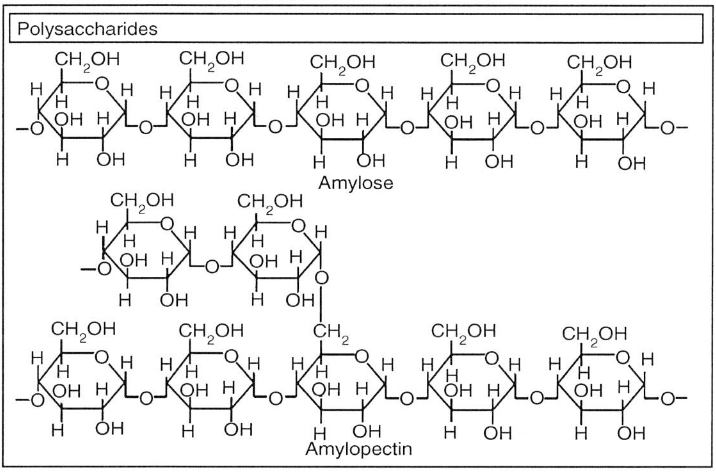

Fig. Common Disaccharides, Trisaccharides and Polysaccharides

As with nucleic acids, primary structure refers to the ordered sequence of the different amino acids in a protein. The carboxyl group and the amino group of amino acids are reactive. As shown in Figure, cells synthesize proteins by attaching the carboxyl group of one amino acid to the amine group the next, with polymerization taking place at the ribosome. This is termed a peptide bond. Since each amino acid has a carboxyl group and an amino group, hundreds of amino acids can be linked together.

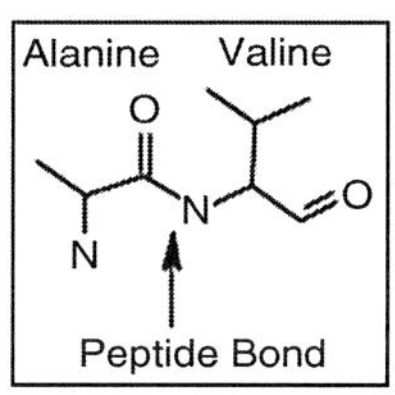

Fig. A Peptide Bond

The peptide bond between an alanine residue and a valine residue is identified by the arrow. Peptide bonds can form between any two of the 20 amino acids and link the carboxyl group of one amino acid and the amine group of the next. There are 20 common amino acids found in proteins and these amino acids can be roughly classified into 3 groups: polar, non-polar and charged. Polar and charged amino acids are hydrophilic and are often found on the surface of a protein, interacting with the surrounding water. In contrast, non-polar (or hydrophobic) amino acids avoid water.

While this categorization is adequate for most purposes, you should recognize that it is a bit simplistic. For example, arginine *does* have a charged hydrophilic group at one end, but the $-CH_2-$ backbone that makes up most of the amino acid is actually quite hydrophobic. Figure shows the chemical structure of all 20 common amino acids.

In the figure, the amino acids are organized into their chemical type and characteristics. Acidic and basic amino acids carry a charge, amino acids with a sulfur or hydroxyl comonent are polar, and aliphatic and aromatic amino acids are non-polar.

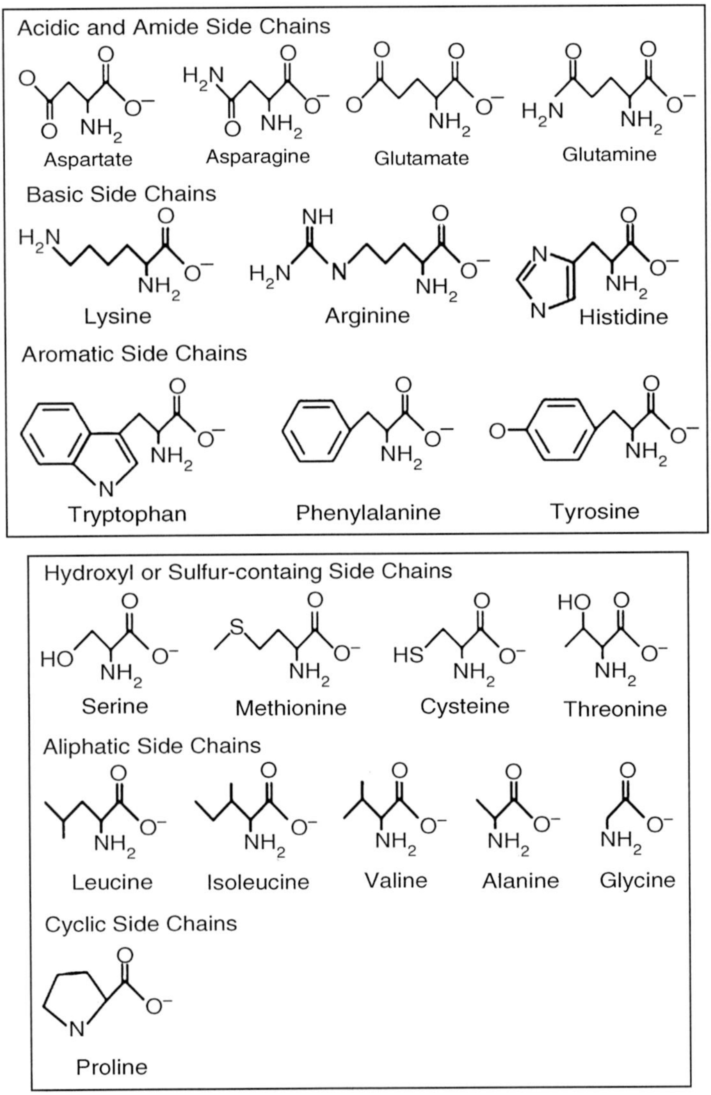

Fig. The Common Amino Acids

Peptides and proteins are formed when a ribosome and the rest of the translation machinery link amino acids together in polymers that range from 10 to 10,000 residues in length. During and after protein synthesis, the residues of the primary sequence dictate how the protein folds. The simplest aspect of protein folding is termed its secondary structure, which refers to the geometry of the local polypeptide chain with respect to their immediate neighbours. How a protein folds is dictated by the primary sequence of amino acids, but predicting the overall structure from the primary sequence remains one of the most important unsolved problems in biology.

Nevertheless, it is clear that the major determinants of this final structure are hydrophobic interactions. During protein folding, hydrophobic amino acids must be hidden from the water interface by being buried in the interior of the protein. This burying defines the protein core which then influences the immediate structure around it and greatly affects the protein's overall structure.

COMMON SECONDARY STRUCTURES

Peptide bonds between adjacent amino acids can rotate and twist to allow a large number of interactions, but two local organization schemes, the α helix and the β sheet, are found in many proteins. Their prevalence is certainly because they happen to form particularly energetically favourable structures. Formation of these structures is driven by favourable hydrogen bonding and hydrophobic interactions between nearby amino acids in the protein. The α helix resembles a ribbon of adjacent amino acids wrapped around a tube to form a staircase–like structure.

Figure shows different representations of an α helix. This structure is very stable, yet cn be flexible in specific cases and is sometimes seen in parts of a protein that may need to bend or move.

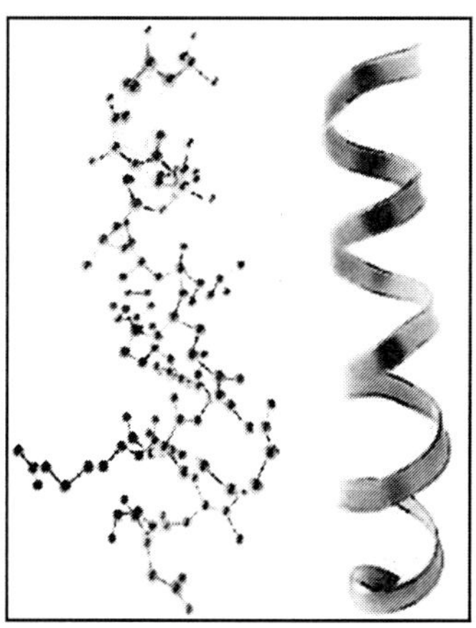

Fig. The α Helix Motif in Proteins

This figure shows two common depictions of an α–helix, an extremely common and important motif in proteins. The ball–and–stick depiction on the left shows the various side chains, but the helical nature of the structure is not so obvious. The ribbon representation on the right only traces the backbone of the peptide bonds and overemphasizes the symmetry a bit, but shows the helical nature of the structure. In the β sheet, two protein chains (perhaps different segments of the same protein) align themselves in a planar structure such that hydrogen bonds can form between facing amino acids in each sheet.

Figure shows different representations of this structure. The β sheet is different from the α helix in that it can involve amino acids from different sections of the protein, which come together to form this structure. Also, the structure tends to be rigid and less flexible than the α helix.

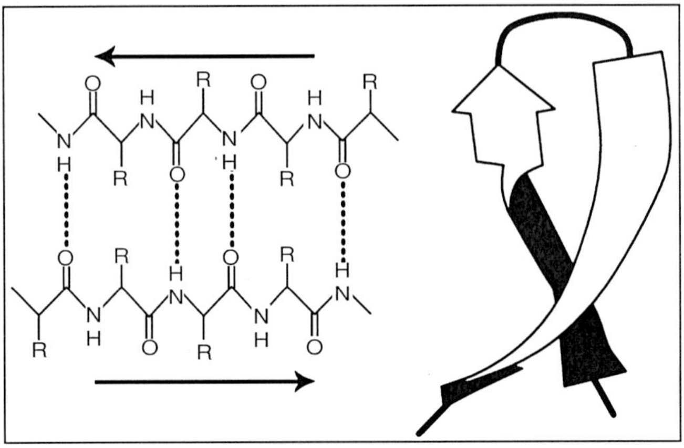

Fig. The β–sheet. Two views of a β–sheet.
(Left) A Diagram of β Sheet Showing Hydrogen Bonding between Protein Strands (Right) A Ribbon Representation of a β–sheet

As noted before, the most important stabilizing force in proteins is burying hydrophobic residues from the surrounding water, but there are other chemical features that are important in creating and stabilizing tertiary structure. These include hydrogen bonds, ionic interactions, and sulfhydryl bonds. Ionic interactions are attractions between groups of opposite charge. There are amino acids with negatively charged side groups (aspartate, glutamate) and amino acids with positively charged side groups (typically lysine and arginine).

If opposite charges are brought close enough together, there will be an attractive force that can contribute to protein structure. It also follows that like charges repel each other and this can also dictate protein structure. Sulfhydryl linkages are covalent bonds between cysteine groups. Cysteine is a unique amino acid in that it has a sulfur group at the end of its variable side group that is available for binding to other groups. Often in proteins, nearby sulfhydryl groups on cysteines form a covalent bond and these are often crucial for the stabilization of the mature protein or for it to perform its function. Figure shows some representations of sulfhydryl bond in proteins.

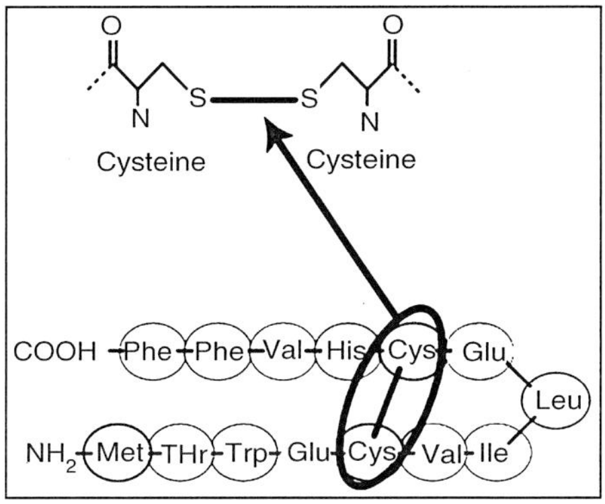

Fig. The Chemical Structure of a Sulfhydryl Bond. Several Views of Sulfhydryl Linkages are Shown. In the Top Panel, the Covalent Sulfhydryl Linkage between two Cysteine Residues is Shown. In the Bottom Panel, this Sort of Cys–Cys Bond is Shown in the Context of a Section of a Protein.

Figure shows the tertiary structure of ribulose bisphosphate carboxylase (RubisCo), one of the most abundant enzymes on this planet. This enzyme takes energy, obtained most often from the sun, and uses it to convert carbon dioxide into carbohydrate. Its abundance reflects the fact that virtually all photosynthetic organisms, from bacteria to plants, use this enzyme to produce much of the cell's carbon. Note the α helices throughout the protein and the β sheet near the bottom. These structures help to hold the protein in its proper conformation so that it can carry out its enzymatic activity. The figure shows the structure of ribulose bisphosphate carboxylase, typically abbreviated Rubisco, from the bacterium *Rhodospirillum rubrum*. As described in the text, this complex structure has alpha helices (shown as cylinders) and beta sheets (shown as the flat ribbons) throughout.

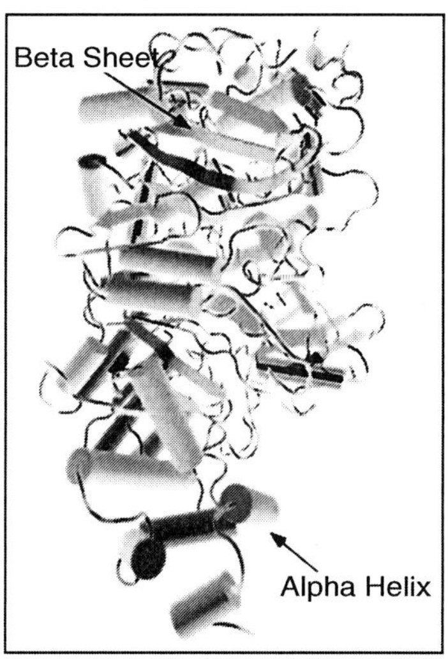

Fig. Tertiary Structure of Ribulose Bisphosphate Carboxylase

Now we have talked about certain secondary structures that recur in many proteins, but it also happens that there are conserved tertiary structures as well. In other words, there are regions of proteins, which are sometimes termed motifs or domains, that are structurally identical in two or more different proteins. Figure shows some examples of protein motifs. By "structurally identical," we mean that they have similarly sized α helices and β sheets that are organized in a similar overall way. Interestingly, these domains typically do not have identical primary structures. Instead, it appears that a variety of different sequences can fold to give the same overall tertiary form.

It is usually clear from the sequence that structurally similar domains from different proteins are evolutionarily related. That is, at one time there was a single protein with the domain and through evolution, an extra copy of the gene was created. Subsequent random mutations in each gene altered the primary sequences of the proteins, but the only changes that were tolerated were those that maintained the structure, and therefore the function, of each copy of the domain. This implies that these motifs often have a specific function. For example, there is a very common motif that proteins use for binding ATP. A second class of examples would be the set of motifs that are used to bind DNA. Through genetic changes, these motifs have been grafted onto other protein regions, so that the same ATP–binding (or the DNA–binding) domain performs a similar function for proteins with completely different overall activities. Of course the unique functions are defined by those portions of the protein that are not in common.

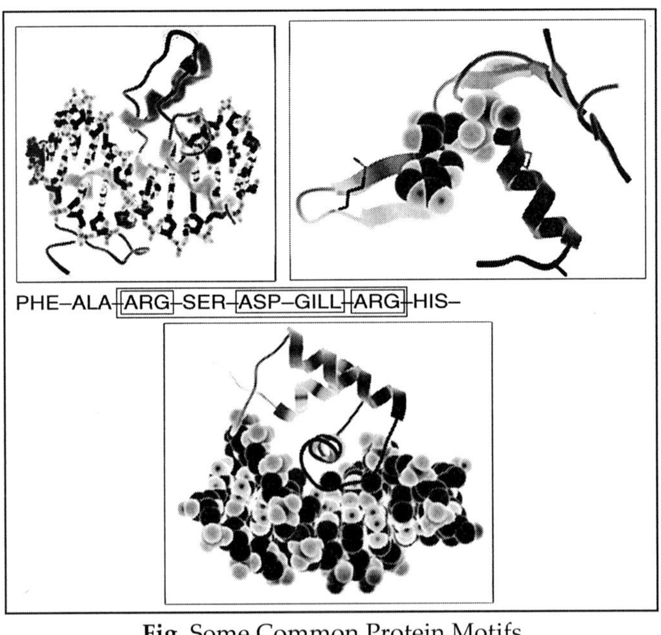

Fig. Some Common Protein Motifs

Three examples of protein motifs are shown. The top left panel shows two zinc–finger motifs, each bound to a Zinc atom shown in pale blue. This is a common motif found in DNA–binding proteins and the interaction is shown here (the ball–and–stick figure running from left to right in this panel is a section of DNA). Note that the Zinc does not interact with the DNA, but instead stabilizes the protein motif to allow it to form the precise shape to bind to DNA. The top right panel shows an ATP–binding domain, with the ATP displayed as the space–filling model and the protein shown in a ribbon format. The bottom panel shows another common DNA–binding motif, termed the helix–turn–helix, as found in the *lac* repressor. Here the DNA is shown in space–filling form and the two helices after which the structure is named are shown in black and white (the latter is shown end–on and is interacting with the DNA).

The existence of these conserved motifs was revealed by the analysis of the crystal structures of many different proteins. One saw similar structural regions performing similar functions in proteins that otherwise were quite dissimilar. Not surprisingly, since there are some residues that need to be conserved to form these structures, we can now predict some of these motifs based on the sequence of the proteins as well. The explosion of DNA sequence information and tools that can find these motifs has allowed the prediction of protein functions solely from sequences to be done with ever greater accuracy.

From this description you can see that proteins can often be looked at as a collection of "protein pieces," each of which serves a function and has an evolutionary history. We refer to these as functional domains. The enzyme pyruvate kinase, which generates ATP by removing it from phosphoenol pyruvate, has three domains, termed A, B and C. The A domain is responsible for binding substrate (phosphoenolpyruvate and ADP) while the C domain serves a regulatory role, binding fructose–6–bisphosphate and stimulating activity.

Up to this point, we have talked as if proteins are static rigid structures, but this is certainly wrong. Instead, most proteins are flexible, dynamic structures that respond to the conditions around them. Movement is typically essential to their function, especially for enzymes that need to grab a substrate, react with it and release product. In some cases, the flexibility is over very small distances; moving the residues of an active site just a few tenths of an angstrom (10^{-10} meters) would have a major effect on binding and reacting with a substrate that is itself only a few angstroms in size. However, some protein movements are huge, where one domain changes its position relative to that of the rest of the protein. This obviously changes the tertiary structure of that protein. Thus, when we talk about tertiary structures, typically determined by crystallography, we are really referring to only one of the possible structures that the protein can exist in while performing its biological function. The structure we see through crystallography is simply the form of the protein that happened to form crystals and whose structure was then solved. This flexibility has many important implications for a protein. At too high a temperature the protein might become too dynamic. This might overwhelm the stabilizing energy that maintains the protein, so that it unfolds or denatures. At too low a temperature, a protein might be so stable that it locks into a single structure and lacks the flexibility necessary to carry out its function. Interestingly, thermophilic microbes have been found to have more rigid proteins that can tolerate the turbulent environment of higher temperatures.

TOOLS FOR STUDYING CELLS

Cell biologists would love to do what you just did—shrink down and actually see, touch, and hear the inner workings of cells. Because that's impossible, they've developed an ever–growing collection of approaches to study cellular innards from the outside. Among them are biochemistry, physical analysis, microscopy, computer analysis, and molecular genetics. Using these techniques, researchers can exhaustively inventory the individual molecular bits and pieces that make up cells, eavesdrop on cellular communication, and spy on cells as they adapt to changing environments. Together, the approaches provide vivid details about how cells work together in the body's organs and tissues. We'll start by discussing the traditional tools of the trade—microscopes—then touch on the new frontiers of quantum dots and computational biology.

LIGHT MICROSCOPES: THE FIRST WINDOWS INTO CELLS

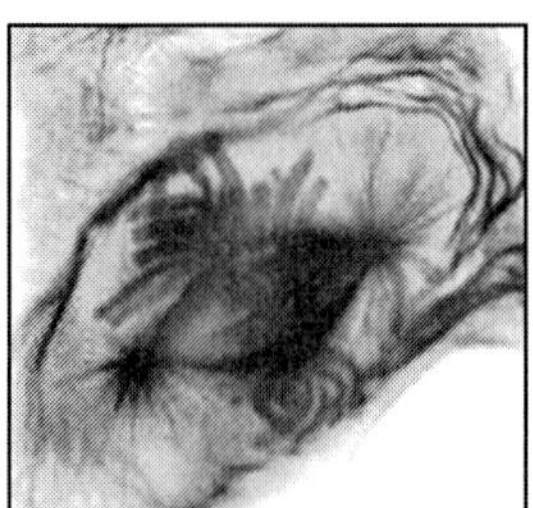

Scientists first saw cells by using traditional light microscopes. In fact, it was, looking through a microscope at a thin slice of cork, who coined the word "cell." He chose the word to describe the boxlike holes in the plant cells because they reminded him of the cells of a monastery. Scientists gradually got better at grinding glass into lenses and at whipping up chemicals to selectively stain cellular parts so they could see them better. By the late 1800s, biologists already had identified some of the largest organelles (the nucleus, mitochondria, and Golgi).

Researchers using high–tech light microscopes and glowing molecular labels can now watch biological processes in real time. The scientists start by chemically attaching a fluorescent dye or protein to a molecule that interests them. The coloured glow then allows the scientists to locate the molecules in living cells and to track processes—such as cell movement, division, or infection—that involve the molecules. Fluorescent labels come in many colours, including brilliant red, magenta, yellow, green, and blue. By using a collection of them at the same time, researchers can label multiple structures inside a cell and can track several processes at once. The technicolour result provides great insight into living cells—and is stunning cellular art.

ELECTRON MICROSCOPES: THE MOST POWERFUL OF ALL

In the 1930s, scientists developed a new type of microscope, an electron microscope that allowed them to see beyond what some ever dreamed possible. The revolutionary concept behind the machine grew out of physicists' insights into the nature of electrons. As its name implies, the electron microscope depends not on light, but on electrons. The microscopes accelerate electrons in a vacuum, shoot them out of an electron gun, and focus them with doughnut–shaped magnets. As the electrons bombard the sample, they are absorbed or scattered by different cell parts, forming an image on a detection plate.

Although electron microscopes enable scientists to see things hundreds of times smaller than anything visible through light microscopes, they have a serious drawback: They can't be used to study living cells. Biological tissues don't survive the technique's harsh chemicals, deadly vacuum, and powerful blast of electrons. Electron microscopes come in two main flavours:

transmission and scanning. Some transmission electron microscopes can magnify objects up to 1 million times, enabling scientists to see viruses and even some large molecules. To obtain this level of detail, however, the samples usually must be sliced so thin that they yield only flat, two–dimensional images. Photos from transmission electron microscopes are typically viewed in black and white. Scanning electron microscopes cannot magnify samples as powerfully as transmission scopes, but they allow scientists to study the often intricate surface features of larger samples. This provides a window to see up close the three–dimensional terrain of intact cells, material surfaces, microscopic organisms, and insects. Scientists sometimes use computer drawing programmes to highlight parts of these images with colour.

STUDYING SINGLE MOLECULES: CONNECTING THE QUANTUM DOTS

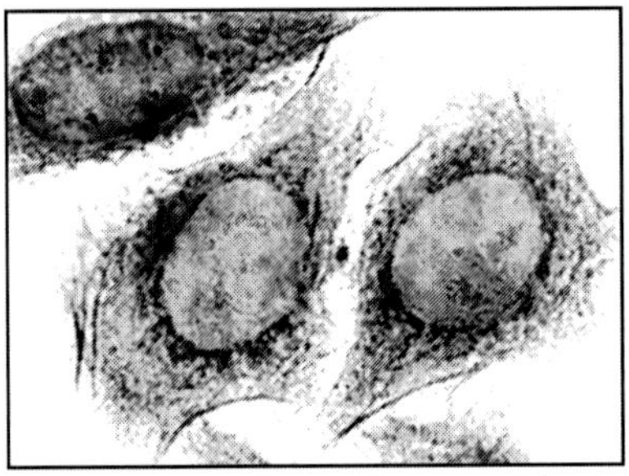

Dyes called quantum dots can simultaneously reveal the fine details of many cell structures. Someday, the technique may be used for speedy disease diagnosis, DNA testing, or analysis of biological samples. Whether they use microscopes, genetic methods, or any other technique to observe specific molecules, scientists typically flag every molecule of a certain type, then study these molecules as a group. It's rather like trying to understand a profession—say, teaching, architecture, or medicine—by tagging and observing all the workers in that profession simultaneously. Although these global approaches have taught us a lot, many scientists long to examine individual molecules in real time—the equivalent of following individual teachers as they go about their daily routines.

Now, new techniques are beginning to allow scientists to do just that. One technology, called quantum dots, uses microscopic semiconductor crystals to label specific proteins and genes. The crystals, each far less than a millionth of an inch in diameter, radiate brilliant colours when exposed to ultraviolet light. Dots of slightly different sizes glow in different fluorescent colours—larger dots shine red, while smaller dots shine blue, with a rainbow of colours in between. Researchers can create up to 40,000 different labels by mixing quantum dots of different colours and intensities as an artist would mix paint. In addition to coming in a vast array of colours, the dots also are brighter and more versatile than more traditional fluorescent dyes: They can be used to visualize individual molecules or, like the older labeling techniques, to visualize every molecule of a given type. Quantum dots promise to advance not only cell biology but also a host of other areas. Someday, the technology may allow doctors to rapidly analyse thousands of genes and proteins from cancer patients and tailor treatments to each person's molecular profile. These bright dots also could help improve the speed, accuracy, and affordability of diagnostic tests for everything from HIV infection to allergies. And, when hitched to medicines, quantum dots might deliver a specific dose of a drug directly to a certain type of cell.

COMPUTERS CLARIFY COMPLEXITY

Say you're hungry and stranded in a blizzard: If you eat before you seek shelter, you might freeze to death, but if you don't eat first, you might not have the strength to get yourself out of the storm. That's analogous to the decisions cells have to make every day to survive. For years, scientists have examined cell behaviours—like the response to cold or hunger—one at a time. And even that they did bit by bit, labouriously hammering out the specific roles of certain molecules. This approach made it difficult or impossible to study the relative contributions of—and the interplay between—genes that share responsibility for cell behaviours, such as the 100 or so genes involved in the control of blood pressure. Now, computers are allowing scientists to examine many factors involved in cellular behaviours and decisions all at the same time.

The field of computational biology blossomed with the advent of high-end computers. For example, sequencing the 3.2 billion base pairs of the human genome, which was completed in 2003, depended on computers advanced enough to tackle the challenge. Now, state-of-the-art equipment and a wealth of biological data from genome projects and other technologies are opening up many new research opportunities in computer analysis and modeling. So, much as microscopes and biochemical techniques revolutionized cell biology centuries ago, computers promise to advance the field just as dramatically in this new century.

SCIENCE SCHISMS

In living cells, material moves both forward (above arrows in figure) and backward (below arrows in figure) from each tip of the Golgi. For simplicity, we have illustrated forward movement only on the top and backward movement only on the bottom of the Golgi cartoon. One great thing about science is that you're allowed to argue about your work.

To gain new information, scientists ask a lot of questions. Often, the answers spur more questions. The never–ending circle not only feeds curiosity; it also can lead to important and sometimes unexpected breakthroughs. Sometimes, scientists studying the same topic but using different experimental approaches come up with different conclusions.

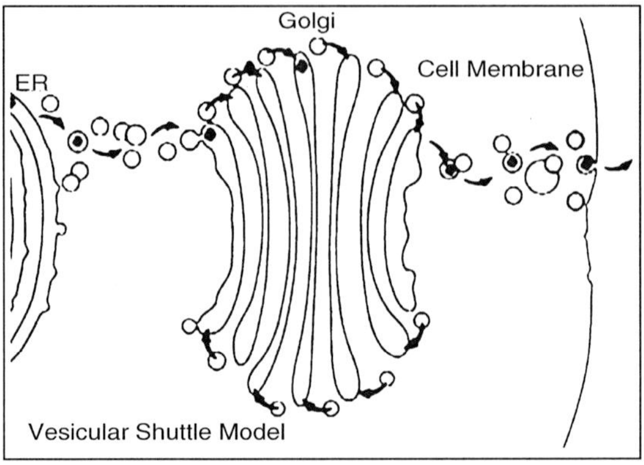

Take the Golgi, for example. Think it's noncontroversial? The details of how this organelle forms inside your cells have kept two camps of researchers in a lively battle. On one side of the debate is Graham Warren of Yale University School of Medicine in New Haven, Connecticut, who argues that the Golgi is an architectural structure that cannot be made from scratch. He believes that newly made proteins are packaged in the rough ER and are sent for further processing to a pre–existing structure (the Golgi) that is made up of different compartments. This is called the vesicular shuttle model.

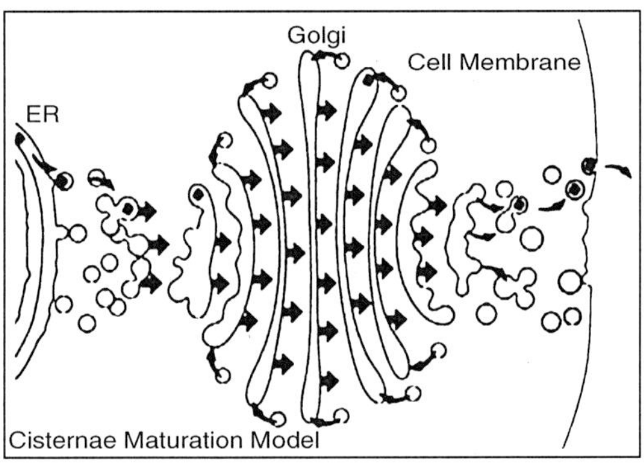

On the other side is Jennifer Lippincott–Schwartz of the National Institute of Child Health and Human Development (part of the National Institutes of Health) in Bethesda, Maryland. She says that the Golgi makes itself from scratch. According to her theory, packages of processing enzymes and newly made proteins that originate in the ER fuse together to form the Golgi. As the proteins are processed and mature, they create the next Golgi compartment. Intriguing new data suggest that perhaps neither model is completely correct. This will likely lead to yet another model. You may not see what all the fuss is about, but the differing Golgi theories say very different things about how cells function. Understanding basic cellular processes, such as how the Golgi works, ultimately can have a profound impact on the development of methods to diagnose, treat, and prevent diseases that involve those processes.

7

Molecular Biology

Molecular Biology continues to make significant progress toward understanding the fundamental molecular events that are at the core of all life. Its goal is to combine the knowledge gained in the many areas of scientific inquiry employed here — from molecular genetics to structural biology to computational biology to name a few — to obtain new insights into the functioning of biomolecules and their role in disease causality. Members of the department utilize the full range of contemporary research tools, including nuclear magnetic resonance spectroscopy, x-ray crystallography, electron microscopy, mass spectrometry, computer modeling, bioinformatics, and laser spectroscopy, as well as the techniques of modern molecular biology and genetics.

One of the most notable recent achievements in the department is the determination of the three-dimensional structure of the T-cell receptor, a key component of the immune response. Understanding its structure and function may enable scientists to enhance the effectiveness of the immune system through the development of new, highly targeted therapeutics. This same research team determined the structure of the erythropoietin receptor bound to a small cyclic peptide that mimics the function of the natural ligand for the receptor. In addition to providing insight into the mechanism of signal transduction by erythropoietin, this work has implications for the rational design of small-molecule drugs that stimulate the development of red blood cells. Other significant accomplishments include the structural determination of the active site of the enzyme that regulates the activity of nitric oxide, the discovery of a novel cortical neuropeptide — cortistatin — that induces sleep, and the solution of the structure of superoxide dismutase, a discovery that has shed light on the mechanism by which mutations cause amyotrophic lateral sclerosis, or Lou Gehrig's disease.

Protein engineering is a major activity of research activity in the department. Improved methods for the generation of catalytic antibodies have been developed, and new antibodies that efficiently catalyze disfavored chemical reactions or facilitate enantiomeric synthesis have been obtained. Directed evolution of proteins is being investigated as a means to produce novel therapeutic agents and to provide deeper understanding of the principles of molecular recognition and catalysis. Additionally, antibodies that bind cocaine have been generated and their potential use in the treatment of cocaine addiction is being explored. In the area of metalloenzyme engineering, scientists have introduced new activities into enzymes by manipulating the protein structure or by introducing new metal-binding centers. Directed evolution methods are being used to generate novel RNA and DNA enzymes.

The department continues its investigation into the theoretical and experimental understanding of protein folding. In addition, progress continues unabated in areas of molecular modeling and computational drug design. On the experimental side, novel compounds are being developed that can be designed to target specific DNA sequences and could eventually lead to a new class of therapeutic agents. Numerous advances have been made in the elucidation of the fundamental molecular processes involved in cell cycle regulation. New insights have been obtained into the mechanism by which mitosis is regulated and the role of various cyclins and kinases in these events. Ongoing research is providing a new understanding of transcriptional regulation during cell growth and the molecular basis for DNA and RNA recognition by zinc-finger proteins. Also, the molecular neurobiology group has identified and characterized several new genes with important functions in the central nervous system, including a receptor protein implicated in mediating circadian rhythms and a new neuropeptide hormone homologous to somastostatin.

CELLS

Cells are fundamental building blocks of living organisms. Cells contain a nucleus, mitochondria and chloroplasts, endoplasmatic reticulum, ribosomes, vacuoles, etc. The nucleus is important organelle because it houses chromosomes which include the DNA. The DNA is in essence a blueprint of the organism as it encodes information needed to synthesize proteins. Molecular biologists would like to understand how human biology works with the hope to treat diseases like cancer. One can look at simpler organisms such as yeasts to understand how human biology works. Admittedly, unicellular yeasts are very different from humans who have approximately 10^{14} cells. However, the DNA is similar across all living organisms. For example, humans share 99 per cent of DNA with chimps. Naturally, we would like to know what information contained in that 1 per cent of DNA is so critical to determine all the distinguishing features of humans, and we will try to answer this question.

DNA

DNA stands for deoxyribonucleic acid. DNA is an extremely long molecule that forms a double-helix. The double-helix backbone of the molecule consists of sugars and phosphates, and there is one base attached to each sugar. There are four types of bases: cytosine (C), guanine (G), adenine (A) and thymine (T). The DNA consists of two strands, and each base attached to one strand forms a bond with a corresponding base on the other strand. A only links with T and C links with G. A triplet of bases encodes an amino acid. Protein is a sequence of amino acids, and the functional subunit of DNA that encodes a protein is called a gene

GENE EXPRESSION - FROM DNA TO PROTEIN

Gene expression is a two-step process in which DNA is converted into a protein it encodes. The first step is DNA transcription. In this step, the information from the archival copy of DNA is imprinted into short-lived mRNA. The structure of RNA is a little different, it contains ribose instead of deoxyrybose, and the four bases that bind to it are cytosine (C), guanine (G), adenine (A) and uracil (U). During transcription, DNA unfolds, and mRNA is created by pairing mRNA bases with the bases of RNA. In this process C in DNA translates to G, G to C, A to U, and T to A. After mRNA is translated, it is transported to the ribosome.

The second step, protein translation occurs at the ribosome. During translation, the sequence of codons (triplets of bases) of mRNA is, with the help of tRNA, translated into a sequence of aminoacids. Gene expression seems to be a straightforward process, but it is the control of gene expression that causes most phenotypic differences in organisms. Since many diseases result from complex changes on the molecular level, we need to observe and model these processes on the system level. Gene regulatory circuits are an example of machinery that allows us to depict gene expression graphically. In this model, arrow from A and B to D indicates that if genes A and B are made into protein, so is gene D.

Data Analysis in Genomics

Recent advances in Biology, such as sequencing of the genome, have lead to creation of enormous datasets. However, our knowledge of the inter-relationships that lead to function lags behind. Data analysis in genomics is really challenging for several reasons. First, the data are intrinsically noisy because they are results of measurements and observations. Second, the data are heterogeneous and vary process by process. Finally, our coverage and accuracy of measurements varies process by process. To overcome these challenges and extract information about biological processes from the data, integrated analysis that uses probabilistic methods to deal with the noise are used.

Predicting Function of Unknown Proteins

The genome is just a sequence of letters. How do we determine what the corresponding proteins do? We already have a lot of information about some proteins, and we know about some of the interactions. This information has been captured in the Gene Ontology. The ontology is a structured vocabulary that describes proteins in terms of the associated biological processes, cellular components, molecular functions, etc. Machine learning techniques are used to take advantage of the new advanced datasets. One possible method uses individual classifiers for each class. We note that the resulting predictions may be inconsistent. These predictions are subsequently combined and inconsistencies are resolved. This method is illustrated in the accompanying slides. 10 linear SVMs were used as the weak classifiers, and the median of the results was used. Finally, hierarchical consistency was enforced. Hierarchical consistency improved the accuracy of the predictions for a majority of the nodes. The AUC of the original classifiers on the x-axis, and the AUC after the consistency enforcement on the y-axis. Since AUC is the area under the precision recall curve, the higher the value the better, AUC increased.

The improvement in AUC is not uniform. It was observed that the increase is largest for leaves of the network. Validation is, of course, important. Validation is done as usual by holding some examples out. If a particular gene is predicted to be involved in DNA replication, an experiment can be performed. A copy of yeast that is missing the gene is created, and mitosis is observed. If both the mother and daughter cell get a copy of DNA, the gene could not have been involved in DNA replication and vice versa.

TECHNIQUES OF MOLECULAR BIOLOGY

Since the late 1950s and early 1960s, molecular biologists have learned to characterize, isolate, and manipulate the molecular components of cells and organisms. These components include DNA, the repository of genetic information; RNA, a close relative of DNA whose functions range from serving as a temporary working copy of DNA to actual structural and enzymatic functions as well as a functional and structural part of the translational apparatus; and proteins, the major structural and enzymatic type of molecule in cells.

EXPRESSION CLONING

One of the most basic techniques of molecular biology to study protein function is expression cloning. In this technique, DNA coding for a protein of interest is cloned (using PCR and/or restriction enzymes) into a plasmid (known as an expression vector). This plasmid may have special promoter elements to drive production of the protein of interest, and may also have antibiotic resistance markers to help follow the plasmid. This plasmid can be inserted into either bacterial or animal cells. Introducing DNA into bacterial cells is called transformation, and can be completed with several methods, including electroporation, microinjection, passive uptake and conjugation.

Introducing DNA into eukaryotic cells, such as animal cells, is called transfection. Several different transfection techniques are available, including calcium phosphate transfection, liposome transfection, and proprietary transfection reagents such as Fugene. DNA can also be introduced into cells using viruses or pathenogenic bacteria as carriers. In such cases, the technique is called viral/bacterial transduction, and the cells are said to be transduced. In either case, DNA coding for a protein of interest is now inside a cell, and the protein can now be expressed.

A variety of systems, such as inducible promoters and specific cell-signaling factors, are available to help express the protein of interest at high levels. Large quantities of a protein can then be extracted from the bacterial or eukaryotic cell. The protein can be tested for enzymatic activity under a variety of situations, the protein may be crystallized so its tertiary structure can be studied, or, in the pharmaceutical industry, the activity of new drugs against the protein can be studied.

POLYMERASE CHAIN REACTION (PCR)

The polymerase chain reaction is an extremely versatile technique for copying DNA. In brief, PCR allows a single DNA sequence to be copied (millions of times), or altered in predetermined ways. For example, PCR can be used to introduce restriction enzyme sites, or to mutate (change) particular bases of DNA. PCR can also be used to determine whether a particular DNA fragment is found in a cDNA library. PCR has many variations, like reverse transcription PCR (RT-PCR) for amplification of RNA, and, more recently, real-time PCR (qPCR) which allow for quantitative measurement of DNA or RNA molecules.

GEL ELECTROPHORESIS

Gel electrophoresis is one of the principal tools of molecular biology. The basic principle is that DNA, RNA, and proteins can all be separated using an electric field. In agarose gel electrophoresis, DNA and RNA can be separated based on size by running the DNA through an agarose gel. Proteins can be separated based on size using an SDS-PAGE gel. Proteins can also be separated based on their electric charge, using what is known as an isoelectric gel.

SOUTHERN BLOTTING

Named after its inventor, biologist Edwin Southern, the Southern blot is a method for probing for the presence of a specific DNA sequence within a DNA sample. DNA samples before or after restriction enzymedigestion are separated by gel electrophoresis and then transferred to a membrane by blotting via capilliary action. The membrane can then be probed using a DNA probe labeled using a complement of the sequence of interest. Most original protocols used radioactive labels, however now non-radioactive alternatives are available. Southern blotting is less commonly used in laboratory science due to the capacity of using PCR to detect specific DNA sequences from DNA samples. However, these blots are still used for some applications, such as measuring transgene copy number in transgenic mice, or in the engineering of gene knockout embryonic stem cell lines.

NORTHERN BLOTTING

The Northern blot is used to study the expression patterns a specific type of RNA molecule as relative comparison among of a set of different samples of RNA. It is essentially a combination of denaturing RNA gel electrophoresis, and a blot. In this process RNA is separated based on size and is then transferred to a membrane that is then probed with a labeled complement of a sequence of interest. The results may be visualized through a variety of ways depending on the label used, however, most result in the revelation of bands representing the sizes of the RNA detected in sample. The intensity of these bands is related to the amount of the target RNA in the samples analyzed. The procedure is commonly used to study when and how much gene expression is occurring by measuring how much of that RNA is present in different samples. It is one of the most basic tools for determining at what time certain genes are expressed in living tissues.

WESTERN BLOTTING

Antibodies to most proteins can be created by injecting small amounts of the protein into an animal such as a mouse, rabbit, sheep, or donkey (polyclonal antibodies)or produced in cell culture (monoclonal antibodies). These antibodies can be used for a variety of analytical and preparative techniques. In Western blotting, proteins are first separated by size, in a thin gel sandwiched between two glass plates in a technique known as SDS-PAGE (for Sodium Dodecyl Sulphate Poly-Acrylamide Gel Electrophoresis). The proteins in the gel are then transferred to a PVDF, nitrocellulose, nylon or other support membrane. This membrane can then be probed with solutions of antibodies. Antibodies that specifically bind to the protein of interest can then be visualized by a variety of techniques, including coloured products, chemiluminescence, or autoradiography.

Analogous methods to western blotting can also be used to directly stain specific proteins in cells and tissue sections. However, these *immunostaining* methods are typically more associated with cell biology than molecular biology. The terms "Western" and "Northern" are jokes: The first blots were with DNA, and since they were done by Ed Southern, they came to be known as Southerns. I don't think Patricia Thomas, inventor of the RNA blot, which became known as a "Northern", actually used the term.To carry the joke further, one can find reference in the literature to "southwesterns" (Protein-DNA interactions) and "far-westerns" (Protein-Protein interactions).

ARRAYS

A DNA array is a collection of spots attached to a solid support such as a microscope slide; each spot contains one or more DNA molecules. Arrays make it possible to put down a large number of very small (100 micrometre diameter) spots on a single slide; if each spot has a DNA molecule that is complementary to a single gene, one can analyze the expression of every gene in an organism in a single experiment. For instance, the common baker's yeast, *Saccharomyces cerevisiae,* contains about 7000 genes; with a microarrray, one can measure quantitatively, how each gene is expressed, and how that expression changes, for example, with a change in temperature.

There are many different ways to fabricate microarrays; the most common are silicon chips, microscope slides with spots of ~ 100 micrometre diameter, custom arrays, and arrays with larger spots on porous membranes (macroarrays). Arrays can also be made with molecules other than DNA. For example, an antibody array can be used to determine what proteins or bacteria are present in a blood sample

ABANDONED TECHNOLOGY

As new procedures and technology become available, the older technology is rapidly abandoned. A good example is methods for determining the size of DNA molecules. Prior to gel electrophoresis, with agarose and polyacrylamide, DNA was sized with rate sedimentation in sucrose gradients, a slow and labour intensive technology requiring expensive instrumentation; prior to sucrose gradients, viscometry was used. Aside from their historical interest, it is worth knowing about older technology as it may be useful to solve a particular problem.

HISTORY

Molecular biology was established in the 1930s, the term was first coined by Warren Weaver in 1938 however. Warren was director of Natural Sciences for the Rockefeller Foundation at the time and believed that biology was about to undergo a period of significant change given recent advances in fields such as X-ray crystallography. He therefore channeled significant amounts of (Rockefeller Institute) money into biological fields.

MOLECULAR BIOLOGY OF CELL

The cell membrane or plasma membrane is a biological membrane that separates the interior of all cells from theoutside environment. The cell membrane is selectively permeable to ions and organic molecules and controls the movement of substances in and out of cells. The basic function of the cell membrane is to protect the cell from its surroundings. It consists of the lipid bilayer with embedded proteins. Cell membranes are involved in a variety of cellular processes such as cell adhesion, ion conductivity and cell signaling and serve as the attachment surface for several extracellular structures, including the cell wall, glycocalyx, and intracellular cytoskeleton. Cell membranes can beartificially reassembled. Microorganisms typically face the world as single cells rather than as the multi-cellular assemblies of higher organisms. Each single cell must therefore contain all the structures necessary for managing its internal state and dealing with the outside environment. Not surprisingly, this evolutionary process results in the use of rather similar structures and processes to solve similar need in different microorganisms. However, prokaryotes have been on this Earth for a long period of time and this has allowed them to differentiate into a dizzying number of different species. Eukaryotic microbes are not quite so diverse, but they still display a remarkable range of properties.

The diversity of the microbial organisms also means that this survey of structures is not exhaustive. No one cell contains all the structures that we describe here, but we will explore the more common structures that have been observed by scientists in the past 150 years. A distinction in this discussion must be made between the two types of prokaryotes: the Archaea and their cousins, the Bacteria. We will initially focus on the Bacteria, since that is what we know the most about. Many of the structures we will examine are found in both the Bacteria and the Archaea, but there are some significant differences and these will be covered at the end of the chapter. Finally, we will talk about the features that are distinctive among the microbial eukaryotes. So how did scientists find out so much about such very small organisms? As you might guess, many techniques come into play when tackling a subject as complex as bacterial structure. Electron microscopes have been important, of course, but so have genetics, molecular biology and biochemistry. Microscopes help scientists to visualize where these structures are located and how they are arranged spatially in the microbe. Bacterial genetics and molecular biology identify and analyse the genes necessary for the synthesis and regulation of these structures. Biochemistry permits the detailed examination of each part separately, with implications for its role in the living bacterium. The powerful combination of these disciplines has provided a deep understanding of how a bacterium is put together, but there is still much to learn.

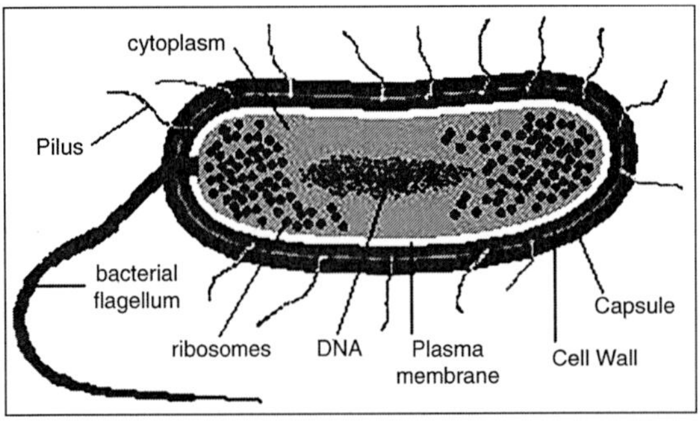

Fig. The Generalized Bacterium

This chapter on microbial structure is separated into two sections. In the first section, we describe the chemical nature of the types of molecules and polymers that are important in carrying out the business of biology. In the second section, we examine the functional units in the cell, describing how the various chemical structures in the cell interact to carry out important cellular functions. In this discussion we assume that the student has had an introductory chemistry course (at least in high school) and is somewhat familiar with chemical notation. First and foremost, it is important to point out that there are some universal structures that all living cells contain. They are the basic building blocks of life: DNA, RNA, protein and cellular membranes. Most, but not all, bacteria also possess cell walls. Beyond these essentials, the frequency of the rest of the structures we mention here ranges in the bacterial world from quite common to very rare.

Sugars are Common in the Cell

Sugars serve three basic purposes in the cell: as carbon and energy sources, as reservoirs of carbon and energy, and as parts of cellular structures. Large amounts of energy can be extracted from sugars by processes referred to as catabolism. This may explain why many microorganisms show a preference for sugars if given a choice of energy sources. The term carbohydrate is often used to refer to them because their chemical formula can be broken down into $[C(H_2O)]x$ where x is any number greater than three.

Sugar Monomers

Single sugar molecules are typically 3 to 7 carbons long and are termed Figure shows that each carbon on the sugar molecule is decorated with a hydroxyl group (OH) except for one carbon that forms a carbonyl group. If the carbonyl group is at the end of the molecule, it forms an aldehyde; if the carbonyl group is in the middle of the molecule, it forms a ketone as indicated in Figure.

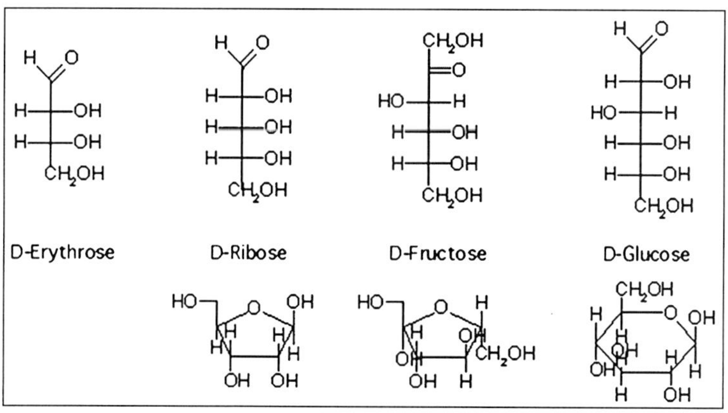

Fig. Common Monosaccharides

All sugars can exist as linear molecules in solution and those greater than 5 carbons long can also circularize with the carbonyl group attacking a hydroxyl on one of the other carbons. The circular sugar contains 5 to 7 members in the ring with one of the members being oxygen. Glucose, fructose and ribose are some of the more common sugars found in the cell.

Sugar Polymers

Sugars readily polymerize: the combination of two sugars is called a disaccharide, while the combination of three is called a tri saccharide, and polymers of greater than three sugars are referred to as polysaccharides. Sugars are connected by α or β linkages. If the hydrogen is pointing up, it is a β linkage, if the hydrogen is pointing down, it is a α linkage as shown in Figure. This distinction may seem trivial, but it greatly influences the properties of the molecule. (The name of the linkage further depends upon the orientation of the hydrogen on the lowest numbered carbon that forms the bond. The lowest numbered carbon is decided by the rules of organic chemistry, but this is really not important for our discussion.

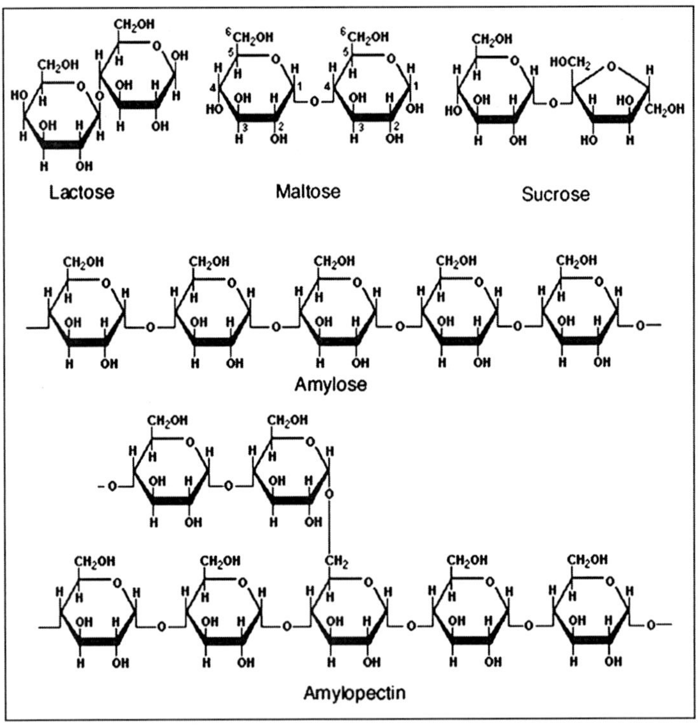

Fig. Common Disaccharides, Trisaccharides and Polysaccharides

Just realise there is a set pattern of numbering for each organic molecule and this helps determines the α or β linkage.) For example, starch is made up of glucose linked by α-1,4 bonds. Starch is also water-soluble and can serve as a food source for many organisms. In contrast, cellulose, containing glucose linked by β-1,4 bonds, is insoluble in water and is not as readily degraded by most microbes. Polysaccharides might contain only one type of sugar monomer or a variety of different ones, sometimes in repeating units. Polymers of sugars can serve as storage products for the cell. The breakdown of sugar yields a huge amount of energy, which means that they are terrific molecules for effectively storing energy for later utilization. Starch and glycogen are two examples of sugar polymers that are used in this way. Polysaccharides also serve as structural components of many different molecules in the cell including nucleic acids and the cell wall.

Nucleic Acids Store Information and Process it

- DNA and RNA are composed of two parts, the sugar phosphate backbone and nucleotide bases.
- The bases in DNA are purines and pyrimidines (cytosine and thymine). RNA contains the same bases, except uracil is substituted for thymine and they contain an extra 5'-hydroxyl.
- DNA polymers can be thousands of base pairs long. RNA polymers are hundreds to thousands of bases long.
- RNA is single-stranded. DNA is double-stranded and forms a double helix in the cell.
- RNA molecules have significant secondary and tertiary structure that is important in their functions.

Deoxyribonucleic acid (DNA) and ribonucleic acid (RNA) are involved in information storage and processing. DNA serves as the cell's hereditary information, while RNA is involved in converting that information into functional products, such as proteins. RNA and DNA are long polymers of only 4 nucleotides: adenine, guanine, cytosine and thymine (or uracil for RNA). Figure lists the structure of the five nucleotides found in nucleic acids.

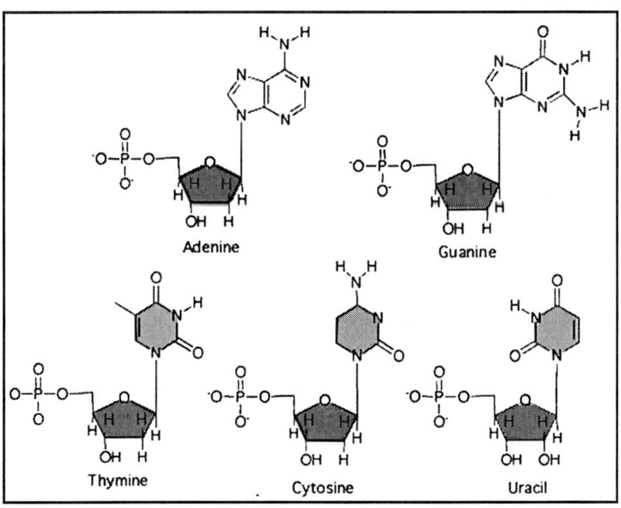

Fig. The Structure of Nucleotides

The nucleotides for adenosine (A), guanine (G), cytosine (C) and thymine (T) as found in DNA are shown. The first three are also found in RNA, but when incorporated into that polymer, the associated sugar has two hydroxyls, as shown in the model of uracil (U). Uracil is the RNA equivalent of the DNA nucleotide thymine.

The Figure Many thousands of DNA nucleotides are strung together to form genes and chromosomes. In this picture the bases of the two anti-parallel (running in opposite direction) strands of a DNA double helix are shown. Note the sugar-phosphate backbone from which the bases extend and pair with matching bases from the other strand. The dashed lines show hydrogen bonds between the base pairs.

The bases of the four nucleotides are different, but there is also a pattern. Adenine (A) and guanine (G) are purines, and therefore have a distinctive two-ring structure; they differ in the chemical groups attached to the rings. Likewise, cytosine (C), thymine (T) and uracil (U) are all pyrimidines and share a single-ringed structure, but also differ in their attached groups.

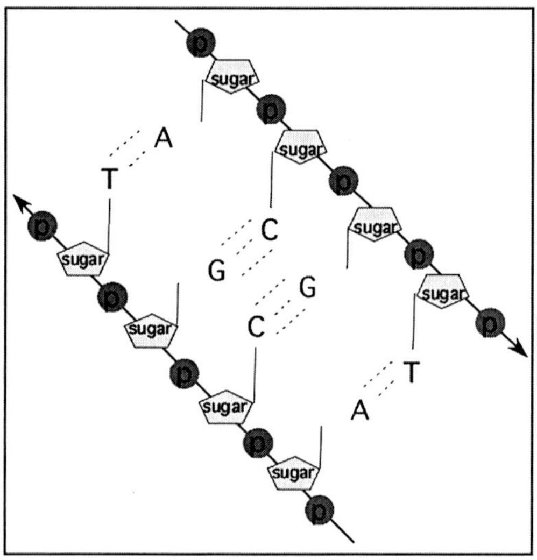

Fig. A Schematic of the Nucleic Acid Polymer

Not surprisingly, as these extra chemical groups distinguish the different purines and pyrimidines structurally, they are also responsible for their important functional differences as well. The bases in a nucleic acid polymer are capable of forming hydrogen bonds with neighbouring bases on a second strand of nucleic acid, a process termed base pairing. However, there are rules to this association. Adenine is capable of forming two hydrogen bonds with thymine (or uracil) and cytosine can base pair with guanine, forming three hydrogen bonds. Some of these bonds require the extra chemical groups mentioned above. If two single strands of nucleic acid have sequences that can base pair along the polymer (such sequences are sometimes said to complement), they will generate a long double-stranded polymer that has a staircase topology as shown in Figure. This structure is termed a double helix, since two strands form the molecule and it spirals around an axis in a regular pattern.

Such a reaction between two complementary DNA strands is spontaneous: if you mix two complementary single strands of nucleic acid together in a test tube at a reasonable temperature, pH and salt concentration, they will find each other and anneal to form a double-stranded polymer.

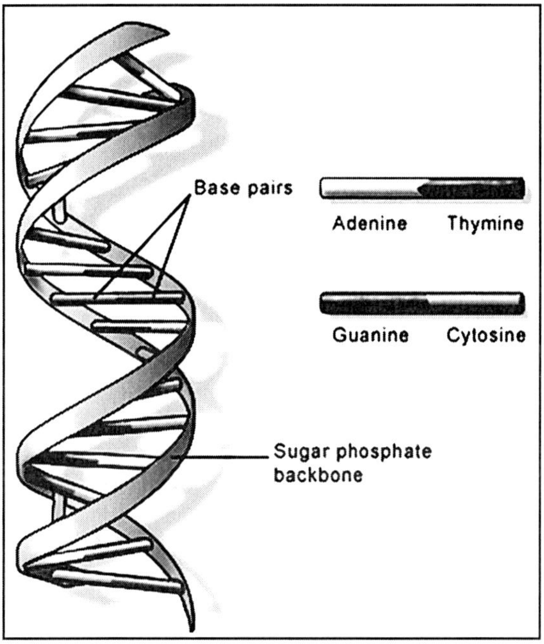

Fig. The Double Helix

Note how the two nucleic acid strands spiral around each other in a regular repeating pattern. There are 10 bases per turn of the double helix. Bases pair with one another in the centre of the helix cylinder and form what some have likened to a spiral staircase.

Secondary and Tertiary DNA Structures

DNA almost always exists in cells as a double-stranded structure of complementing strands. It happens that this double-stranded form is rather stable, and resists tight twists and turns. It is often assumed that the stability is due to the hydrogen bonding between the bases, but this is not the case (the bases would hydrogen bond just as well to water). Instead, it is largely due to the interaction between adjacent base pairs along the helix, which is termed "base stacking" Figure for an example. Finally, the larger organization of the DNA strands with respect to each other, termed the tertiary structure, is also fairly similar in all DNA molecules. One implication is that proteins that want to distinguish between different DNA molecules must do so by reading different primary structure sequences by interacting with the outer surface of the base pairs.

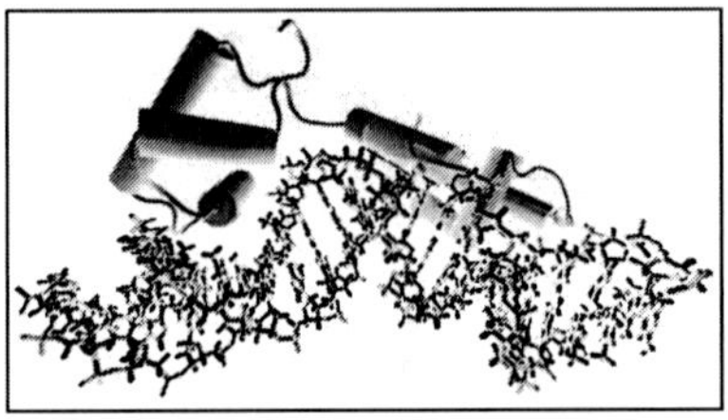

Fig. A Molecular model of the Lac Repressor

The repressor binding to its recognition sequence. Any DNA-binding protein that recognize specific DNA sequences must make specific contacts with that recognition sequence by reading the pattern of bases in the DNA. These proteins do this by making very specific contacts between themselves and those atoms on DNA that are different in the different bases. In this figure, the DNA helix is shown by the ball-and-stick figure running along the bottom and the protein is depicted by cylinders (which refer to alpha helices) and other "squiggles" (which refer to any other structure in the protein). The point of the figure are that the DNA is not opened up and that the recognition regions of the proteins are very often alpha helices that lay parallel with or perpendicular to the DNA helix.

Not shown are the side chains of the amino acids that actual make the specific contacts with the DNA. In this and subsequent molecular figures, of three different displays will be shown: ball and stick, which shows all atoms larger than hydrogen; ribbon models, in which a ribbon that runs along the carbon backbone of the protein is displayed; and space-filling models, where the actual atomic surface of the molecule is shown. Sometimes, as in this figure, the ribbon form is modified slightly be showing cylinders for alpha helices instead of an actual helix as in Figure. Most of the structures shown in this text are the product of X-ray crystallography, which essentially displays the position of every atom (larger than hydrogen) in a molecule. Increasingly all biological questions are being thought about in terms of these structures at the atomic scale.

Structures of RNA

In composition and therefore primary structure, RNA is similar to DNA, except that uracil (U) takes the place of thymine in the molecule and the ribose unit on each sugar contains a additional hydroxyl group. However, most RNA in cells exists as single-stranded molecules and not a complex of two different strands as with DNA. Now if complementary base sequences are present in an RNA molecule, it can fold back upon itself and base pair, so that many

RNA molecules have at least some double-stranded regions. However, this bending and folding means that RNA molecules typically have much more complicated tertiary structures than does DNA. Both the single-stranded loops and the double-stranded stems are critical for the function of most RNA molecules. Many are involved in creating physical structures, such as ribosomes, that are involved in processing information. The other general class of RNAs are messenger RNAs, which represent a version of the DNA primary structure that is suitable for translation into protein.

CELL THEORY

All living organisms are composed of cells, and all cells arise from other cells. These simple and powerful statements form the basis of the cell theory, first formulated by a group of European biologists in the mid-1800s. So fundamental are these ideas to biology that it is easy to forget they were not always thought to be true.

EARLY OBSERVATIONS

The invention of the microscope allowed the first view of cells. English physicist and microscopist Robert Hooke (1635–1702) first described cells in 1665. He made thin slices of cork and likened the boxy partitions he observed to the cells (small rooms) in a monastery. The open spaces Hooke observed were empty, but he and others suggested these spaces might be used for fluid transport in living plants. He did not propose, and gave no indication that he believed, that these structures represented the basic unit of living organisms.

Marcello Malpighi (1628–1694), and Hooke's colleague, Nehemiah Grew (1641–1712), made detailed studies of plant cells and established the presence of cellular structures throughout the plant body. Grew likened the cellular spaces to the gas bubbles in rising bread and suggested they may have formed through a similar process. The presence of cells in animal tissue was demonstrated later than in plants because the thin sections needed for viewing under the microscope are more difficult to prepare for animal tissues. The prevalent view of Hooke's contemporaries was that animals were composed of several types of fibres, the various properties of which accounted for the differences among tissues.

At the time, virtually all biologists were convinced that organisms were composed of some type of fundamental unit, and it was these "atomistic" preconceptions that drove them to look for such units. While improvements in microscopy made their observations better, it was the underlying belief that there was some fundamental substructure that made the microscope the instrument of choice in the study of life. In 1676 the Dutch microscopist Antony van Leeuwenhoek (1632–1723) published his observations of single-cell organisms, or "little animalcules" as he called them. It is likely that Leeuwenhoek was the first person to observe a red blood cell and a sperm cell. Leeuwenhoek made numerous and detailed observations on his microorganisms, but more than one hundred years passed before a connection was made between the obviously cellular structure of these creatures and the existence of cells in animals or plants.

DEVELOPMENT OF THE CELL THEORY

In 1824 Frenchman Henri Milne-Edwards suggested that the basic structure of all animal tissues was an array of "globules," though his insistence on uniform size for these globules puts into question the accuracy of his observations. Henri Dutrochet (1776–1847) made the connection between plant cells and animal cells explicit, and he proposed that the cell was not just a structural but also a physiological unit: "It is clear that it constitutes the basic unit of the organized state; indeed, everything is ultimately derived from the cell". Dutrochet proposed that new cells arise from within old ones, a view that was echoed by his contemporary François Raspail (1794–1878). Raspail was the first to state one of the two major tenets of cell theory: *Omnis cellula e cellula,* which means "Every cell is derived from another cell." However, despite this ringing and famous phrase, his proposed mechanism of cell generation was incorrect. Raspail was also the founder of cell biochemistry, making experiments on the chemical composition of the cell and their response to changing chemical environments.

In 1832 Barthelemy Dumortier (1797–1878) of France described "binary fission" (cell division) in plants. He observed the formation of a mid-line partition between the original cell and the new cell, which, Dumortier noted, "seems to us to provide a perfectly clear explanation of the origin and development of cells, which has hitherto remained unexplained". These observations led him to reject the idea that new cells arise from within old ones, or that they form spontaneously from noncellular material. The discovery of cell division is usually attributed to Hugo von Mohl (1805–1872), but Dumortier proceeded him in this regard.

Von Mohl did coin the word "protoplasm" for the material contained in the cell. The first unequivocal description of the cell nucleus was made by a Czech, Franz Bauer, in 1802 and was given its name in 1831 by Robert Brown (1773–1858) of Scotland, who is best remembered for discovering the random "Brownian" motion of molecules. The first accurate description of the nucleolus was made in 1835. Schleiden and Schwann, who are usually given credit for elucidating the cell theory, made their marks in 1838 and 1839. In 1838 Matthais Schleiden (1804–1881) proposed that every structural element of plants is composed of cells or the products of cells. However, Schleiden insisted on priority for several ideas that were not his and clung to the idea that cells arise by a crystallization-like process either within other cells or from outside, which Dumortier had dispensed with some years earlier. (In Schleiden's defence, it should be remembered that drawing incorrect conclusions from limited observations is a risk inherent in science, especially when working on the frontier of a new field.) In 1839 a fellow German, Theodor Schwann (1810–1882), proposed that in animals too every structural element is composed of cells or cell products. Schwann's contribution might be regarded as the more groundbreaking, since the understanding of animal structure lagged behind that of plants. In addition, Schwann made the explicit claim that the fundamental laws governing cells were identical between plants and animals: "A common principle underlies the development of all the individual elementary subunits of all organisms".

A special word should be said here about the Czech Jan Purkyòe (1787–1869), or Purkinje, as his name is usually given. Purkinje was the premiere cytologist of his day, and one of the most influential formulators of the cell theory. He gave his name to structures throughout the body, including the Purkinje cells of the cerebellum. Purkinje, in fact, deserves much of the credit that usually goes to Schwann, for in 1837 he proposed not only that animals were composed principally of cells and cell products (though he left room for fibres) but also that the "basic cellular tissue is again clearly analogous to that of plants". Unfortunately, Schwann did not credit Purkinje in his influential publication.

REPRODUCTION AND INHERITANCE

Despite the work of Dumortier, the origins of new cells remained controversial and confused. In 1852 a German, Robert Remak (1852–1865), published his observations on cell division, stating categorically that the generation schemes proposed by Schleiden and Schwann were wrong. Based on his observations of embryos, Remak stated instead that binary fission was the means of reproduction of new animal cells. This view was widely publicized not by Remak but by Rudolf Virchow (1821–1902), unfortunately without crediting Remak. Virchow is also usually given the credit for the phrase *Omnis cellula e cellula,* indicating the importance of cell division in the creation of new cells.

The understanding of the central importance of chromosomes lagged well behind their discovery. In 1879 Walther Flemming (1843–1905) noted that the chromosomes split longitudinally during mitosis (a term he introduced). Wilhelm Roux (1850–1924) proposed that each chromosome carried a different set of hereditable elements and suggested that the longitudinal splitting observed by Flemming ensured the equal division of these elements. This scheme was confirmed in 1904 by Theodor Boveri (1862–1915). Combined with the rediscovery of Gregor Mendel's 1866 paper on heritable elements in peas, these results highlighted the central role of the chromosomes in carrying genetic material. The chemical nature of the gene was determined in a series of experiments over the next fifty years, culminating in the determination of the structure of deoxyribonucleic acid (DNA) in 1953 by James Watson and Francis Crick.

FUNCTION OF NERVE FIBRES

The neuron is differentiated for the conduction and transmission of nerve impulses. After embryonic life, neurons do not divide, but remain in a permanent interphase throughout the entire life of the organism. During this time a neuron undergoes changes in volume and in the number and complexity of its processes and functional contacts, but the number of neurons is not increased by cell division. This fact may be of paramount significance, since, in addition to conducting and transmitting impulses, nerve cells store instinctive and learned *information* (*e.g.*, conditioned reflexes, memory) – a property that would be best served by a more permanent system of structures.

Neurons are adapted to their specialized functions by means of different types of outgrowths. The cell body (perikaryon) may emit one or more short outgrowths, or *dendrites*, which carry nerve impulses centripetally, and a longer one, the *axon*, which carries the impulse centrifugally to the next neuron or the effector. An axon is also called a *nerve fibre* when it is wrapped in the different sheaths after emerging from the neuron. The axon terminates, ramifying in the *telodendrons*, or endings. Some neurons have only one dendrite and the axon (*i.e., bipolars*) and others have only the axon (*i.e., monopolars*), in addition to the most common *multipolars*. In invertebrates most neurons are monopolar. The different types of neurons and neuronal interconnections are discussed in histology and neuroanatomy textbooks. Only a few general considerations will be made here.

NEUROFIBRILS AND NEUROTUBULES

In fixed and stained preparations observed under the light microscope fine filaments, called *neurofibrils,* can be demonstrated in the cytoplasm of the neuron. These neurofibrils run in all directions and continue into the dendrites, axon, and nerve fibre. Better information about the structure of the axon can be obtained by polarization and electron microscopy. The axoplasm of myelinated nerve fibres may be extruded and separated from the myelin sheath, permitting its study with the electron microscope and with polarization microscopy. In these studies the partial volume occupied by the axially oriented material has been found to be less than 1 per cent. In the extruded axoplasm a fibrillar material was detected. In addition, mitochondria, strands of canaliculi, and vesicles of the endoplasmic reticulum have been observed. The fibrillar material present in the axon, dendrites, and perikaryon is formed principally of long, tubular elements (the *neurotubules* or *neuronal microtubules*), 20 to 30 nm in diameter, which were previously seen in nerve homogenates.

Unmyelinated nerve fibres show numerous neurotubules in cross section. Among the neurotubules, there are also cross sections of neurofilaments. The walls of the neurotubules show a subunit structure similar to that generally found in microtubules. Neurotubules containing dense core have been described. The electron-dense material of the cores has been interpreted as a result of endoluminal migration of some products. In the neurofilaments some lateral outgrowths or side-arms have been observed. Neurotubules have been isolated from mammalian brain cells. Neurotubules and neurofilaments, when clumped together under the action of the fixatives and with the addition of colloid silver, form the neurofibrils of classic histology. Although neurofibrils were described more than a century ago, their significance remained practically unknown. The hypothesis that they are involved in nerve conduction has been disproved. There is now no doubt that nerve conduction takes place at the surface membrane of the axon.

FUNCTIONS OF THE NEURON

The perikaryon is characterized by the presence of considerable amounts of basophilic material – the Nissl substance – which, as in other cells, is composed of ribosomes and endoplasmic reticulum. A well-developed Golgi complex is also characteristic of the neuron.

The great abundance of ribosomes is related to the biosynthetic functions of the perikaryon, which has to maintain a volume of cytoplasm in its outgrowths that may be considerably greater than its own. (In mammals axons may be as long as 1 metre or longer). If a nerve fibre is cut, the distal part degenerates (wallerian degeneration), and the proximal stump may regenerate later on by a growing process that is dependent on the perikaryon. There is also experimental evidence that the axon is continuously growing and being used at the ending.

Orthograde Directed Macromolecular

The experiments just mentioned –on degeneration –should be correlated with the fact that the axon and the nerve endings are generally devoid of polysomes, and are unable to undertake a significant local protein synthesis. In fact, local synthesis by nerve endings may account for only 2 per cent of the rapidly renewed proteins. For this reason most axonal and synaptic proteins must be manufactured in the perikaryon and transported to the nerve endings. The study of the axonal transport (*axon flow*) of macromolecules in the *orthograde direction* (*i.e.*, from the perikaryon to the nerve endings) has become one of the most intense fields of research in neurobiology.

Macromolecules (such as proteins, glycoproteins, and enzymes) which are soluble in the axoplasm or integrate the various axonal and synaptic structures (*i.e.*, axonal membrane, neurotubules, neurofilaments, mitochondria, synaptic vesicles, nerve-ending membranes, etc.), are first synthesized in the cell body and then find their way into the dendritic arborizations. They may also pass into the axon and nerve endings, at a distance that may vary from millimeters to meters. The proteins present in nerve endings have half-lives which range from 12 hours to 50 to 100 days; to be replaced, therefore, new proteins must reach the nerve endings to compensate for their loss.

Experimental work that has used colchicine and vinblastine has demonstrated that axoplasmic transport is related to the neurotubules.[13] In fact, the integrity of these axonal organelles is critical in maintaining a normal axoplasmic flow. One of the best approaches to the study of axonal transport is the use of radiolabeled precursors for proteins or glycoproteins (*e.g.*, ^{3}H-lysine, ^{3}H-fucose) and the monitoring of the final product by radioautography at the electron microscopic level. Experiments of this kind have been made on the ciliary ganglion of the chicken because it has definite advantages. In fact, the axons originate from the Edinger-Westphal nuclei, situated near the cerebral aqueduct, where the precursor can be injected.

The axons are 10 mm long and terminate in a single giant nerve ending (the presynaptic calyx), which makes contact with the postsynaptic cell body of the ciliary ganglion. A radioautograph obtained after the injection of ^{3}H-lysine. In the lower portion of the postsynaptic ganglion cell is shown to be completely devoid of silver grains, indicating that there is no transynaptic migration of proteins. In the upper part, the presynaptic ending is covered with silver grains corresponding to proteins that have migrated from the perikaryon. The radioactive proteins are distributed over synaptic vesicles, mitochondria, and axoplasm.

Orthograde Axonal Transport

The studies just described, as well as others, have revealed that the proteins may be transported in an orthograde direction at different rates, and that there are generally two definite groups of proteins – those of fast or those of slow transport.

Rapid Axonal Flow

Axonal flow of about 280 mm per day is found in proteins that are synthesized in polysomes bound to the endoplasmic reticulum (ergastoplasm), that are transferred to the Golgi complex, and then penetrate the axon and dendrites, mainly by way of tubules and vesicles of the smooth endoplasmic reticulum. On their way through the Golgi apparatus, some proteins are converted into glycoproteins by the addition of carbohydrates indicates that the proteins of fast transport contribute to the renewal of the axon membrane, the turnover of macromolecules associated with synaptic vesicles, and the presynaptic membrane. Another fraction is related to axonal and nerve ending mitochondria. The fast moving proteins are thus assigned mainly to the renewal of various membrane components; by cell fractionation they are recovered in the particulate fractions. Practically all glycoproteins are also transferred by this fast phase of axonal flow. Since these fast moving proteins and glycoproteins can cover 280 mm per day, it is understandable that they begin to appear in the ciliary ganglion within one hour of the injection of the precursors into the aqueduct.

Slow Axonal Flow

In proteins present in the soluble fraction of the nervous tissue, and to a lesser degree in mitochondria, axonal flow is slow. The proteins are synthesized mainly by free polysomes and probably bypass the Golgi complex. Soluble enzymes, such as choline acetyltransferase, protein subunits of tubulin that assemble to form the neutrotubules, or the protein of microfilaments, are transported by slow axonal flow.

These enzymes enter the axon at a speed of 1 to 1.5 mm a day and exhibit maximal accumulation in the nerve ending at 6 days. In general, these protein turn over at a slow rate in the axon, and only a minor fraction of them enters the nerve ending.

The fast and slow moving proteins that enter the axon and nerve endings may compensate for the local breakdown by proteolytic enzymes, as well as the release of proteins that may be related to the function of synaptic vesicles. A possible postsynaptic transfer of macromolecules of small magnitude should also be contemplated.

Retrograde Axonal Transport

We have described earlier the axon flow of macromolecules in the orthograde direction, but there is also an axon transport that goes the opposite way, *i.e.*, a *retrograde axon flow*. Although by 1917 Levi, observing the axons of cultured nerve cells, had detected movement of particles in both orthograde and retrograde directions, the latter type of transport had been neglected until recently. By injecting high concentrations of horseradish peroxidase or serum albumin, coupled to a fluorescent dye, in the vicinity of nerve terminals, it was found that these proteins were transferred through the axon to the corresponding cell bodies.

For example, if peroxidase is injected into the visual cortex, at the occipital pole of the brain, it is possible to trace the neurons into the lateral geniculate nucleus. While this technique has become of great importance in tracing neuroanatomic connections in the central nervous system and is now widely used, the physiologic importance of the retrograde transport was questionable. In fact, peroxidase is not a protein normally found in the nervous system and had to injected at rather high concentrations.

The physiologic importance of this mechanism was recognized when it was found that the *nerve growth factor* (NGF) (a protein of low molecular weight that is indispensable for the growth of adrenergic neurons) is transported by a retrograde mechanism. For example, NGF is injected into the anterior chamber of the eye of a rat, the radioactivity in taken up by the adrenergic terminals of the irregular and transported to the superior cervical ganglia where the adrenergic perikarya are present. These findings have suggested that NGF could act as a retrograde "trophic" messenger between the effector cells and the neurons.

At present the following two types of retrograde transport are postulated:

- A *non-specific* type which depends, as indicated above, on the injection of a macromolecule at high concentration
- A *specific retrograde axonal transport* which depends on the binding of the macromolecule to specific binding sites present at the nerve-ending membrane. The typical example of specific transport is that of the NGF.

However, other macromolecules, such as tetanus and cholera toxins and several lectins, may also be transported by a selective mechanism. The transport velocity of NGF in sympathetic axons was estimated at 2 to 3 mm/hr. and in sensory neurons, at 10 to 13 mm/hr. By coupling NGF to peroxidase it has been possible to follow the subcellular components that are involved in the retrograde axonal transport under the electron microscope. They follow essentially the same route for fast orthograde transport.

The macromolecules are found in the axon in tubules of smooth endoplasmic reticulum and then in the perikaryon, in tubules, multivesicular bodies, and dense bodies that may correspond to secondary lysosomes. We mentioned that the NGF could be considered as a retrograde trophic factor, and it is well known that it is indispensable for the growth of the adrenergic neuron. If newborn rats are treated with 6-hydroxydopamine (precursor analogue of the neurotransmitter norepinephrine) the adrenergic nerve terminal and the cell bodies are destroyed.

However, the degeneration of the cell bodies can be prevented with NGF. Similar results are obtained by surgical axotomy or by the action of *vinablstine*, which acts on the assembly of microtubules and blocks both orthograde and retrograde axonal transport. The NGF is indispensable at the early stages of neuronal growth, and with increasing age (after 3 weeks) the neuronal perikaryon becomes resistant to axotomy, to the action of vinblastine, and to 6-hydroxydopamine. In the adult the last-mentioned neurotoxic drug affects only the adrenergic nerve endings.

The NGF not only produces the growth of the adrenergic neuron but also stimulates the key enzymes involved in the metabolism of the neurotransmitter, *tyrosine hydroxylase* and *dopamine-b-hydroxylase*. Since the specific retrograde transport involves binding to specific receptor sites present at the nerve-ending membrane, by using different ligands, (*i.e.*, NGF, toxins, lectins, and others) it is possible to obtain qualitative and quantitative information about the composition of this membrane.

Trans-synaptic Transport

Tetanus toxin, one of the most potent neurotoxins, is transported by the axons of the motoneurons to the anterior horn of spinal cord by a retrograde transport similar to that of NGF. However, the toxin does not remain confined to the perikarya and dendrites of the motoneurons but is also found in the presynaptic nerve terminals that are attached to the neuron. From the medical viewpoint it is interesting that the trans-synaptic transfer of the toxin coincides with the first symptoms of local tetanus. In addition to tetanus toxin, a series of neurotropic viruses (*e.g.*, polio virus) reach the nervous system by binding to the peripheral nerve terminals and then undergoing retrograde transport, possibly followed by trans-synaptic transfer.

In this way viruses may gain access to the central nervous system in spite of the blood-brain barrier that blocks entry from the blood-stream. More detailed information about the molecular mechanism of the interaction of toxins and viruses with the nerve-ending membrane would have medical importance in developing specific therapeutic approaches to cope with these generally lethal infections.

Nerve Conduction Velocity

Nerve fibres are *non-myelinated* when wrapped only in Schwann cells. *Myelinated* nerve fibres also have a myelin sheath that consists of a multi-layer lipoprotein system. In the autonomic system of vertebrates, most nerve fibres are unmyelinated and are contained within invaginations of the plasma membrane of the Schwann cells. The myelin sheath is interrupted at the *nodes of Ranvier*. The distance between nodes varies with the diameter of the fibre. The *internode, i.e.*, the distance between successive nodes, is the segment of myelin that is produced and contained within a single Schwann cell. The internode is 0.2 mm in a bull frog fibre of 4 μm, about 1.5 mm in a fibre of 12 μm, and 2.5 mm in one of 15μm.

Table. Properties of Neurons of Different Sizes (Cat And Rabbit Saphenous Nerves)

	Group		
Properties	A	B	C
Diameter of fibre (μm)	20-1	3	–
Conduction velocity (m/sec)	100-5	14-3	2
Duration of action potential (msec)	0.4-0.5	1.2	2.0
Absolute refractory period (msec)	0.4-1.0	1.2	2.0

Within the internode, obliquitous (conic) *incissures* go across the myelin sheath where the myelin leaflets have a looser disposition. At the node the myelin lamellae are loosely arranged, and a small zone of axon is in direct contact with the extracellular fluid.

The myelin sheath acts as an insulator and, as a consequence, myelinated fibres conduct nerve impulses at a much faster rate than unmyelinated fibres. The diameter of the fibre also influences the conduction rate. As nerve fibres can be classified according to their diameters into groups A, B, and C. C fibres are unmyelinated. The diameter may vary from 20 μm in A fibres to less than 1μm in C fibres, and the conduction velocity varies from 100 to 2 meters or less per second. The rate of conduction of the nerve impulse follows a linear relationship with the fibre diameter in mammalian myelinated fibres, and it is also related to the internode distance.

Action Potential

When a nerve fibre is stimulated, a profound change is produced in the electrical properties of the surface membrane and in the steady potential, and a wave of depolarization is conducted. For the study of the physicochemical phenomena underlying the conduction of the nerve impulse, consult general physiology Here, the subject will be discussed superficially as a continuation of the of active transport and membrane using intracellular recover was demonstrated that upon excitation potential suddenly changes. At the stimulation there is not only a dipolar but also an overshoot, and the potential becomes positive. With radioactive tracers it has been found at the point of stimulation there is a suddenly several hundredfold increase in permeable Na^+. At the end of this period the again becomes essentially increases, and leaks out of the cell, repolarizing the nervous. In other words, during the rising phase spike N^+ enters, and in the descending K^+ is extruded. Complete restoration of the balance takes a longer time after the element event.

The action potential that develops in the fibre has several other:

- Stimulus produces a slight local depolarization in the fibre, which, after reaching a *threshold of activation,* produces spikes same amplitude. If the intensity of the is increased, the height of the spike always mains the same. This is called an *all-or response.*
- The nerve impulse is *decremental; i.e.,* the amplitude of the spike not decrease and is the same all along the of the nerve fibre. The action potential is well adapted to conduction over long distance without losses.
- On nerve impulse has passed over any point of fibre, there is a *refractory period* during which cannot react to another stimulus.

The *propagation* of the nerve impulse is generally explained by the so-called *local circuit.* At the point of stimulation the becomes depolarized (negative outside) and as a sink towards which the current flows for the adjacent areas. This we of depolarization advances along the nerve for at a rate of conduction that is characteristic each fibre. While this wave of depolarization advances, repolarization is so rate that only a fraction of the nerve fibre (a millimeters or centimeters, depending on conduction rate) is depolarized at a time. In recovery period. Sodium leaves the cell by action of the sodium pump a potassium re-enters to restore the steady state. This recovery is probably produced at the expense of high-energy phosphate bonds. However, impulses continue to discharge for some time in the absence of oxygen, and even when glycolysis is inhibited, which indicates that high energy bonds are stored at the membrane.

Myelinated Nerve Fibre

The preceding theory of nerve conduction applies to unmyelinated nerve fibres. In myelinated fibres it is thought that the local circuits occur only at the nodules. According to this so-called *saltatory theory,* at the internode the impulse is conducted electrotonically, and at each node the action potential is boosted to the same height by ionic mechanisms. In this way the amount of Na^+ and K^+ exchanged is greatly reduced and the net work required is much less. Stimulation of myelinated nerve fibres with fine electrodes has shown that at the nodes the threshold of stimulation is much lower than at the internode. It has been found that the nerve impulse can jump across one anesthetized node, but not two of them.

Non-propagated Potentials

Physiologic studies have demonstrated that in addition to the all-or-none response, there is another type of electrical activity in nervous tissue. This is, by far, the most frequent in the central nervous system and is referred to as a *graded response.* In the graded response the impulse is not *propagated* and the *amplitude* varies with the intensity of the stimulus. This type of response is characteristic of the physiologic receptors and synapses. Both the *genetator potentials* found at the receptors and the *synoptic potentials* are graded responses.

If a peripheral recepotor, such as a Pacini corpuscle or a stretch receptor (neuromuscular spindle), is mechanically stimulated, a local graded, and decremental potential is recorded the amplitude and duration of with depend on the intensity and duration of the stimulus. In the Pacini corpuscle it is possible to remove most of the connective lamellae that surround the nerve ending without impairing the generator potential.

It appears that in this case the *biological transducer* capable of transforming the mechanical energy (pressure) into the electrical energy (generator potential) is localized at the sensory part of the ending. Probably the mechanical deformation of the ending produces a change in permeability with entrance of ions and partial depolarization. In the Pacini corpuscle it has been observed that the nerve impulse starts at the first node of Ranvier.

The intensity of the sensory stimulus is reflected in the amplitude of the generator potential and this in turn in the *frequency* of the propagated signal- the stronger the generator potential, the higher the frequency. In this way the information received is coded for conduction along the nerve fibre in the form of a train or volley of impulses.

Axon Membrane

The introduction of the concepts of Na^+ and K^+ channels, and we described the $Na^+ - K^+$ pump ($Na^+ - K^+$ ATPase) the function of which is to extrude Na^+ from the axon and to allow the entrance of K^+. These three components (*i.e.*, Na^+ and K^+channels and the Na^+ - K^+ pump) are essential to explaining the molecular basis of the propagation of action potential along the axonal the axonal membranes, as well those of the muscle. An ion channel can be envisioned as a hydrophilic channel formed by the interstice between protein subunits embedded in the axon membrane.

It is now generally admitted that the ion channel consists of two functional elements:

1. A *selectivity filter,* which determines the kind of ion that will be translocated,
2. *A gate,* which by opening and closing the channel, regulates the ion flow.

In the Na^+ and K^+ channels of nerve (and muscle) membranes this gate is regulated by the membrane potential; in other words, the gating mechanism is electrically driven, the voltage dependence of Na^+ and K^+ channels is indicated diagrammatically. It may be observed that in the resting (polarized)condition both the Na^+ and K^+ channels are closed. With depolarization, the Na^+ channel is opened and during repolarization it closes again, and the K^+ channel opens. We shall see later that another type of channel, present at the postsynaptic membrane, is regulated chemically by the interaction of the neurotransmitter with specific receptor molecules.

Sodium Channel

The sodium channel is better known than the potassium channel be cause there are several drugs that affect it in a different manner. For example. *tetrodoxin* (TTX), a toxin derived from the Japanese puffer fish, at concentrations well below I μM completely blocks the Na^+ channels. Other toxion.

By using ^{3}H-TTX the number of channels per μm^2 has been calculated. In the axon membrane, this is rather small number, *i.e.*, between 50 and 500 per μm, but there are many more at the nodes of Ranvier. The TTX binding site appears to be glycoprotein embedded in the lipid phase. The Na^+ channel is blocked by protons and, by comparing its permeability to various ions, a diameter of 0.3 × 0.5 nm has been estimated. [36] A single channel transports about 10^8 Na^+/sec, with a conductivity of $4 \times 10^{-12} Q^{-1}$ which is about one third that of the K^+ channel.

Since TTX has no effect when applied internally, it is generally accepted that the selectivity filter of the Na^+channel is located on the outer side of the membrane. This selectivity filter is independent of the gating mechanism, which operates even after blocking with TTX. This mechanism has been explained by voltage-and time-dependent conformational changes in protein subunits, and at least three of these subunits are postulated. A four-barrier model in which the limiting step is the dehydration of the cation has been postulated for the sodium channel.

Potassium Channel

The potassium conductance is selectively inhibited by tetraethylammonium ions (TEA); however, the affinity is too low to carry out binding assays. The affinity can be increased by replacing one of the ethyl groups of TEA with a more hydrophobic side chain. In the squid giant axon, TEA blocks only the K^+ channel from the inside. This molecule has the same diameter as the hydrated potassium ion. This and other findings have led to the postulation of a funnel shape for the axonal K^+ channel. The large opening of the funnel is towards the interior, and it has a diameter of about ion, as well as TEA. The selectivity filter, which corresponds to the narrow end of the funnel, is only about 0.3 nm in diameter, *i.e.*, large enough to let only the dehydrated potassium ion pass. A hydrophobic site is thought to be present the large entrance of the channel. The gating mechanism, sensitive to changes in the membrane potential, could be located on the inside of the channel. The number of sites for the squid axon has been calculated to be *70/μm* 2 27.

8

Enzyme Structure of DNA

Before a cell can begin to divide or differentiate, the genetic information within the cell's DNA must be copied, or "transcribed," onto complementary strands of RNA. RNA polymerase II (pol II) is an enzyme that, by itself, can unwind the DNA double helix, synthesize RNA, and proofread the result. When combined with other molecules that regulate and control the transcription process, pol II is the key to successful interpretion of an organism's genetic code. However, the size, complexity, scarcity, and fragility of pol II complexes have made analysis of these macromolecules by x-ray crystallography a formidable challenge. A team of structural biologists has met this challenge using data obtained from both the Stanford Synchrotron Radiation Laboratory and the Macromolecular Crystallography Facility at the ALS.

The resultant high-resolution model of a 10-subunit pol II complex suggests roles for each of the subunits and will allow researchers to begin unraveling the intricacies of DNA transcription and its role in gene expression. In this work, the researchers studied the pol II enzyme from the yeast *Saccharomyces cerevisiae,* which is likely to be an excellent model for the human enzyme in light of its highly similar gene sequences. It is also the best-characterized form of the pol II enzyme, having been the subject of many biochemical and low-resolution structural studies in the past. To obtain a high-resolution structure, the research team drew on its considerable expertise in the preparation of protein crystals: two-dimensional crystals of pol II (minus two small subunits found to impede crystal growth) were used as seeds for growing three-dimensional crystals.

These crystals, when produced in an inert atmosphere to prevent oxidation, enabled the collection of data to 3.5-angstrom resolution. The addition of a final soaking procedure to produce uniform crystals, combined with high-brightness x-ray sources, resulted in a resolution of 3.0 angstroms. The current results bring into focus the somewhat fuzzy features previously observed in or inferred from earlier experiments. More importantly, the structural details suggest possible explanations for some of the unusual characteristics of this enzyme, which include a high processivity (the ability to synthesize very long strands of RNA) and the tendency to work in periodic spurts separated by pauses. While it is known that additional proteins (transcription factors) play a role in controlling the activity of pol II, scientists have yet to understand how such proteins interact with pol II binding sites to perform their various functions. The pol II model reported here establishes the positions of the various subunits and provides detailed information about the DNA/RNA binding domains. The data reveal two main subunits separated by a deep cleft where DNA can enter the complex. At the end of the cleft is the active site, where the DNA can be unwound for a short distance (the "transcription bubble") and a DNA/RNA hybrid can be produced. Two prominent grooves lead away from the active site, either of which could accommodate the exiting RNA transcript. An opening below the active site may allow the entry of nucleotides (for manufacturing RNA) and transcription factors (for regulating the process).

The same opening may provide room for the leading end of the RNA strand during "backtracking" maneuvers, which are important for proofreading and for traversing obstacles such as DNA damage. Other notable features that might help account for the great stability of this transcribing complex include a pair of "jaws" that appear to grip the DNA strands as they enter the complex and, closer to the active site, a clamp on the DNA that could possibly be locked in the closed position by the presence of RNA. The high-resolution pol II structure reported here is a landmark achievement, pulling together threads from numerous diverse research efforts into a cohesive whole. Further study should yield many new insights into the detailed mechanisms of pol II and its transcription factors. Construction of an atomic model is already well underway.

UNRAVELING DNA

Encoded into the double-helical strands of DNA, the human genome is the complete set of instructions required to make a human being. While the mapping of the human genome (roughly three billion components) is certainly a Herculean accomplishment, it is only the first step towards realizing the full potential of genomic medicine in the diagnosis, monitoring, and treatment of disease. Beyond knowing what the genetic blueprint says, scientists must understand how that blueprint gets interpreted, or "expressed" as an individual with unique traits. The pol II enzyme is the catalyst for a major step in this process.

As a pol II molecule slides along a DNA molecule, it "unzips" the strands of the DNA double helix, synthesizes a complementary strand of RNA (which will carry the genetic information to where it is needed), and verifies that no mistakes have occurred. This process is regulated by transcription factors—separate molecules that bind to pol II and determine which genes are expressed, at what stage of development, and in which tissue. Done correctly, this process results in healthy cell growth and differentiation; otherwise, aberrations such as cancer can be the result. Thus, details of the structure of pol II, including information about its binding sites and how they interact with transcription factors, will provide valuable insight into the detailed mechanisms underlying the flow of genetic information from DNA to RNA to protein, which is necessary for life and health.

Without hydrogen bonds there could be no life because they hold the double helix of DNA together, and this they do by charge attractions. A hydrogen bonded to oxygen, or nitrogen, becomes slightly positively charged which enables it to attract a centre of negative charge on another molecule, such as another oxygen or nitrogen atom. The hydrogen bond is then written, *e.g.*, O–H…N, with the dotted line signifying the hydrogen bond. There are also O–H…O, N–H…O and N–H…N bonds, the last being among the weakest. The secondary effects they have on structures, molecular vibrations, etc., can be used to infer hydrogen bonding, but there is no primary way of observing them because of their inherent weakness. NMR appears to be the least useful technique because neither of the common isotopes, oxygen-16 or nitrogen-14, has a magnetic nucleus. However, nitrogen-15 has a magnetic moment, and by replacing ^{14}N by N, Grzesiek has opened up a new area of investigating these enigmatic bonds.

Paper reports for the first time the direct observation by NMR of an N–H…N hydrogen bond between nucleic acids enriched with ^{15}N, by measuring the coupling of the nitrogen atoms. Grzesiek, working with Andrew Dingley of the Heinrich-Heine University in Düsseldorf, has been able to do this and show that the coupling is surprisingly large. Normally atomic nuclei only couple with each other if they are linked by normal chemical bonds, and in theory hydrogen bonds have neither the strength nor stability for this to occur. The German researchers studied an ^{15}N enriched sample of the T1 domain of the potato spindle tuber viroid and were able to prove that N–H...N hydrogen bonding was present between the base pairs, uridine…adenosine and guanosine…cytidine, with couplings of approximately 7 Hz. How could they be certain that the signals they were observing are due to N–H…N hydrogen bonds?

The answer was to use triple resonance techniques to examine base pairs that hydrogen bond only via O–H…N hydrogen bonds, and show that the signal they had previously observed was absent. Grzesiek's second paper on hydrogen bonds, coauthored by Florence Cordier, Heinrich-Heine University, extends the work in an even more remarkable way by measuring the NMR coupling between nitrogens and carbons in the backbone hydrogen bonds of the human protein ubiquitin. The carbons are part of a carbonyl (C=O) group, so are one removed from the hydrogen bond, *i.e.*, N–H…O=C. This time they used material enriched with ^{15}N and ^{13}C (normal ^{12}C has no nuclear magnet), and there too was the evidence for these hydrogen bonds, albeit with an interaction an order of magnitude weaker (at –0.25 to –0.9 Hz) than the N–

H...N coupling. Nevertheless, the couplings correlate with the strength of the hydrogen bond, being stronger in the stronger bonds. Grzesiek's third paper, was done in conjunction with Dingley and researchers at UCLA. Together they studied not only the hydrogen bonding of Watson-Crick base pairs but also of Hoogsten base pairs within a DNA triplex consisting of one purine and two pyrimidine strands.

Four different base pairs were identified, their various couplings distinguished–including those of the weaker interactions at the "frayed ends" of the DNA chains–and relationships with other hydrogen bonding parameters, such as bond length, were established. In addition they were able to show that density functional computer simulations by computer could reproduce these findings exactly.

PROKARYOTIC DNA REPLICATION

Duplication of DNA is one of the fundamental properties of "molecule of life". Duplication or replication takes place once in a cell cycle. The duration and initiation point differs from one system to another. When conditions are favorable cell cytoplasmic mass increases and when cytoplasmic mass reaches a ratio to that of the cell size to 2L, bacterial cell division is triggered. To complete its cell cycle it requires hardly 30 -40 minutes. Several genes involved in different steps of cell division have been characterized. Par excellent system to understand molecular process of prokaryotic cell division is E.coli. The size of the E.coli chromosome is 4.6×10^6 base pairs, but it is compacted by the binding of histone like proteins. First the nucleoid i.e the chromosome frees from the membrane and unwinds. Such DNA initiates replication and completes in a matter of 15-20 minutes.

Then the cytoplasm starts dividing almost in the centre of the cell by central septal ring formation. It is at this time the daughter DNA molecules are partitioned and segregated to the two compartments. The central periseptal ring that started 1-2 minutes after DNA replication now furrows inwards and divides the cytoplasm in the middle region. Then mid-wall splits in the middle and daughter cell are set free, hence the Schizomycetes to bacteria (E.coli) Bacterial cell division requires the participation of the following genes and gene products. Eukaryotic cell division is more complex and highly regulated. The paradigm for it is yeast, for it is a unicellular system, grows fast and requires hardly 60 minutes for one cell cycle. One can obtain different kinds of mutation like conditional and null or auxotrophic kinds. In culture cells, the duration of one cell cycle is about 12 hrs. Onion root tip cells divide once in 12 hrs.

Whereas embryonic cells in its early stages require just 30 minutes for one cell cycle and they go through 6 to 8 such cell divisions very fast. Replication of DNA in cell division is an important phase. The replication is very accurate and it is nucleotide by nucleotide and there is no provision for making any mistakes, even if mistakes are made they are immediately corrected or repaired, otherwise consequences are serious and deleterious. In general DNA replication is precise, exact and regulated and involves initiation, elongation and termination steps, but mechanisms and rate of replication vary from one system to the other.

REPLICATION IS SEMI CONSERVATIVE

Meselson and Stahl showed that DNA replicates in semi conservative mode. There were some doubts about the mode whether it is conservative, dispersive or semi conservative. But Meselson and Stahl used nonradioactive heavy Nitrogen isotope (^{15}N), as the source of Nitrogen. When cells were grown on such source, ^{15}N incorporated nitrogen bases get incorporated into DNA, thus it becomes heavier than the DNA that is grown in normal (^{14}N) nitrogen media.

When two such DNAs, one grown in the presence of ^{15}N source and the other in ^{14}N (normal), are extracted separately and stained with ethidium Bromide, then subjected to equilibrium density centrifugation at ultra speed of 42000rpm for about 24 –36 hrs, the DNA, separates as two distinct bands, the one that is heavier (^{15}N) is the lower band than the normal (^{14}N) upper DNA. If cells grown in heavy isotopes are shifted to normal Nitrogen source and grown for one or two cell generation and if DNA is isolated from it and if one combine this with the first two DNA sources and subject them to equilibrium density ultra centrifugation (using cesium chloride or cesium sulphate), one can observe three bands, the top DNA band is from cells grown in ^{14}N without isotope, the second band is the hybrid band with one strand ^{15}N and the other with normal 14N-Nitrogen and the third band is with heavy nitrogen where both the strands are labeled with ^{15}N.

This experiment virtually and unambiguously proved that DNA replicates in semi conservative mode. Taylor using root tips, labeled with ^{32}P radioisotopes, of Vicia faba demonstrated semi conservative mode of replication at chromosomal level. Semi conservative means that the two daughter molecules produced at the end of replication, each of the molecules retains one parental strand and the other one is the newly made one. So it is semi conservative. The mode of replication is absolutely semi conservative irrespective the kind of DNA, whether it is ds DNA or ssDNA, whether it is linear or circular, and the basic mechanism is more or less similar; each of them uses their own specialized mechanisms. Experimentally semiconservative mode of replication has been shown by density gradient centrifugation of non-radioactive isotopes such as ^{15}N and ^{14}N sources.

DNA SYNTHESIS

DNA Replication (also known as DNA synthesis) is a process where the double stranded Deoxyribonucleic Acid (DNA) is copied. Replication is an important first step in cell division, as cells must duplicate their entire genetic constitution before they can divide into two daughter cells. DNA replication is also a critical requirement for DNA repair.

REQUIREMENTS OF DNA REPLICATION

Replication requires three important components in order for this process to work.

Template Strand

The DNA serves as a template to guide the incoming nucleotides. The template strand attaches to the RNA primer strand in order to replicate the strand of DNA. The template just guides the nucleotides to the place that they need to go in. This is another component that is needed in order for DNA replication to even take place. If the DNA only had the template strand and none of the rest of the things then the strand could not replicate. The template starnd is very thin. Through an enzyme called primase the DNA template strand is synthesized one nucleotide at a time.

DNA Polymerase

There are many DNA polymerases within a cell, but only one of which is used in the replication of DNA. In order for DNA to replicate, it must have a primer, which is a small strand of RNA. The DNA template strand synthesizes one nucleotide at a time by the use of primase. The adding of nucleotides to the 3′ end, which is continued until that section of DNA is replicated. The other DNA polymerases that are within the cell are used in other ways besides replication. DNA polymerase III is one certain polymerase, which aids in replication of what is known as the "lagging strand". DNA ligase is the final enzyme, which combines the lagging strand

Free 3' Hydroxyl

The free 3' hydroxyl is the starter strand called a primer which is required in order for DNA to be replicated. The primer strand is not very big, it is shorter then a single strand of RNA. The primer is the complementary strand to the template strand of DNA. This is another one of the things that are needed in order to make the process of DNA replication even possible. The DNA polymerase is added to the 3' end of the primer, which keeps growing until the section is finally complete. After a while the RNA primer is degraded and disappears.

Types of Replication

During the discovery of DNA replication they also discovered that there are three modes of replication that a DNA strand can take. The three types of DNA replication are semiconservative, conservative, and dispersive. These three types of replication were tested on in order to see the possible patterns that would result in the complementary base pairing.

Semiconservative Replication

Through this type of replication the parent strand serves as a template for the new strand. The two offsprings would have one of the parent strand and one new strand.

Conservative Replication

Through this type of replication the double helix serves as a template. This although does not contribute to the new double helix.

Dispersive Replication

In this type of replication the fragments from the parent DNA molecule. The DNA serves as a template for the assembly of the new molecule. The new double helix contains old and new parts of the DNA strands.

DNA Repairing

After the DNA is replicated, there are three DNA repair mechanisms. They are to be used when they feel necessary. This will lessen the chance of human mutation.

The rate of error in which the DNA polymerases makes a mistake is very high. This is why the DNA has mechanisms to cause fewer mutations. The three mechanisms are:

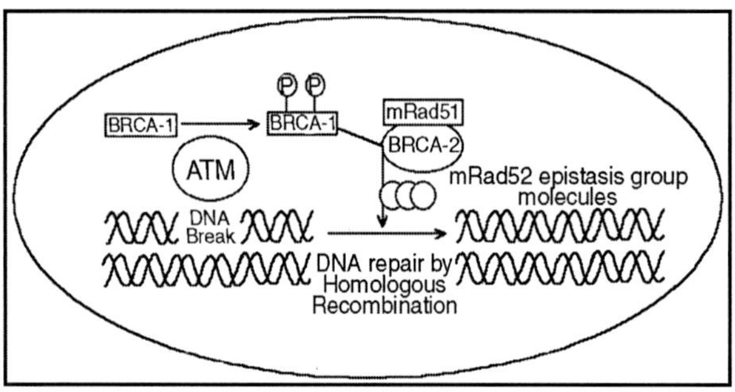

Fig. DNA Repairing

Proofreading

Every time a nucleotide is introduced to an already growing chain, the DNA polymerase checks the connection. If the pair is mismatched, it will be removed and redone. This lowers the overall rate of mutations.

Mismatch Pair

After the replication takes place, another set of proteins check to make sure that there are no more mismatched base pairs and also which strand is the wrong one.

Excision Pair

There can still be damage while the cell is living. Because of this, there are enzymes constantly checking the cell. If they detect a problem, the enzyme cuts the strand and also cuts away the opposite base. DNA polymerase and ligase then fix this base sequence.

LABORATORY TECHNIQUE

Scientists use two techniques in a laboratory that involve DNA replication. They use these techniques to observe genes and genomes. One of the techniques can make multiple copies of DNA from only a short piece. Another technique allows scientists to determine the base sequence of DNA. The two techniques are:

Polymerase Chain Reaction (PCR)

Through this process, a short strand of DNA is copied repetitively. There are three steps that are repeated in the process. The DNA strands are denatured or heated in order to separate the strand. Next, they add a primer, which was artificially made. They also add DNA polymerase and the four deoxyribonucleotide triphosphates. These three are needed in order to replicate DNA. The final step is then the DNA polymerase catalyzes the production of a complementary strand.

DNA Sequencing

This allows scientists to determines the base sequence of a DNA. The process is the DNA is first denatured and a single strand is placed in a test tube. They then add DNA polymerase, primer, the four dNTP's, and small amounts of ddNTP's. The DNA replicates within the tube and after the DNA fragments are denatured. The fragments then go through electrophoresis, which sorts the lengths of DNA fragments. The fragments then pass through a laser beam which excites the fluorescent tags. This information is fed into a computer at which tells the sequence.

Mendelson-Stahl Experiment

Through this experiment Meselson- Stahl convinced the scientists that the correct model for DNA replication was the semiconservative replication. They used density labeling in order to distinguish the old and new strands of DNA. In their experiment they used "heavy" isotopes of nitrogen. This nitrogen is a nonradioactive isotope that made the molecule more dense. The chemically identical molecules containing an isotope known as ^{14}N. They needed a way to distinguish the different DNA densities. Because of this they needed to come up with something to measure the densities of solutions.

Meselson, Stahl, and Jerome Vanguard came up with centrifuge, which is a procedure that involves the spinning of solutions at high speed, which causes the particles to separate and form gradient according to the different densities of the solution. The first part of their experiment they grew cultures of Escherichia coli. One of the cultures was grown in ^{15}N, which made the DNA "heavy". The other culture was placed in a medium using ^{14}N rather than ^{15}N, which made all the DNA "light". In order to see if the use of centrifuge worked they combined the two cultures and when centrifuged it formed two DNA bands. This showed them the different densities with in the two cultures. The next part of their experiment they grew E. Coli in the ^{15}N medium and then transferred the bacteria from that medium to the ^{14}N. Through this they came up with the conclusion that through E. Coli DNA replicates every 20 minutes. They then took DNA samples from each generation through, which they found that the density gradient was different each generation. Their observations could only be explained with the semiconservative model of DNA replication. Their results showed that when the DNNA first underwent replication, it was in 15N, which caused the DNA to be heavy. Because in the semiconservative model of DNA replication, the one strand acts as a template for a second strand which the DNA was in a ^{14}N DNA strand and a ^{15}N and they were of intermediate density.

DNA REPLICATION IS SEMI-CONSERVATIVE

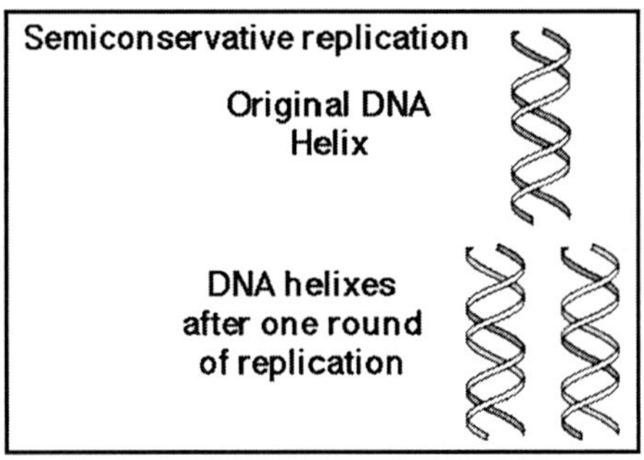

Fig. Semi-Conservative Nature of DNA Replication

DNA replication of one helix of DNA results in two identical helices. If the original DNA helix is called the "parental" DNA, the two resulting helices can be called "daughter" helices. Each of these two daughter helices is a nearly exact copy of the parental helix (it is not 100 per cent the same due to mutations). DNA creates "daughters" by using the parental strands of DNA as a template or guide. Each newly synthesized strand of DNA (daughter strand) is made by the addition of a nucleotide that is complementary to the parent strand of DNA. In this way, DNA replication is semi-conservative, meaning that one parent strand is always passed on to the daughter helix of DNA.

THE LAGGING STRAND

Whereas the DNA polymerase on the leading strand can simply follow the replication fork, because DNA polymerase must move in the 5' to 3' direction, on the lagging strand the enzyme must move away from the fork. But if the enzyme moves away from the fork, and the fork is uncovering new DNA that needs to be replicated, then how can the lagging strand be replicated at all? The problem posed by this question is answered through an ingenious method.

The lagging strand replicates in small segments, called Okazaki fragments. These fragments are stretches of 100 to 200 nucleotides in humans (1000 to 2000 in bacteria) that are synthesized in the 5' to 3' direction away from the replication fork. Yet while each individual segment is replicated away from the replication fork, each subsequent Okazaki fragment is replicated more closely to the receding replication fork than the fragment before. These fragments are then stitched together by DNA ligase, creating a continuous strand. This type of replication is called discontinuous. The first synthesized Okazaki fragment on the lagging strand is the furthest away from the replication fork, which is itself receding to the right. Each subsequent Okazaki fragment starts at the replication fork and continues until it meets the previous fragment. The two fragments are then stitched together by DNA ligase. Because synthesis on the lagging strand takes place in a "backstitching" mechanism, its replication is slightly delayed in relation to synthesis on the leading strand.

The lagging strand must wait for a patch of the parent helix to open up a short distance in front of the newly synthesized strand before it can begin its synthesis back to the end of the daughter strand.This "Lag" time does not occur in the leading strand because it synthesizes the new strand by following right behind as the helix unwinds at the replication fork. Another complication to replication on the lagging strand is the initiation of replication. Whereas the RNA primer on the leading strand only has to trigger the initiation of the strand once, on the lagging strand each individual Okazaki fragment must be triggered. On the lagging strand, then, an enzyme called primase that moves with the replication fork synthesizes numerous RNA primers, each of which triggers the growth of an Okazaki fragment. The RNA primers are eventually removed leaving gaps that are filled by the replication machinery.

THE LEADING STRAND

Since DNA replication moves along the parent strand in the 5' to 3' direction, replication can occur very easily on the leading strand. The nucleotides are added in the 5' to 3' direction. Triggered by RNA primase, which adds the first nucleotide to the nascent chain, the DNA polymerase simply sits near the replication fork, moving as the fork does, adding nucleotides one after the other, preserving the proper anti-parallel orientation.

This sort of replication, since it involves one nucleotide being placed right after another in a series, is called continuous.

DNA STRUCTURE

DNA usually exists in a double-stranded structure, with both strands coiled together to form the characteristic double-helix. Each single strand of DNA is a chain of four types of nucleotide: adenine, cytosine, guanine, and thymine. A nucleotide consists of a phosphate and a deoxyribose sugar forming the backbone of the DNA double helix plus a base that points inwards. Nucleotides are matched between strands through hydrogen bonds to form base pairs. Adenine pairs with thymine and cytosine pairs with guanine. The physical pairing of bases in DNA means that the information contained within each strand is redundant. The nucleotides on a single strand can be used to reconstruct nucleotides on a newly synthesized partner strand. DNA strands have a directionality, and the different ends of a single strand are called the "3' end" and the "5' end" (these refer to the carbon atom in ribose that the next phosphate in the chain attaches to). In addition to being complementary, the two strands of DNA are antiparallel: they are orientated in opposite directions. This directionality has consequences in DNA synthesis, because DNA polymerase can only synthesize DNA in one direction by adding nucleotides to the 3' end of a DNA strand.

DNA POLYMERASE

DNA polymerases are a family of enzymes critical for all forms of DNA replication. A DNA polymerase synthesizes a new strand of DNA by extending the 3' end of an existing nucleotide chain, adding new nucleotides matched to the template strand one at a time. Some DNA polymerases may also have some proofreading ability, removing nucleotides from the end of a strand in order to remove any mismatched bases. DNA polymerases are generally extremely accurate, making less than one error for every million nucleotides added. The energy for the process of DNA polymerization comes from the two additional phosphates attached to each of the unincorporated nucleotides.

These free nucleotides, also known as nucleoside triphosphates, contain a total of three phosphates. When a nucleotide is being added to a growing DNA strand, two of the phosphates are removed and the energy produced is used to attach the remaining phosphate to the growing chain. The energetics of this process may also explain the directionality of synthesis - if DNA were synthesized in the 3' to 5' direction, the energy for the process would come from the 5' end of the growing strand rather than from free nucleotides. During proofreading, if the 5' nucleotide needed to be removed this triphosphate end would be lost, losing the energy source required to add a new nucleotide to the end. DNA polymerase can only extend an existing DNA strand paired with a template strand, it cannot begin the synthesis of a new strand. To do this a short fragment of DNA or RNA, called a primer, must be created and paired with the template strand before DNA polymerase can synthesize new DNA.

DNA REPLICATION WITHIN THE CELL

For a cell to divide, it must first replicate its DNA. This process is initiated at particular points within the DNA, known as "origins", which are targeted by proteins that separate the two strands and initiate DNA synthesis. Origins contain DNA sequences recognized by replication initiator proteins (eg. dnaA in *E coli'* and the Origin Recognition Complex in yeast). These initiator proteins recruit other proteins to separate the two strands and initiate replication forks. Initiator proteins recruit other proteins to separate the DNA strands at the origin, forming a bubble. Origins tend to be "AT-rich" (rich in adenine and thymine bases) to assist this process because A-T base pairs have two hydrogen bonds (rather than the three formed in a C-G pair)—strands rich in these nucleotides are generally easier to separate. Once strands are separated, RNA primers are created on the template strands and DNA polymerase extends these to create newly synthesized DNA. As DNA synthesis continues, the original DNA strands continue to unwind on each side of the bubble, forming replication forks. In bacteria, which have a single origin of replication on their circular chromosome, this process eventually creates a "theta structure". In contrast, eukaryotes have longer linear chromosomes and initiate replication at multiple origins within these.

ENZYMES AS BIOLOGICAL CATALYSTS

A fundamental task of proteins is to act as enzymes—catalysts that increase the rate of virtually all the chemical reactions within cells. Although RNAs are capable of catalyzing some reactions, most biological reactions are catalysed by proteins. In the absence of enzymatic catalysis, most biochemical reactions are so slow that they would not occur under the mild conditions of temperature and pressure that are compatible with life. Enzymes accelerate the rates of such reactions by well over a million-fold, so reactions that would take years in the absence of catalysis can occur in fractions of seconds if catalysed by the appropriate enzyme. Cells contain thousands of different enzymes, and their activities determine which of the many possible chemical reactions actually take place within the cell.

THE CATALYTIC ACTIVITY OF ENZYMES

The enzymes are characterized by two fundamental properties.

- First, they increase the rate of chemical reactions without themselves being consumed or permanently altered by the reaction.
- Second, they increase reaction rates without altering the chemical equilibrium between reactants and products.

These principles of enzymatic catalysis are illustrated in the following example, in which a molecule acted upon by an enzyme (referred to as a substrate [*S*]) is converted to a product (*P*) as the result of the reaction. In the absence of the enzyme, the reaction can be written as follows:

$$S \rightarrow P$$

The chemical equilibrium between *S* and *P* is determined by the laws of thermodynamics and is represented by the ratio of the forward and reverse reaction rates ($S \rightarrow P$ and $P \rightarrow S$, respectively). In the presence of the appropriate enzyme, the conversion of *S* to *P* is accelerated, but the equilibrium between *S* and *P* is unaltered. Therefore, the enzyme must accelerate both the forward and reverse reactions equally. The reaction can be written as follows:

$$S \xrightarrow{E} P$$

Note that the enzyme (*E*) is not altered by the reaction, so the chemical equilibrium remains unchanged, determined solely by the thermodynamic properties of *S* and *P*.

The catalytic activity of enzymes involves the binding of their substrates to form an enzyme-substrate complex (*ES*). The substrate binds to a specific region of the enzyme, called the active site. While bound to the active site, the substrate is converted into the product of the reaction, which is then released from the enzyme. The enzyme-catalysed reaction can thus be written as follows:

$$S + E = ES = E + P$$

Note that *E* appears unaltered on both sides of the equation, so the equilibrium is unaffected. However, the enzyme provides a surface upon which the reactions converting *S* to *P* can occur more readily. This is a result of interactions between the enzyme and substrate that lower the energy of activation and favour formation of the transition state.

Enzymatic Catalysis Process

The binding of a substrate to the active site of an enzyme is a very specific interaction. Active sites are clefts or grooves on the surface of an enzyme, usually composed of amino acids from different parts of the polypeptide chain that are brought together in the tertiary structure of the folded protein. Substrates initially bind to the active site by noncovalent interactions, including hydrogen bonds, ionic bonds, and hydrophobic interactions. Once a substrate is bound to the active site of an enzyme, multiple mechanisms can accelerate its conversion to the product of the reaction. Enzymes accelerate reactions also by altering the conformation of their substrates to approach that of the transition state. The simplest model of enzyme-substrate interaction is the lock-and-key model, in which the substrate fits precisely into the active site. In many cases, however, the configurations of both the enzyme and substrate are modified by substrate binding—a process called induced fit. In such cases the conformation of the substrate is altered so that it more closely resembles that of the transition state. The stress produced by such distortion of the substrate can further facilitate its conversion to the transition state by weakening critical bonds. Moreover, the transition state is stabilized by its tight binding to the enzyme, thereby lowering the required energy of activation.

Coenzymes

In addition to binding their substrates, the active sites of many enzymes bind other small molecules that participate in catalysis. Prosthetic groups are small molecules bound to proteins in which they play critical functional roles. For example, the oxygen carried by myoglobin and hemoglobin is bound to heme, a prosthetic group of these proteins. In many cases metal ions (such as zinc or iron) are bound to enzymes and play central roles in the catalytic process. In addition, various low-molecular-weight organic molecules participate in specific types of enzymatic reactions. These molecules are called coenzymes because they work together with enzymes to enhance reaction rates. In contrast to substrates, coenzymes are not irreversibly altered by the reactions in which they are involved. Rather, they are recycled and can participate in multiple enzymatic reactions.

Coenzymes serve as carriers of several types of chemical groups. A prominent example of a coenzyme is nicotinamide adenine dinucleotide (NAD^+), which functions as a carrier of electrons in oxidation-reduction reactions. NAD^+ can accept a hydrogen ion (H^+) and two electrons (e^-) from one substrate, forming NADH. NADH can then donate these electrons to a second substrate, re-forming NAD^+. Thus, NAD^+ transfers electrons from the first substrate (which becomes oxidized) to the second (which becomes reduced). Several other coenzymes also act as electron carriers, and still others are involved in the transfer of a variety of additional chemical groups (*e.g.*, carboxyl groups and acyl groups. The same coenzymes function together with a variety of different enzymes to catalyse the transfer of specific chemical groups between a wide range of substrates. Many coenzymes are closely related to vitamins, which contribute part or all of the structure of the coenzyme. Vitamins are not required by bacteria such as *E. coli* but are necessary components of the diets of human and other higher animals, which have lost the ability to synthesize these compounds.

Regulation of Enzyme Activity

An important feature of most enzymes is that their activities are not constant but instead can be modulated. That is, the activities of enzymes can be regulated so that they function appropriately to meet the varied physiological needs that may arise during the life of the cell. One common type of enzyme regulation is feedback inhibition, in which the product of a metabolic pathway inhibits the activity of an enzyme involved in its synthesis. Feedback inhibition is one example of allosteric regulation, in which enzyme activity is controlled by the binding of small molecules to regulatory sites on the enzyme. The term "allosteric regulation" derives from the fact that the regulatory molecules bind not to the catalytic site, but to a distinct site on the protein. Binding of the regulatory molecule changes the conformation of the protein, which in turn alters the shape of the active site and the catalytic activity of the enzyme.

The activities of enzymes can also be regulated by their interactions with other proteins and by covalent modifications, such as the addition of phosphate groups to serine, threonine, or tyrosine residues. Phosphorylation is a particularly common mechanism for regulating enzyme activity; the addition of phosphate groups either stimulates or inhibits the activities of many different enzymes.

For example, muscle cells respond to epinephrine (adrenaline) by breaking down glycogen into glucose, thereby providing a source of energy for increased muscular activity. The breakdown of glycogen is catalysed by the enzyme glycogen phosphorylase, which is activated by phosphorylation in response to the binding of epinephrine to a receptor on the surface of the muscle cell. Protein phosphorylation plays a central role in controlling not only metabolic reactions but also many other cellular functions, including cell growth and differentiation.

ENZYME INHIBITION

If the velocity of an enzymatic reaction is decreased or inhibited, the kinetics of the reaction obviously have been perturbed. Systematic perturbations are a basic tool of experimental scientists; much can be learned about the normal workings of any system by inducing changes in it and then observing the effects of the change. The study of enzyme inhibition has contributed significantly to our understanding of enzymes.

The binding of an inhibitor can stop a substrate from entering the enzyme's active site and/or hinder the enzyme from catalysing its reaction. Inhibitor binding is either reversible or irreversible. Irreversible inhibitors usually react with the enzyme and change it chemically. These inhibitors modify key amino acid residues needed for enzymatic activity. In contrast, reversible inhibitors bind non-covalently and different types of inhibition are produced depending on whether these inhibitors bind the enzyme, the enzyme-substrate complex, or both. Many drug molecules are enzyme inhibitors, so their discovery and improvement is an active area of research in biochemistry and pharmacology. A medicinal enzyme inhibitor is often judged by its specificity (its lack of binding to other proteins) and its potency (its dissociation constant, which indicates the concentration needed to inhibit the enzyme). A high specificity and potency ensure that a drug will have few side effects and thus low toxicity. Enzyme inhibitors also occur naturally and are involved in the regulation of metabolism. For example, enzymes in a metabolic pathway can be inhibited by downstream products. This type of negative feedback slows flux through a pathway when the products begin to build up and is an important way to maintain homeostasis in a cell. Other cellular enzyme inhibitors are proteins that specifically bind to and inhibit an enzyme target.

This can help control enzymes that may be damaging to a cell, such as proteases or nucleases; a well-characterised example is the ribonuclease inhibitor, which binds to ribonucleases in one of the tightest known protein–protein interactions. Natural enzyme inhibitors can also be poisons and are used as defences against predators or as ways of killing prey.

Reversible Inhibitors

Types of Reversible Inhibitors

Reversible inhibitors bind to enzymes with non-covalent interactions such as hydrogen bonds, hydrophobic interactions and ionic bonds. Multiple weak bonds between the inhibitor and the active site combine to produce strong and specific binding. In contrast to substrates and irreversible inhibitors, reversible inhibitors generally do not undergo chemical reactions when bound to the enzyme and can be easily removed by dilution or dialysis. There are four kinds of reversible enzyme inhibitors.

They are classified according to the effect of varying the concentration of the enzyme's substrate on the inhibitor:

1. In competitive inhibition, the substrate and inhibitor cannot bind to the enzyme at the same time. This usually results from the inhibitor having an affinity for the active site of an enzyme where the substrate also binds; the substrate and inhibitor *compete* for access to the enzyme's active site. This type of inhibition can be overcome by sufficiently high concentrations of substrate, *i.e.*, by out-competing the inhibitor. Competitive inhibitors are often similar in structure to the real substrate.
2. In uncompetitive inhibition, the inhibitor binds only to the substrate-enzyme complex, it should not be confused with non-competitive inhibitors. Both Vmax and Km decrease (maximum velocity decreases as a result of removing activated complex while binding efficiency increases as a result of Le Chatelier's principle).
3. In mixed inhibition, the inhibitor can bind to the enzyme at the same time as the enzyme's substrate. However, the binding of the inhibitor affects the binding of the substrate, and vice versa. This type of inhibition can be reduced, but not overcome by increasing

concentrations of substrate. Although it is possible for mixed-type inhibitors to bind in the active site, this type of inhibition generally results from an allosteric effect where the inhibitor binds to a different site on an enzyme. Inhibitor binding to this allosteric site changes the conformation (*i.e.*, tertiary structure or three-dimensional shape) of the enzyme so that the affinity of the substrate for the active site is reduced.

4. Non-competitive inhibition is a form of mixed inhibition where the binding of the inhibitor to the enzyme reduces its activity but does not affect the binding of substrate. As a result, the extent of inhibition depends only on the concentration of the inhibitor.

Quantitative Description of Reversible Inhibition

Reversible inhibition can be described quantitatively in terms of the inhibitor's binding to the enzyme and to the enzyme–substrate complex, and its effects on the kinetic constants of the enzyme. In the classic Michaelis–Menten scheme below, an enzyme (E) binds to its substrate (S) to form the enzyme–substrate complex ES. Upon catalysis, this complex breaks down to release product P and free enzyme.

The inhibitor (I) can bind to either E or ES with the dissociation constants K_i or K_i', respectively:

- Competitive inhibitors can bind to E, but not to ES. Competitive inhibition increases K_m (*i.e.*, the inhibitor interferes with substrate binding), but does not affect V_{max} (the inhibitor does not hamper catalysis in ES because it cannot bind to ES).
- Non-competitive inhibitors have identical affinities for E and ES (K_i = K_i'). Non-competitive inhibition does not change K_m (*i.e.*, it does not affect substrate binding) but decreases V_{max} (*i.e.*, inhibitor binding hampers catalysis).
- Mixed-type inhibitors bind to both E and ES, but their affinities for these two forms of the enzyme are different (K_i '" K_i'). Thus, mixed-type inhibitors interfere with substrate binding (increase K_m) and hamper catalysis in the ES complex (decrease V_{max}).

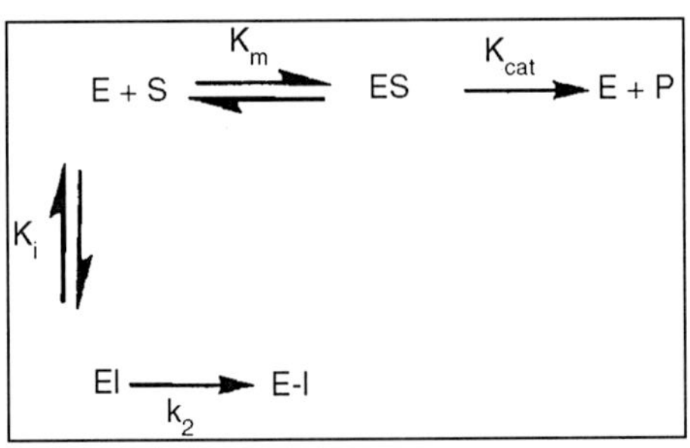

Fig. Kinetic Scheme for Reversible Enzyme Inhibitors

When an enzyme has multiple substrates, inhibitors can show different types of inhibition depending on which substrate is considered. This results from the active site containing two different binding sites within the active site, one for each substrate. For example, an inhibitor might compete with substrate A for the first binding site, but be a non-competitive inhibitor with respect to substrate B in the second binding site.

Measuring the Dissociation Constants of a Reversible Inhibitor

An enzyme inhibitor is characterised by its two dissociation constants, K_i and K_i', to the enzyme and to the enzyme-substrate complex, respectively. The enzyme-inhibitor constant K_i can be measured directly by various methods; one extremely accurate method is isothermal titration calorimetry, in which the inhibitor is titrated into a solution of enzyme and the heat released or absorbed is measured.

However, the other dissociation constant K_i' is difficult to measure directly, since the enzyme-substrate complex is short-lived and undergoing a chemical reaction to form the product. Hence, K_i' is usually measured indirectly, by observing the enzyme activity under various substrate and inhibitor concentrations, and fitting the data to a modified Michaelis–Menten equation

$$V = \frac{V_{max}[S]}{\alpha K_m + \alpha'[S]} = \frac{(1/\alpha')V_{max}[S]}{(\alpha/\alpha')K_m + [S]}$$

where the modifying factors á and á′ are defined by the inhibitor concentration and its two dissociation constants,

$$\alpha = 1 + \frac{[I]}{K_i}$$

$$\alpha' = 1 + \frac{[I]}{K_i'}.$$

Thus, in the presence of the inhibitor, the enzyme's effective K_m and V_{max} become $(\alpha/\alpha')K_m$ and $(1/\alpha')V_{max}$, respectively. However, the modified Michaelis-Menten equation assumes that binding of the inhibitor to the enzyme has reached equilibrium, which may be a very slow process for inhibitors with sub-nanomolar dissociation constants. In these cases, it is usually more practical to treat the tight-binding inhibitor as an irreversible inhibitor however, it can still be possible to estimate K_i' kinetically if K_i is measured independently.

The effects of different types of reversible enzyme inhibitors on enzymatic activity can be visualized using graphical representations of the Michaelis–Menten equation, such as Lineweaver–Burk and Eadie-Hofstee plots. For example, in the Lineweaver–Burk plots at the right, the competitive inhibition lines intersect on the *y*-axis, illustrating that such inhibitors do not affect V_{max}. Similarly, the non-competitive inhibition lines intersect on the *x*-axis, showing these inhibitors do not affect K_m.

However, it can be difficult to estimate K_i and K_i' accurately from such plots, so it is advisable to estimate these constants using more reliable nonlinear regression methods, as described above.

SPECIAL CASES

- The mechanism of partially competitive inhibition is similar to that of non-competitive, except that the EIS complex has catalytic activity, which may be lower or even higher (partially competitive activation) than that of the enzyme–substrate (ES) complex. This inhibition typically displays a lower V_{max}, but an unaffected K_m value.
- Uncompetitive inhibition occurs when the inhibitor binds only to the enzyme–substrate complex, not to the free enzyme; the EIS complex is catalytically inactive. This mode of inhibition is rare and causes a decrease in both V_{max} and the K_m value.
- Substrate and product inhibition is where either the substrate or product of an enzyme reaction inhibit the enzyme's activity. This inhibition may follow the competitive, uncompetitive or mixed patterns. In substrate inhibition there is a progressive decrease in activity at high substrate concentrations. This may indicate the existence of two substrate-binding sites in the enzyme. At low substrate, the high-affinity site is occupied and normal kinetics are followed. However, at higher concentrations, the second inhibitory site becomes occupied, inhibiting the enzyme. Product inhibition is often a regulatory feature in metabolism and can be a form of negative feedback.
- Slow-tight inhibition occurs when the initial enzyme–inhibitor complex EI undergoes isomerisation to a second more tightly held complex, EI*, but the overall inhibition process is reversible. This manifests itself as slowly increasing enzyme inhibition. Under these conditions, traditional Michaelis–Menten kinetics give a false value for K_i, which is time–dependent. The true value of K_i can be obtained through more complex analysis of the on (k_{on}) and off (k_{off}) rate constants for inhibitor association.

Examples of Reversible Inhibitors

As enzymes have evolved to bind their substrates tightly, and most reversible inhibitors bind in the active site of enzymes, it is unsurprising that some of these inhibitors are strikingly similar in structure to the substrates of their targets. An example of these substrate mimics are the protease inhibitors, a very successful class of antiretroviral drugs used to treat HIV. The structure of ritonavir, a protease inhibitor based on a peptide and containing three peptide bonds, is shown on the right. As this drug resembles the protein that is the substrate of the HIV protease, it competes with this substrate in the enzyme's active site.

Enzyme inhibitors are often designed to mimic the transition state or intermediate of an enzyme-catalysed reaction. This ensures that the inhibitor exploits the transition state stabilising effect of the enzyme, resulting in a better binding affinity (lower K_i) than substrate-based designs. An example of such a transition state inhibitor is the antiviral drug oseltamivir; this drug mimics the planar nature of the ring oxonium ion in the reaction of the viral enzyme neuraminidase.

Irreversible Inhibitors

Types of Irreversible Inhibition

Irreversible inhibitors usually covalently modify an enzyme, and inhibition cannot therefore be reversed. Irreversible inhibitors often contain reactive functional groups such as nitrogen mustards, aldehydes, haloalkanes, alkenes, Michael acceptors, phenyl sulfonates, or fluorophosphonates. These electrophilic groups react with amino acid side chains to form covalent adducts. The residues modified are those with side chains containing nucleophiles such as hydroxyl or sulfhydryl groups; these include the amino acids serine (as in DFP, right), cysteine, threonine or tyrosine.

Irreversible inhibition is different from irreversible enzyme inactivation. Irreversible inhibitors are generally specific for one class of enzyme and do not inactivate all proteins; they do not function by destroying protein structure but by specifically altering the active site of their target. For example, extremes of pH or temperature usually cause denaturation of all protein structure, but this is a non-specific effect.

Similarly, some non-specific chemical treatments destroy protein structure: for example, heating in concentrated hydrochloric acid will hydrolyse the peptide bonds holding proteins together, releasing free amino acids. Irreversible inhibitors display time-dependent inhibition and their potency therefore cannot be characterised by an IC_{50} value. This is because the amount of active enzyme at a given concentration of irreversible inhibitor will be different depending on how long the inhibitor is pre-incubated with the enzyme.

Instead, $k_{obs}/[I]$ values are used, wherek_{obs} is the observed pseudo-first order rate of inactivation (obtained by plotting the log of per cent activity vs. time) and $[I]$ is the concentration of inhibitor. The $k_{obs}/[I]$ parametre is valid as long as the inhibitor does not saturate binding with the enzyme (in which case $k_{obs} = k_{inact}$).

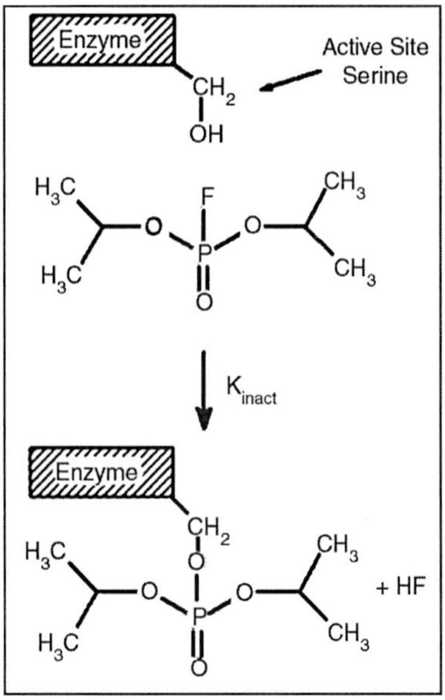

Fig. Reaction of the Irreversible Inhibitor

Analysis of Irreversible Inhibition

Irreversible inhibitors form a reversible non-covalent complex with the enzyme (EI or ESI) and this then reacts to produce the covalently modified "dead-end complex" EI*. The rate at which EI* is formed is called the inactivation rate or k_{inact}. Since formation of EI may compete with ES, binding of irreversible inhibitors can be prevented by competition either with substrate or with a second, reversible inhibitor. This protection effect is good evidence of a specific reaction of the irreversible inhibitor with the active site.

The binding and inactivation steps of this reaction are investigated by incubating the enzyme with inhibitor and assaying the amount of activity remaining over time. The activity will be decrease in a time-dependent manner, usually following exponential decay. Fitting these data to a rate equation gives the rate of inactivation at this concentration of inhibitor. This is done at several different concentrations of inhibitor. If a reversible EI complex is involved the inactivation rate will be saturable and fitting this curve will give k_{inact} and K_i.

Another method that is widely used in these analyses is mass spectrometry. Here, accurate measurement of the mass of the unmodified native enzyme and the inactivated enzyme gives the increase in mass caused by reaction with the inhibitor and shows the stoichiometry of the reaction. This is usually done using a MALDI-TOF mass spectrometre. In a complementary technique, peptide mass fingerprinting involves digestion of the native and modified protein with a protease such as trypsin. This will produce a set of peptides that can be analysed using a mass spectrometre. The peptide that changes in mass after reaction with the inhibitor will be the one that contains the site of modification.

IMMOBILIZED ENZYMES IN BIOPROCESS

Immobilization of biocatalysts helps in their economic reuse and in the development of continuous bioprocesses. Biocatalysts can be immobilized either using the isolated enzymes or the whole cells. Immobilization often stabilizes structure of the enzymes, thereby allowing their applications even under harsh environmental conditions of pH, temperature and organic solvents, and thus enable their uses at high temperatures in nonaqueous enzymology, and in the fabrication of biosensor probes. In the future, development of techniques for the immobilization of multienzymes along with cofactor regeneration and retention system can be gainfully exploited in developing biochemical processes involving complex chemical conversions. The present review outlines some of the above aspects, and delineates the present status and future potentials of immobilized enzymes and nonviable cells in the emerging biotech industries.

Biotechnology is currently considered as a useful alternative to conventional process technology in industrial and analytical fields. This is mainly because, unlike the chemical catalysts, the biological systems have the advantages of accomplishing complex chemical conversions under mild environmental conditions with high specificity and efficiency. Biological systems help in ingredient substitution, processing aid substitution, more efficient processing, less undesirable products, increased plant capacity, increased product yields, and improved or unique products. The variety of chemical transformations catalysed by enzymes has made these catalysts a prime target of exploitation by the emerging biotech industries.

Despite these advantages, the use of enzymes in industrial applications has been limited by several factors, mainly the high cost of the enzymes, their instability, and availability in small amounts. Also the enzymes are soluble in aqueous media and it is difficult and expensive to recover them from reactor effluents at the end of the catalytic process. This restricts the use of soluble enzymes to batch operations, followed by disposal of the spent enzyme-containing solvent.

Over the last few decades, intense research in the area of enzyme technology has provided many approaches that facilitate their practical applications. Among them, the newer technological developments in the field of immobilized biocatalysts can offer the possibility of a wider and more economical exploitation of biocatalysts in industry, waste treatment, medicine, and in the development of bioprocess monitoring devices like the biosensor. Immobilization means associating the biocatalysts with an insoluble matrix, so that it can be retained in a proper reactor geometry for its economic reuse under stabilized conditions. Immobilization thus allows, by essence, to decouple the enzyme location from the flow of the liquid carrying the reagents and products. Immobilization helps in the development of continuous processes allowing more economic organization of the operations, automation, decrease of labour, and investment/capacity ratio. Immobilized biocatalysts offer several other advantages, notable among them is the availability of the product in greater purity.

Purity of the product is very crucial in food processing and pharmaceutical industry since contamination could cause serious toxicological, sensory, or immunological problems. The other major advantages include greater control over enzymatic reaction as well as high volumetric productivity with lower residence time, which are of great significance in the food industry, specially in the treatment of perishable commodities as well as in other applications involving labile substrates, intermediates or products.

IMMOBILIZED NONVIABLE CELLS AS AN ECONOMICAL SOURCE OF ENZYMES

Biocatalysts can be immobilized using either the isolated enzymes or the whole cells or cellular organelles. Immobilization of whole cells has been shown to be a better alternative to immobilization of isolated enzymes. Doing so avoids the lengthy and expensive operations of enzyme purification, preserves the enzyme in its natural environment thus protecting it from inactivation either during immobilization or its subsequent use in continuous system. It may also provide a multipurpose catalyst, specially when the process requires the participation of number of enzymes in sequence. The major limitations which may need to be addressed while using such cells are the diffusion of substrate and products through the cell wall, and unwanted side reactions due to the presence of other enzymes.

The cells can be immobilized either in a viable or a nonviable form. Immobilized nonviable cell preparations, which are normally obtained by permeabilizing the intact cells, for the expression of intracellular activity are useful for simple processes that require single-enzyme with no requirement for cofactor regeneration, like hydrolysis of sucrose or lactose. On the other hand immobilized viable cells, which serve as 'controlled catalytic biomass', have opened new avenues for continuous fermentation on heterogeneous catalysis basis by serving as self-proliferating biocatalysts.

The present review will focus mainly on immobilized isolated enzyme systems, and the use of immobilized nonviable cells as a source of enzymes. For the aspects on the immobilized viable cell systems, the readers can refer to another article in this issue. Some of the aspects on the applications of immobilized viable cells in biochemical process development and monitoring have also been reviewed recently. Most of the enzymes used at industrial scale are normally the extracellular enzymes produced by the microbes. This has been mainly due to their ease of isolation as crude enzymes from the fermentation broth. Moreover, the extracellular enzymes are more stable to external environmental perturbations compared to the intracellular enzymes. However, over 90 per cent of the enzymes produced by a cell are intracellular. The economic exploitation of these, having a variety of biochemical potentials, has been limited in view of the high cost involved in their isolation. Also, compared to extracellular enzymes, the intracellular enzymes are more labile. Delicate and expensive separation methods are required to release the enzymes undamaged from the cell, and to isolate them.

This increases the labour and the cost of the enzyme. These problems could now be obviated by the use of permeabilized cells as a source of enzyme. Permeabilization of the cells removes the barrier for the free diffusion of the substrate/product across the cell membrane, and also empties the cell of most of the small molecular weight cofactors, etc., thus minimizing the unwanted side reactions. Side reactions, which can occur due to the presence of other enzymes in a cell, can also be minimized by inactivating such enzymes prior to or after immobilization.

Such permeabilized cells, which are often referred to as nonviable or nongrowing cells, can be exploited in an immobilized form as a very economical source of intracellular enzyme for simple bioconversions like hydrolysis, isomerization and oxidation reactions that do not need a cofactor-regeneration system. The decision to immobilize cells either in a viable or nonviable form is very important and depends on their ultimate application. Thus the permeabilized *K. fragilis* cells, which are nonviable, convert lactose in milk to glucose and galactose' whereas the viable (nonpermeabilized) cells convert the lactose to ethanol and are useful in the complete desugaration of milk.

A variety of physical, chemical, and enzymatic techniques have been developed for the permeabilization of cells. Some techniques like entrapment in polyacrylamide itself have been shown to result in permeabilization of the cells. A few typical examples of immobilized enzyme preparation obtained using permeabilized cells.

TECHNIQUES AND SUPPORTS FOR IMMOBILIZATION

A large number of techniques and supports are now available for the immobilization of enzymes or cells on a variety of natural and synthetic supports. The choice of the support as well as the technique depends on the nature of the enzyme, nature of the substrate and its ultimate application. Therefore, it will not be possible to suggest any universal means of immobilization. It can only be said that the search must continue for matrices which provide facile, secure immobilization with good interaction with substrates, and which conform in shape, size, density and so on to the use for which they are intended. Care has to be taken to select the support materials as well as the reagents used for immobilization, which have GRAS status, particularly when their ultimate applications are in the food processing and pharmaceutical industries. Macromolecular, colloidal, viscous, sticky, dense or particulate food constituents or waste streams also limit the choice of reactor and support geometries. Commercial success has been achieved when support materials have been chosen for their flow properties, low cost, nontoxicity, maximum biocatalysts loading while retaining desirable flow characteristics, operational durability, ease of availability, and ease of immobilization.

The variety of techniques and supports investigated for the immobilization have been reviewed in a number of articles and books. The objective of the preceding part of the paper is to review briefly, in the light of current developments, some approaches that have been used for the immobilization of biocatalysts with reference to the studies carried out in the author's laboratory as well as few others.

Techniques for immobilization have been broadly classified into four categories, namely entrapment, covalent binding, cross-linking and adsorption. A combination of one or more of these techniques has also been investigated. It must be emphasized that in terms of economy of a process, both the activity and the operational stability of the biocatalysts are important. They determine its productivity, which is the activity integrated over the operational time.

Entrapment

Entrapment has been extensively used for the immobilization of cells, but not for enzymes. The major limitation of this technique for the immobilization of enzymes is the possible slow leakage during continuous use in view of the small molecular size compared to the cells. Biocatalysts have been entrapped in natural polymers like agar, agarose and gelatine through thermoreversal polymerization, but in alginate and carrageenan by ionotropic gelation. A number of synthetic polymers have also been investigated. Notable among them are the photo-crosslinkable resins, polyurethane prepolymers' and acrylic polymers like polyacrylamide.

Among these, the most widespread matrix made from monomeric precursors is the polyacrylamide gel. Polyacrylamide may not be a useful support for use in food industry in view of its toxicity, but can have potentials in the treatment of waste and in the fabrication of analytical devices containing biocatalysts.

One of the major limitations of entrapment technique is the diffusional limitation as well as the steric hindrance, especially when the macromolecular substrates like starch and proteins are used. Diffusional problems can be minimized by entrapment in fine fibres of cellulose acetate or other synthetic materials or by using an open pore matrix. Recently, the development of so-called hydrogels and thermoreactive water-soluble polymers, like the albumin-poly (ethylene glycol) hydrogel, have attracted attention in the field of biotechnology. In the area of health care, they offer new avenues for enzyme immobilization. Such gels with a water content of about 96 per cent provide a microenvironment for the immobilized enzyme close to that of the soluble enzyme with minimal diffusional restrictions.

COVALENT BINDING

Covalent binding is an extensively used technique for the immobilization of enzymes, though it is not a good technique for the immobilization of cells. The functional groups extensively investigated are the amino, carboxyl, and the phenolic group of tyrosine. Enzymes are covalently linked to the support through the functional groups in the enzymes, which are not essential for the catalytic activity.

It is often advisable to carry out the immobilization in the presence of its substrate or a competitive inhibitor so as to protect the active site. The covalent binding should also be optimized so as not to alter its conformational flexibility. Some of these problems however, can be obviated by covalent bonding through the carbohydrate moiety when a glycoprotein is concerned. A number of industrially useful enzymes are glycoproteins wherein the carbohydrate moiety may not be essential for its activity. In general, functional aldehyde group can be introduced in a glycoprotein by oxidizing the carbohydrate moiety by periodate oxidation without significantly affecting the enzyme activity. The enzyme could then be covalently linked to a support containing an alky amine group through Schiffs base reaction. Enzymes like glucose oxidase, peroxidase, invertase, etc. have been immobilized using this technique.

Covalent binding has been extensively investigated using inorganic supports. Enzymes covalently bound to inorganic supports have been used in the industry. Enzymes have also been bound to synthetic membranes, thus integrating biconversion and downstream processing. Large-scale processes using such an approach have been demonstrated for the preparation of invert sugar using invertase.

Cross-linking

Biocatalysts can also be immobilized through chemical cross-linking using homo- as well as heterobifunctional cross-linking agents. Among these, glutaraldehyde which interacts with the amino groups through a base reaction has been extensively used in view of its GRAS status, low cost, high efficiency, and stability. The enzymes or the cells have been normally cross-linked in the presence of an inert protein like gelatine, albumin, and collagen. Studies from our laboratory have shown the use of raw hen egg white as an economic, easily available novel proteinic support rich in lysozyme for the immobilization of enzyme or nonviable cells either in a powder, bead or highly porous foam form, using glutaraldehyde as the cross-linker. The unique feature of this support is the large concentration of lysozyme naturally present in hen egg white which gets co-immobilized, thus imparting the bacteriolytic property to the support.

Adsorption followed by cross-linking has also been used for the immobilization of enzymes. The technique of cross-linking in the presence of an inert protein can be applied to either enzymes or cells. The technique can also be used for the immobilization of enzymes by cross-linking the cell homogenates. Osmotic stabilization of cellular organelles or halophilic cells prior to immobilization using cross-linkers has also shown promise.

Adsorption

This is perhaps the simplest of all the techniques and one which does not grossly alter the activity of the bound enzyme. In case of enzymes immobilized through ionic interactions, adsorption and desorption of the enzyme depends on the basicity of the ion exchanger. Moreover, a dynamic equilibrium is normally observed between the adsorbed enzyme and the support which is often affected by pH as well as the ionic strength of the surrounding medium. This property of reversibility of binding has often been used for the economic recovery of the support.

This has been successfully adapted in industry for the resolution of racemic mixtures of amino acids, using amino acid acylase. A variety of commercially available ion exchangers have been investigated for this purpose. One of the techniques, which has gained importance more recently, is the use of polyethylenimine for imparting polycationic characteristics to many of the neutral supports based on cellulose or inorganic materials. Enzymes with low pI, like invertase, urease, glucose oxidase, catalase, and other enzymes have been bound through adsorption followed by cross-linking on polyethylenimine-coated supports.

Immobilization of enzymes through hydrophobic interaction has also shown promise. One of the important features of this technique, which is of great significance, is that, unlike ionic binding, hydrophobic interactions are usually stabilized by high ionic concentrations, thus enabling the use of high concentrations of substrates as desired in an industrial process without the fear of desorption.

Other types of strong interactive binding techniques have also been reported for the reversible immobilization of enzymes. A typical example is the immobilization of soybean b -amylase on phenyl-boronate agarose, which can be reversed for the recovery of the support using sorbitol. Varieties of biospecific interactions have also been investigated for the reversible immobilization of enzymes by adsorption. Enzymes like acetyl choline esterase, ascorbic acid oxidase, invertase, peroxidase, glucose oxidase, etc. have been immobilized by biospecific-reversible immobilization on lectin-bound supports and invertase, using polyclonal antiinvertase antibodies. Traditional enzyme immobilization procedures involve isolation of the enzyme, followed by use of several steps for the immobilization.

Costs of enzyme purification, the immobilization procedure, bioreactor operational stability, and bioreactor regeneration are the major factors that determine the cost of a bioreactor process. Development of techniques for the simultaneous isolation and immobilization of enzymes from crude extracts has obviated these problems. Some typical examples include the immobilization of a streptavidin-b -galactosidase fusion protein expressed in *Escherichia coli* and bioselectively adsorbed from a crude cell lysate to biotin which is covalently immobilized on controlled pore glass, and the simultaneous purification and immobilization of D-amino acid oxidase from *Trigonopsis variabilis* cell lysate adsorbed on phenyl sepharose. The technique can have future potentials especially in the downstream processing and immobilization of enzyme/proteins obtained by recombinant DNA technology.

Techniques for the adhesion of whole cells on polymeric surfaces are also currently gaining considerable importance. The major advantage for the cells immobilized through adhesion is reduction or elimination of the mass transfer problems associated with the commonly used gel entrapment method. The technique of immobilization usually being followed is the microbial colonization by recycling of the cell suspension along with nutrients, such that a biofilm is gradually formed. This often results in the immobilization of cells in a viable form for use in heterogeneous fermentations. Useful techniques have been developed also for the immobilization of nonviable cells to be used as an enzyme source for simple chemical conversions. Notable among them include treating the cells or the support with trivalent metal ions like Al or Fe^{3+} or charged colloidal particles, and use of polycationic polymers like chitosan.

Novel techniques have been developed to adhere cells strongly on a variety of polymeric surfaces including glass, cotton cloth, cotton threads, and other synthetic and inorganic surfaces using polyethylenimine. This technique has also been recently used for the simultaneous filtration and immobilization of cells from a flowing suspension, thus integrating downstream processing with bioprocessing. This technique may have future potentials for the immobilization of cells for food applications. Polyethylenimine is nontoxic and the United States of America Food and Drug Administration has permitted its use as a direct food additive under the Food Drug and Cosmetic Act.

ENZYMES AND LIFE PROCESSES

The living cell is the site of tremendous biochemical activity called metabolism. This is the process of chemical and physical change which goes on continually in the living organism. Build-up of new tissue, replacement of old tissue, conversion of food to energy, disposal of waste materials, reproduction - all the activities that we characterize as "life." This building up and tearing down takes place in the face of an apparent paradox. The greatest majority of these biochemical reactions do not take place spontaneously. The phenomenon of catalysis makes possible biochemical reactions necessary for all life processes. Catalysis is defined as the acceleration of a chemical reaction by some substance which itself undergoes no permanent chemical change. The catalysts of biochemical reactions are enzymes and are responsible for bringing about almost all of the chemical reactions in living organisms.

Without enzymes, these reactions take place at a rate far too slow for the pace of metabolism. The oxidation of a fatty acid to carbon dioxide and water is not a gentle process in a test tube - extremes of pH, high temperatures and corrosive chemicals are required. Yet in the body, such a reaction takes place smoothly and rapidly within a narrow range of pH and temperature. In the laboratory, the average protein must be boiled for about 24 hours in a 20 per cent HCl solution to achieve a complete breakdown. In the body, the breakdown takes place in four hours or less under conditions of mild physiological temperature and pH.

CHEMICAL NATURE OF ENZYMES

All known enzymes are proteins. They are high molecular weight compounds made up principally of chains of amino acids linked together by peptide bonds.

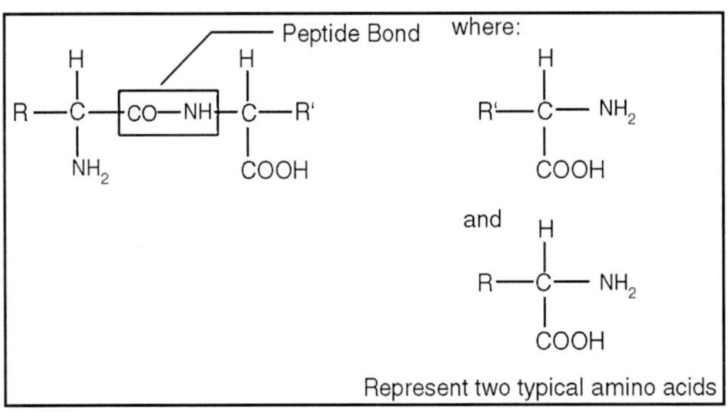

Fig. Typical Protein Structure - Two Amino Acids Joined by a Peptide Bond.

Enzymes can be denatured and precipitated with salts, solvents and other reagents. They have molecular weights ranging from 10,000 to 2,000,000. Many enzymes require the presence of other compounds - cofactors - before their catalytic activity can be exerted. This entire active complex is referred to as the holoenzyme; *i.e.*, apoenzyme (protein portion) plus the cofactor (coenzyme, prosthetic group or metal-ion-activator) is called the holoenzyme.

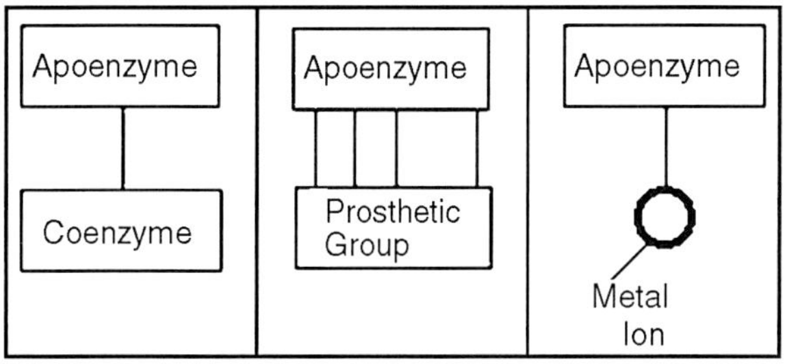

Fig. Holoenzymes- Apoenzymes Plus Various Types of Cofactors.

Apoenzyme + Cofactor = Holoenzyme

According to Holum, the cofactor may be:

- *A coenzyme*: A non-protein organic substance which is dialyzable, thermostable and loosely attached to the protein part.
- *A prosthetic group*: An organic substance which is dialyzable and thermostable which is firmly attached to the protein or apoenzyme portion.
- *A metal-ion-activator*: These include K^+, Fe^{++}, Fe^{+++}, Cu^{++}, Co^{++}, Zn^{++}, Mn^{++}, Mg^{++}, Ca^{++}, and Mo^{+++}.

Specificity of Enzymes

One of the properties of enzymes that makes them so important as diagnostic and research tools is the specificity they exhibit relative to the reactions they catalyze. A few enzymes exhibit absolute specificity; that is, they will catalyze only one particular reaction. Other enzymes will be specific for a particular type of chemical bond or functional group.

In general, there are four distinct types of specificity:

1. *Absolute specificity*: The enzyme will catalyze only one reaction.
2. *Group specificity*: The enzyme will act only on molecules that have specific functional groups, such as amino, phosphate and methyl groups.
3. *Linkage specificity*: The enzyme will act on a particular type of chemical bond regardless of the rest of the molecular structure.
4. *Stereochemical specificity*: The enzyme will act on a particular steric or optical isomer.

Though enzymes exhibit great degrees of specificity, cofactors may serve many apoenzymes. For example, nicotinamide adenine dinucleotide (NAD) is a coenzyme for a great number of dehydrogenase reactions in which it acts as a hydrogen acceptor. Among them are the alcohol dehydrogenase, malate dehydrogenase and lactate dehydrogenase reactions.

Naming and Classification

Except for some of the original enzymes such as pepsin, rennin, and trypsin, most enzyme names end in "ase". The International Union of

Biochemistry (I.U.B.) initiated standards of enzyme nomenclature which recommend that enzyme names indicate both the substrate acted upon and the type of reaction catalyzed. Under this system, the enzyme uricase is called urate: O_2 oxidoreductase, while the enzyme glutamic oxaloacetic transaminase (GOT) is called L-aspartate: 2-oxoglutarate aminotransferase.

Enzymes can be classified by the kind of chemical reaction catalyzed:

- Addition or removal of water
  - *Hydrolases*: These include esterases, carbohydrases, nucleases, deaminases, amidases, and proteases
  - Hydrases such as fumarase, enolase, aconitase and carbonic anhydrase
- Transfer of electrons
  - Oxidases
  - Dehydrogenases
- Transfer of a radical
  - *Transglycosidases*: Of monosaccharides
  - *Transphosphorylases and phosphomutases*: Of a phosphate group
  - *Transaminases*: Of amino group
  - *Transmethylases*: Of a methyl group
  - *Transacetylases*: Of an acetyl group
- Splitting or forming a C-C bond
  - Desmolases
- Changing geometry or structure of a molecule
  - Isomerases
- Joining two molecules through hydrolysis of pyrophosphate bond in ATP or other triphosphate
  - Ligases

9

Cell Biology

THE LIFE CYCLE OF CELLS

A cell in an adult organism can be viewed as a steady-state system. The DNA is constantly read out into a particular set of mRNAs, which specify a particular set of proteins. As these proteins function, they are also being degraded and replaced by new ones, and the system is so balanced that the cell neither grows, shrinks, nor changes its function. This static view of the cell, however, misses the all-important dynamic aspects of cellular life. The dynamics of a cell can best be understood by examining the course of its life. A new cell arises when one cell divides or when two cells, like a sperm and an egg cell, fuse. Either event sets off a cell-replication programme that is encoded in the DNA and executed by proteins.

This programme usually involves a period of cell growth, during which proteins are made and DNA is replicated, followed by cell division, when a cell divides into two daughter cells. Whether a given cell will grow and divide is a highly regulated decision of the body, assuring that an adult organism replaces worn out cells or makes more cells in response to a new need. Examples of the latter are the growth of muscle in response to exercise or damage, and the proliferation of red blood cells when a person ascends to a higher altitude and needs more capacity to capture oxygen. However, in one major and devastating disease — cancer — cells multiply even though they are not needed by the body. To understand how cells become cancerous, biologists have intensely studied the mechanisms that control the growth and division of cells.

THE CELL CYCLE FOLLOWS A REGULAR TIMING MECHANISM

Most eukaryotic cells live according to an internal clock; that is, they proceed through a sequence of phases, called the cell cycle, during which DNA is duplicated during the synthesis (S) phase and the copies are distributed to opposite ends of the cell during mitotic (M) phase. Progress along the cycle is controlled at key checkpoints, which monitor the status of a cell, for instance, the internal amount of DNA or the presence of extracellular nutrients. When certain conditions are met, the cell proceeds to the next checkpoint. The cycle begins after the cell divides into two daughter cells, each containing an identical copy of the parental cell's genetic material. The cell cycle of prokaryotes is simple and fast. Replication of the single chromosome begins at a particular DNA sequence, the replication origin, which is anchored to the cell membrane. Once DNA replication is complete, assembly of new membrane and cell wall forms a septum, which eventually divides the cell in two. Because the origins of the two newly formed chromosomes are anchored to different membrane sites, each daughter cell receives one chromosome. In ideal growth conditions, the bacterial cell cycle is repeated every 30 minutes.

Only a few types of eukaryotic cells can grow and divide as quickly as bacteria. Most growing plant and animal cells take 10 – 20 hours to double in number, and some duplicate at a much slower rate. Many cells in adult animals, such as nerve cells and striated muscle cells, do not divide at all. They have temporarily exited from the cell cycle after mitosis and entered a "paused or quiescent" state called G_0. Because eukaryotic cells are larger and more complex than prokaryotic cells, a specialized mechanism coordinates their replication of genomic DNA, distribution of chromosomes, and cell division.

MITOSIS

Mitosis is the mechanism in eukaryotes for partitioning the genome equally at cell division. To accomplish this complex task, plant and animal cells build a specialized machine, called the mitotic apparatus, which captures the chromosomes and then pushes and pulls them to opposite sides of the dividing cell. Remarkably, the mitotic apparatus is a temporary structure that exists only during mitosis to distribute the genetic material. Although the events of mitosis unfold continuously, they are conventionally divided into four substages representing phases of chromosome movement. During the first substage, prophase, the replicated chromosomes, each comprising two identical chromatids, are condensed into compact packets and then released to the cytoplasm when the nuclear membrane breaks down. During metaphase and anaphase, the chromosomes are sorted, and each chromatid of a pair moves to opposite sides of the cell. The end of mitosis is marked by re-formation of a membrane around each set of chromosomes (telophase). Division of the cytoplasm, called cytokinesis, then yields two daughter cells, each with a $2n$ complement of genetic material.

Cell division in plant and animal cells differs mainly at cytokinesis. Animal cells divide in two by pinching of the cytoplasm. However, because a plant cell is surrounded by a rigid cell wall, daughter cells are formed by building a new cell membrane and cell wall between the two daughter nuclei, thereby cutting the cytoplasm into two portions.

CELL DIFFERENTIATION CREATES NEW TYPES OF CELLS

The most complicated example of cellular dynamics occurs when a cell changes, or *differentiates,* to carry out a specialized function. This process often is marked by a change in the microscopic appearance, or *morphology,* of the cell. For example, the different structures of a nerve cell and a muscle cell reflect their respective functions in long-distance communication and contraction, highlighting the biological principle that "form follows function." Cell differentiation creates the diversity of cell types that arise during the development of an organism from a fertilized egg. This is a process of extensive cell multiplication and differentiation. A mammal that starts as one cell becomes an organism with hundreds of diverse cell types such as muscle, nerve, and skin.

Here we see at its most dramatic the power of DNA to control cellular behaviour: development is a DNA-orchestrated set of cellular changes (easily tens of thousands of them) that occur virtually without fail. The almost perfect resemblance of "identical" twins is a testament to the programme encoded by DNA to reproducibly direct the development of a human being. Nowhere is the variety of cellular activities and responses better illustrated than in the body's immune system. It is there that many cell types come together in organized tissues specifically designed to allow the body to distinguish its own cells from those of foreign invaders.

Within the immune system, we see both development of specialized cells that can recognize invading cells and formation of tissues from cells that originate in various parts of the body. The immune-system cells not only actively survey their environment with surface receptor proteins like antibodies, but also change their properties when they encounter a foreign substance, allowing the body to rid itself of invaders.

CELLS DIE BY SUICIDE

Unchecked cell growth and multiplication produce a mass of cells, a tumor. *Programmemed cell death* plays the very important role of population control by balancing cell growth and multiplication. In addition, cell death also eliminates unnecessary cells. For example, during embryogenesis, the digits of our fingers and toes are sculpted by the death of cells in the intervening spaces. If these cells remained alive, our hands and feet would become webbed. Thus the timing and location of cell death, as well as cell growth and division, must be precisely controlled.

Cell death follows an internal programme of events called apoptosis, in which all traces of a cell vanish. The first visible sign of apoptosis is condensation of the nucleus and fragmentation of the DNA. The cell soon shrivels and is consumed by macrophages. A cell is directed to commit suicide when an essential factor is removed from the extracellular environment or when an internal signal is activated.

Thus, the default state of the cell is to remain alive. The discovery of genes that suppress the growth of tumors by activating cell death stimulated an exciting new line of cancer research that may lead to more effective treatment strategies

INTEGRATING CELLS INTO TISSUES

The evolution of multicellular organisms permitted specialized cells and tissues to form; a flowering plant has at least 15 cell types, and a vertebrate hundreds. In both plants and animals, cells that are specialized to carry out a particular task are found together in the tissues in which the task is performed: a xylem or meristem; a liver, a muscle, or a nerve ganglion. Different types of cells in a tissue are often arranged in precise patterns of staggering complexity. For instance, the hundreds of different types of neurons in the human brain are interconnected to one another through a network of some 10^{15} synaptic connections! The coordinated functioning of many types of cells within tissues, and of multiple specialized tissues, permits the organism as a whole to move, metabolize, reproduce, and carry out other essential activities.

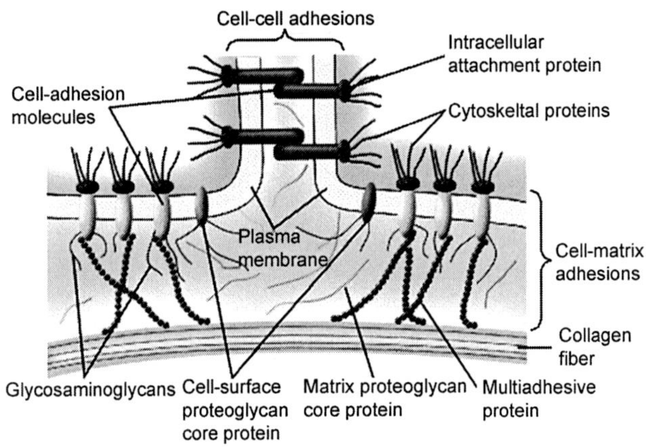

Fig. Integrating Cells into Tissues.

A key step in the evolution of multicellularity must have been the ability of cells to contact tightly and interact specifically with other cells. Various integral membrane proteins, collectively termed cell-adhesion molecules (CAMs), enable many animal cells to adhere tightly and specifically with cells of the same, or similar, type; these interactions allow populations of cells to segregate into distinct tissues. Following aggregation, cells elaborate specialized cell junctions that stabilize these interactions and promote local communication between adjacent cells.

Animal cells also secrete a complex network of proteins and carbohydrates, the extracellular matrix (ECM), that creates a special environment in the spaces between cells. The matrix helps bind the cells in tissues together and is a reservoir for many hormones controlling cell growth and differentiation. The matrix also provides a lattice through which cells can move, particularly during the early stages of differentiation. Defects in these connections lead to cancer and developmental malformations. The extracellular matrix has three major protein components: highly viscous proteoglycans, which cushion cells; insoluble collagen fibres, which provide strength and resilience; and soluble multiadhesive matrix proteins, which bind these components to receptors on the cell surface. Different combinations of these components tailor the strength of the extracellular matrix for different purposes. For example, animals contain many types of extracellular matrices, each specialized for a particular function such as strength (in a tendon), cushioning (in cartilage), or adhesion. In the case of smooth muscle cells that surround an artery, the extracellular matrix must provide strong but flexible connections.

The extracellular matrix is not just an inert framework or cage that supports or surrounds cells. The matrix also communicates directly and indirectly with the intracellular signaling pathways that direct a cell to carry out specific functions. For example, the ability of hepatocytes—the principal cells in the liver—to express liver-specific proteins depends on their association with a matrix of appropriate composition. Specific ECM components can directly activate cytosolic signal-transduction pathways by binding to cell-adhesion protein receptors in the plasma membrane. Alternatively, by binding growth factors and other hormones, the matrix can either sequester these signals from cells or, conversely, present them to cells, thereby indirectly inducing or inhibiting signaling pathways. Morphogenesis—the later stage of embryonic development during which form is achieved by cell movements and rearrangements—also is critically dependent on ECM components, which are constantly being remodeled, degraded, and resynthesized locally. Even in adults—in areas of wounding, for example —degradation and resynthesis of ECM components occurs.

THE ORDERED STEPS IN THE CELL CYCLE

G1: The first gap in the normal cell cycle is called G1 and is the period when the necessary proteins for DNA replication are synthesized. However, this phase of the cell cycle is not only characterized by synthesis of replication machinery. During this period the cell must monitor both the internal and external environments to ensure that all the preparations for DNA synthesis have been completed and that overall conditions for cell division are favourable. As discussed below, there is a major check-point in a normal cell cycle that is critical for ensuring that all is well for the cell to enter S-phase.

S-phase: The duplication of the cellular content of DNA occurs during S-phase, so-called because this is the phase when DNA is synthesized. This phase of the cell cycle is the longest taking 10-12 hours of a typical 24hr eukaryotic cell cycle.

G2: During the second gap phase of the cell cycle the cell undertakes the synthesis of the proteins required to assemble the machinery required for separation of the duplicated chromosomes (the process called mitosis) and ultimately division of the parental cell into two daughter cells (the process termed cytokinesis). Like the G1 phase, the G2 phase is also a stage when the internal and external environments are monitored to ensure that faithful replication of the DNA has occurred and that conditions are favourable for cytokinesis. In addition, as for the G1 phase there is a major check-point at the end of the G2 phase that controls the entry into M-phase.

M-phase: During M-phase there is an ordered series of events that leads to the alignment and separation of the duplicated chromosomes (called sister chromatids) This process is divided into distinct steps that were originally identified and characterized through light microscopic observations of dividing cells. The steps of mitosis are termed prophase, prometaphase, metaphase, anaphase and telophase. Although cytokinesis is the process by which the parental cell is physically separated into two new daughter cells, it actually begins during anaphase. The processes that occur during M-phase require much less time than those of S-phase, generally lasting only 1-2 hrs.

During prophase the duplicated chromosomes condense while outside the nucleus the mitotic spindle assembles between the two centrosomes. The centrosome is an organelle that serves as the main microtubule organizing centre that is involved in the attachment of microtubules to the sister chromatids. During prometaphase the nuclear membrane breaks apart and the chromosomes can attach to spindle microtubules and begin active movement. During metaphase the chromosomes are aligned at the equator of the spindle midway between the spindle poles. The sister chromatids are attached to opposite poles of the spindle.

During anaphase the sister chromatids synchronously separate to form the two sets of daughter chromosomes. Each sister chromatid is slowly pulled towards the spindle pole it faces. During telophase the two daughter chromosomes arrive at the spindle poles and decondense. A new nuclear envelope forms around each set of chromosomes which forms the two new nuclei. This process marks the end of mitosis and sets the stage for cytokinesis.

CHECKPOINTS AND CELL CYCLE REGULATION

It should seem obvious that the processes that drive a cell through the cell cycle must be highly regulated so as to ensure that the resultant daughter cells are viable and each contains the complement of DNA found in the original parental cell. There are many "parts" to the systems that control the transit through a eukaryotic cell cycle. These "parts" include mechanisms to control the timing of events so that each individual process is turned on and off at the appropriate time, mechanisms to initiate each event in the correct order and to also ensure that each event is triggered only once per cell cycle, controls to ensure events occur in a linear, irreversible direction, redundancy, or back-ups to ensure the cycle functions properly even in the context of some malfunctioning parts, and systems that are adaptable so that cell cycle events can be modified in the context of different cell types and/or environmental conditions. Many of the most important discoveries about the mechanisms that control events of the cell cycle were elucidated using yeasts which are single cell eukaryotes. By analysis of various mutants that inactivated genes encoding essential components of cell cycle control systems in yeast many important control genes were identified. These genes were identified as cell division cycle genes or cdc genes. Thus, many cell cycle control genes in mammalian cells are also called cdc genes. Much of the control of the progression through the phases of a cell cycle are exerted at checkpoints. There are many such checkpoints but the two most critical are those that occur near the end of G1 prior to S-phase entry and those near the end of G2 prior to mitosis.

As indicated above, there is the need for cell cycle control mechanisms to exert their influences at specific times during each transit through a cell cycle. The heart of this timing control is the responsibility of a family of protein kinases that are called cyclin-dependent kinases, CDKs. The kinase activity of these enzymes rises and falls as the cell progresses through a cell cycle. Different CDKs operate at different points in the cell cycle. As would be expected, the oscillating changes in the activity of CDKs leads to oscillating changes in the phosphorylation of various intracellular proteins. These phosphorylations alter the activity of the modified proteins which then effect changes in events of the cell cycle. The cyclical activity of each CDK is controlled by a complex series of proteins, the most important of which are the cyclins, hence the name of the enzymes as cyclin-dependent kinases. The CDKs are absolutely dependent upon their interaction with the cyclins for activity.

Unless they are tightly bound CDKs have no kinase activity. The cyclins were originally idenitified because they undergo a cycle of synthesis and degradation at specific points in each cell cycle. Thus, whereas the levels of the various CDKs remain fairly constant throughout the cell cycle, their activities changes in concert with the fluctuations of the cyclins. Four different classes of cyclins have been defined on the basis of the stage of the cell cycle in which they bind and activate CDKs. These four classes are G1-cyclins, G1/S-cyclins, S-cyclins, and M-cyclins.

The cyclin nomenclature and associated CDK in mammalian cells are listed here:

Cyclin-CDK Complex	Cyclin	CDK Partner
G1-CDK	cyclin D	CDK4, CDK6
G1/S-CDK	cyclin E	CDK2
S-CDK	cyclin A	CDK2
M-CDK	cyclin B	CDK1

- There are three D cyclins in mammals: D1, D2, and D3
- CDK1 is the same as CDC2 in fission yeast and CDC28 in budding yeast

The G1-cyclins are not found in all eukaryotic cells but in those where they are synthesized they promote passage through a restriction point in late G1 called Start. The G1/S-cyclins bind to their cognate CDKs at the end of G1 and it is this interaction that is required to commit the cell to the process of DNA replication in S-phase. The S-cyclins bind to their cognate CDKs during S-phase and it is this interaction that is required for the initiation of DNA synthesis. The M-cyclins bind to their cognate CDKs and in so doing promote the events of mitosis.

Although CDKs are inactive unless bound to a cyclin, there is more to the activation process than just the interaction of the two parts of the complex. When cyclins bind to CDKs they alter the conformation of the CDK resulting in exposure of a domain that is the site of phosphorylation by another kinase called CDK-activating kinase (CAK). Following phosphorylation the cyclin-CDK complex is fully active. In addition to control of CDK kinase activity by cyclin binding and CAK phosphorylation, control is exerted to inhibit CDK activity through interaction with inhibitory proteins as well as by inhibitory phosphorylation events.

Thus, there is extremely tight control on the overall activity of each CDK. One of the inhibitory kinases that phosphorylates CDKs is called Wee1. The inhibitory phosphorylations are removed through the action of a phosphatase called CDC25. The action of these two regulatory enzymes on CDK activity is most important at the level of the M-CDK activity at the onset of mitosis. Proteins that bind to and inhibit cyclin-CDK complexes are called CDK inhibitory proteins (CKI, for cyclin-kinase inhibitor). Mammalian cells express two classes of CKI.

These are called CIPs for CDK inhibitory proteins and INK4 for inhibitors of kinase 4. The CIPs bind and inhibit CDK1, CDK2, CDK4, and CDK6 complexes, whereas the INK4s bind and inhibit only the CDK4 and CDK6 complexes. There are at least three CIP proteins in mammalian cells and these are identified as $p^{21Cip1/WAF1}$ (gene symbol=CDKN1A), p^{27KIP1} (gene symbol=CDKN1B), and p^{57KIP2} (gene symbol=CDKN1C). The expression of each of these CIPs is controlled by specific events that may have occurred during cell cycle transit. For example p^{21Cip1} expression is induced in response to DNA damage.

This induction is under the control of the action of the tumor suppressor protein p53. There are at least four INK4 proteins that are each identified by their molecular weights: $p^{15INK4B}$, $p^{16INK4A}$, and $p^{18INK4C}$ (these were the first 3 characterized) as well as p^{19INK4}. The $p^{16INK4A}$ protein is also a tumor suppressor since loss of its function leads to cancer. All the INK4 proteins contain 4 tandem repeats of a sequence of amino acids that were first identified in ankyrin and are thus referred to as ankyrin repeats. As indicated above, many cells reside in a resting or quiescent state but can be stimulated by external signals to re-enter the cell cycle. These external growth promoting signals are the result of growth factors binding to their receptors. Most growth factors induce the expression of genes that are referred to as early and delayed-response genes.

The activation of early response genes occurs in response to growth factor receptor-mediated signal transduction resulting in phosphorylation and activation of transcription factor proteins that are already present in the cell. Many of the induced early response genes are themselves transcription factors that in turn activate the expression of delayed-response genes. In the context of the cell cycle, these delayed-response genes encode proteins of the G1-CDK complexes. One such early response gene is the proto-oncogene MYC. With respect to the cell cycle some of the genes turned on by activation of MYC are cyclin D, proteins of the ubiquitin ligase complex called SCF (Skp1/cullin/F-box protein) and the members of the E2F transcription factor family. There are six members of the E2F family: E2F1 through E2F6). The synthesis of cyclin D will result in the activation of G1-CDK complexes. The synthesis of components of SCF leads to the degradation of p27KIP1 which normally inhibits G1-CDK complexes.

The synthesis of E2F family members results in increased synthesis of proteins involved in DNA synthesis as well as the synthesis of the S-phase cyclins A and E and CDK2. Regulation of E2F activity by the tumor suppressor pRB will be discussed below. The cyclical degradation of the cyclins is effected through the action of several different ubiquitin ligase complexes. The action of ubiquitin ligases in protein turn-over is discussed in more detail in the Protein Modifications page. There are two important ubiquitin ligase complexes that control the turn-over of cyclins and other cell cycle regulating proteins. One is the SCF complex which functions to control the transit from G1 to S-phase and the other is called anaphase promoting complex (APC) which controls the levels of the M-phase cyclins as well as other regulators of mitosis.

One important function of APC is to control the initiation of sister chromatid separation which begins at the metaphase-anaphase transition. The attachment of the sister chromatids to the opposite poles of the mitotic spindles occurs early during mitosis. The ability of the sister chromatids to be pulled apart is initially inhibited because they are bound together by a protein complex termed cohesin complex. The cohesin complex is deposited along the chromosomes as they are duplicated during S-phase. Anaphase can only begin with the disruption of the cohesin complex. The breakdown of the cohesin complex is brought about as a consequence of the activation of the ubiquitin ligase activity of the APC. APC targets a protein called securin. Securin functions to inhibit the protease called separase and the action of separase is to degrade the proteins of the cohesin complex, thus allowing sister chromatid separation.

TUMOR SUPPRESSORS AND CELL CYCLE REGULATION

Tumor suppressors are so called because cancer ensues as a result of a loss of their normal function, *i.e.* these proteins suppress the ability of cancer to develop. It would seem obvious, therefore, that one import function of tumor suppressors would be control of the progression of a cell through a round of the cell cycle. If cells are able to synthesize damaged DNA before it is repaired or to divide when the DNA is damaged then the resulting daughter cells can pass on the resultant DNA damage to their progeny. The result can be catastrophic resulting in cancer.

For this reason, the two most important check points in the eukaryotic cell cycle are the G1-S transition and the entry into mitosis. The former prevents DNA replication prior to repair of damaged DNA and the latter prevents damage that may have occurred to the DNA during replication to propagated into daughter cells during mitosis. Following the isolation and characterization of two tumor suppressor genes in particular it was found that they function to control the ability of cells to progress through these two important checkpoints. The protein encoded by the retinoblastoma susceptibility gene (pRB) and the p53 protein are both tumor suppressors. The function of pRB is to act as a brake preventing cells from exiting G1 and that of p53 is to inhibit progression from S-phase to M-phase. The best understood effect of G1-CDK activity is that exerted on transcription factors of the E2F family, hereafter referred to simply as E2F. In the context of the cell cycle regulation, E2F activates the expression of cyclin A, cyclin E and CDK2. These proteins are components of the S-CDK complexes necessary for progression through S-phase.

The activity of E2F is itself controlled via interaction with pRB. When pRB binds E2F it can no longer function as a transcription factor as it is sequestered in the cytosol. Interaction of pRB and E2Fcorrelates to the state of phosphorylation of pRB and the affinity between the two proteins is highest when pRB is hypophosphorylated. Phosphorylation of pRB is maximal at the start of S phase and lowest after mitosis and entry into G1. Stimulation of quiescent cells with mitogen induces phosphorylation of pRB, while in contrast, differentiation induces hypophosphorylation of pRB. One of the most significant substrates for phosphorylation by the G1 cyclin-CDK complexes is pRB.

When pRB is phosphorylated by G1 cyclin-CDK complexes it releases E2F allowing E2F to transcriptionally activate its target genes. MDM2 is so named since it was isolated as an amplified gene in the tumorigenic mouse cell line 3T3DM. In response to DNA damage, *e.g.* as a result of uv-irradiation or irradiation, cells activate several kinases including checkpoint kinase 2

(CHK2) and ataxia telangiectasia mutated (ATM). One target of these kinases is p53. ATM also phosphorylates MDM2. When p53 is phosphorylated it is released from MDM2 and can carry out its transcriptional activation functions. One target of p53 is the cyclin inhibitor p21Cip1 gene.

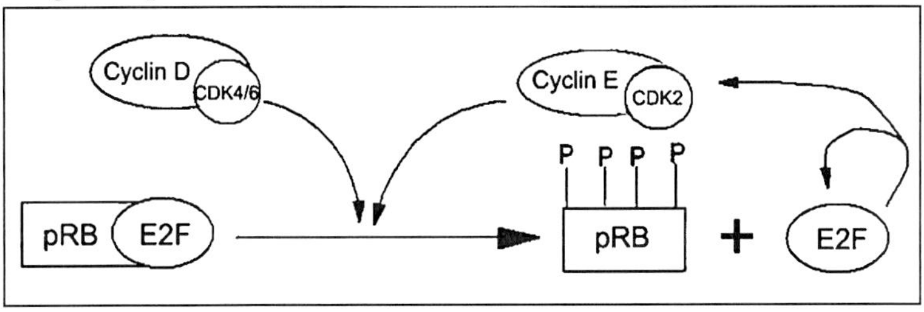

Fig. Regulation of E2F by pRB

Activation of p21Cip1 leads to increased inhibition of the cyclin D1-CDK4 and cyclin E-CDK2 complexes thereby halting progression through the cell cycle either prior to S-phase entry or during S-phase. As a consequence of p53-induced synthesis of p21 expression, there is a convergence between the roles of p53 and pRB in regulation of cyclin-CDK complexes. In either case the aim is to allow the cell to repair its damaged DNA prior to replication or mitosis. Even given the limited discussion of the functions of pRB and p53 it is still easy to understand how loss of either protein function can lead to aberrant cell cycle progression and the potential for the development of cancer.

THE MECHANICS OF CELL DIVISION

The process of cell division must occur in a highly ordered and accurate manner so as to ensure that each daughter cell receives an identical copy of the parental cells genome. The processes described above relate to the regulation of the progressive steps taken as a cell progresses to cytokinesis. This section will discuss the biochemical processes undertaken to effect accurate separation of the duplicated DNA (the sister chromatids) and cytokinesis.

Following duplication of the chromosomes the sister chromatids are held together through multisubunit protein complexes referred to as cohesins. These complexes are found all along the length of each chromatid as the DNA is replicated. Following DNA replication the chromosomes condense and this is the role of proteins called condensins. Chromosome condensation is the first easily identifiable sign that a cell is about to enter M-phase. Cohesins and condensins are structurally related and act in concert to prepare the chromosomes for mitosis.

The task of separating the sister chromatids such that each daughter cell receives one copy of each chromosome is carried out by the mitotic spindle which is composed of microtubules and several proteins that interact with them. An additional cytoskeletal structure is required for the actual separation of the cell into two new cells. This structure is referred to as the contractile ring. The process of mitosis occurs through a highly ordered series of five steps as outlined above. The actual separation of the parental cell into two daughter cells can be considered the sixth step in mitosis. Whereas, prophase, prometaphase, metaphase, anaphase, and telophase occur in a strictly controlled sequential fashion, cytokinesis begins in anaphase and continues until the cell divides. During interphase the microtubule machinery is in a constant state of dynamic instability. Individual microtubules are growing or shrinking at any given moment.

During prophase the activation of M-CDK complexes initiates a change in the microtubule structures to one where there are a large number of shorter microtubules surrounding each centrosome. The centrosomes are cytoplasmic nucleation sites for the mitotic spindles. M-CDK complexes initiate these changes via the phosphorylation of microtubule motor proteins and microtubule-associated proteins (MAPs). During prometaphase the nuclear envelope abruptly breaks down as a consequence of M-CDK complexes phosphorylating the nuclear lamina. The dissolution of the nuclear membrane allows the microtubules to access the mitotic spindles.

When microtubules attach to the mitotic spindle they become stabilized. The microtubules eventually become attached at the kinetochore which is a complex protein structure that assembles onto the highly condensed DNA at the centromere. The chromosomes are pulled back and forth by the microtubules eventually becoming aligned equidistant from the two spindle poles. The alignment of the chromosomes forms the metaphase plate. The chromosomes oscillate about the metaphase plate awaiting the signal that will induce the sister chromatids to separate. This phase of mitosis is referred to as the spindle-attachment checkpoint and it ensures that cells do not enter anaphase until all of the chromosomes are attached to both poles of the mitotic spindles. As described above, sister chromatids begin to separate with the activation of the APC. As each sister chromatid is pulled towards one of the poles of the mitotic spindle the kinetochore microtubules depolymerize.

By the end of anaphase, the daughter chromosomes have separated to opposite ends of the cells and have begun to decondense which signals the onset of telophase. Telophase is denoted by the reassembly of the nuclear envelope around each group of daughter chromosomes. During this process the lamins that make of the nuclear lamina are dephosphorylated allowing them to re-associate with the nuclear envelope. Following the formation of the new nuclear envelopes, the chromosomes decondense into their interphase state and transcriptional activity begins anew. The cell is now ready for the final process, complete separation into two daughter cells.

MICROTUBULAR ORGANELLES

Many cells organs can be considered to be derived from special assemblies of microtubules. In fact, cilia, flagella, basal bodies, and centrioles have, as one of their main components, groups of microtubules arranged in a special fashion. For this reason and in spite of the complexity of their structure, these organelles will be considered at this point, along with some of their physiological aspects.

Cillary and Flagellar Motions

Cillary and flagellar all motility is adapted to liquid media and is executed by minute, specially differentiated appendices that vary in size and number. They are called *flagella* if they are few and long, and *cilia,* if short and numerous. In Protozoa, especially the Infusoria, each cell has hundreds or thousands of minute cilia, and their movement permits a rapid procession of the organism in the liquid medium.

In some special regions of the Infusoria, several cilia fuse and form larger conical appendices, *the cirri,* or membranes observed as *undulating membranes*. Out of entire class of Protozoa, the Flagellata, is characterized by the presence of flagella. The spermatozoa of metazoans move as isolated cells by means of flagella. On the other hand, epithelial cells that possess vibratile cilia and constitute true ciliated sheets are relatively common. These may cover large areas of the external surface of the body and determine the motion of the animal. Such is the case with some Platyhelminthes and Nemertea and also with larvae of Echinodermata, Mollusca, and Annelida.

More often, the ciliated epithelial sheets line cavities or internal tubes, such as the air passages of the respiratory system or various parts of the genital tract. In these organs all the cilia move simultaneously in the same direction, and fluid currents are thus produced. In some cases, the currents serve to eliminate solid particles in suspension (*e.g.,* in the respiratory system). The eggs of amphibians and mammals are driven along the oviduct with the aid of vibratile cilia.

The Cillary Apparatus

Some of the essential components of cillary apparatus are:

- The *cilium,* which is the slender cylindroid process, limited by the plasmalemma (cell membrane), that projects from the free surface of the cell,
- The *basal body,* or granule, the intracellular organelle similar to the centriole from which it originates,
- In some cells fine fibrils—called *ciliary rootlets*—that arise from the basal granule and converge into a conical bundle, the pointed extremity of which ends at one side of the nucleus.

Regarding the basal bodies they are embedded in the ectoplasmic layer beneath the cell surface. In general, they are spaced uniformly and in parallel rows. Cilia and flagella are extremely delicate filaments whose thickness is often at the limit of the resolving power of the light microscope.

Various epithelia have appendices similar in shape to cilia, but immobile; these are called *stereocilia.* Examples are the processes of the epithelial cells of the epididymis. In the macula and crista of the inner ear, there are stereocilia in addition to motile cilia, or *kynocilia*. Stereocilia do not contain microtubules.

The Axoneme

Regarding the term Axoneme the some is applied to the axial basic microtubular structure of cilia and flagella, and is it thought to be the essential motile element. In various flagella *accessory structures* made of fibres, microtubules, crystalline bodies, or mitochondrial derivatives may be found. The axoneme of a cilium or flagellum may range from a few microns to 1 to 2 millimeters in length, but the outside diameter is about 0.21 μm. It is generally surrounded by an *outer ciliary membrane,* which is continuous with the plasma membrane. All the components of the axoneme are embedded within the *ciliary matrix.*

With the fundamental 9 + 2 microtubular pattern and the other essential structural components of the axoneme. A plane perpendicular to the line joining the two central tubules divides the axoneme into a right and a left symmetrical half. It is generally admitted that the plane of the ciliary beat is perpendicular to this plane of symmetry.

The pairs of peripheral microtubules have an ellipsoidal profile, whereas the central ones are circular. The diameter of the peripheral tubules varies from 18 to 25 nm with the central tubules being slightly larger. The two microtubules of each peripheral pair can be clearly distinguished by several morphologic features. The doublets are skewed at about 10 degrees so that one tubule, designated subfibre A, lies closer to the axis than the other (*i.e.,* subfibre B). The microtubule of subfibre A is smaller, but complete, whereas that of subfibre B is larger and incomlete, since it lacks the wall adjacent to A. In fact, while A has 13 tubulin subunits, B has only 11. Furthermore, subfibre.

A has processes—the so-called *dyein arms* —that are oriented in the same direction in all microtubules. This orientation is clockwise when the axoneme is viewed from base to tip.

Dynein Arms

The arms of subfibre A which are generally called *dynein arms* because they contain dynein, a higher molecular weight ATPase (300,000 to 400,000 daltons). This is a Mg^{2+}-and Ca^{2+} -activated enzyme, which after solubilization can recombine at the same position on the A microtubules. Two isoenzymatic forms of the ATPase, dynein I and II, have been separated by electrophoresis.[16] Dynein I constitutes the largest proportion of the axonemal ATPase and is present in the arms of subfibre A. As will be mentioned later, the interaction between tubulin and dynein is thought to underlie the basic mechanism of ciliary and flagellar motion.

Nexin Links

The doublets are linked by *interdoublet* or *nexin links*. This name is derived from the fact that another protein called *nexin* has been isolated from these links. This protein has a molecular weight of 150,000 to 160,000 daltons. The function of these links is unknown, but they may serve as structures that maintain the geometric integrity of the axoneme during the sliding motion.

Radial Spokes

The spokes terminate in a dense knob or head, which may have a forklike structure. Certain observations that the spokes are attached perpendicularly to the ciliary axis where it is straight and that they are relatively detached in bent or tilted regions of the axis has led to the hypothesis that they may be active in the conversion of active sliding between the outer doublets into local axial being.

Basal Bodies (Kinetosomes) and Centrioles

After the classic works of Hennegay and Lenhosesk in the year 1897, it has been suggested that basal bodies (or kinetosomes) of cilia and flagella are homologous with the centrioles found in mitotic spindles. In some cells it was discovered that a centriole engaged in mitosis could carry a cilium at the same time. This homology was fully confirmed with the introduction of electron microscopy. Centrioles are cylinders that measure, on the average, 0.2 μm–0.5 μm; at times they may be as long as 2 μm. This cylinder is open on both ends, unless it carries a cilium. In the later case, it is separated from the cilium by a *ciliary plate*. In freeze-fractured cilia it has been possible to observed the so-called ciliary necklace. This consists of two to six rows of membrane particles present just below the ciliary plate where the central pair of microtubules end.

In the wall of the centriole there are nine groups of microtubules arranged in a circle. Each group is a triplet formed of three tubules (rather than two, as in cilia) that are skewed towards the centre. Since the angle made by the set of tubules with the tangent to the cylinder varies from 70 degrees in the proximal end to 20 to 30 degrees in the distal end, the tubules twist from one end to the other or describe a helical course. The tubules are designated *subfibre A, B,* and C. from the centre towards the periphery.

Both subfibres A and B cross the ciliary plate and are continuous with the corresponding subfibres in the axonemene; subfibre C terminates near the *ciliary* or *basal plate*. There are no central microtubules in the centrioles and no central microtubules in the centrioles and no special arms; however, they are linked by connectives. The proximal portion of the centriole has a cartwheel appearance with spokes that radiate from a central hub and connect to subfibre of each triplet. The presence of the cartwheel structure at the proximal end provides the centriole with a structural and functional polarity.

The growth of the centriole is from the distal end, and in the case of kinetosomes, it is from this end that the cilium is formed. Further more, the *procentrioles*, which are formed at right angles to the centriole, are located near the proximal end. Both the above given centrioles may give rise to a cilium but in most cases only one does, and the sister centriole remains at a right angle to the kinetosome. In most multiciliated cells there are single basal bodies. It will be shown later that in sensory cells having nonmotile ciliary derivatives and a 9 + 0 configuration, there are two basal bodies. The two centrioles of a pair are always located at a distance from each other of at least 0.8 μm and are perpendicular to each other.

Ciliary Rootlets

In some of the cells, ciliary rootlets originate from the basal body. Most rootlets are striated, having a regular cross-banding with a repeating period of 55 to 70 nm. Five intraperiodic sub-bands have been observed in the cross-striated rootlets. The striated fibres are composed of parallel microfilaments, 3 to 7 nm in diameter, which in turn are formed of globular subunits. These fibres and filaments may serve a structural role such as anchoring the kinetosomes. By analogy with other microfilaments, a contractile role has been postulated for the ciliary rootlets. Furthermore, ATPase has been found associated with the cross-bands of certain rootlets.

Basal Feet and Satellites

Basal feels are dense processes that are arranged perpendicularly to the basal body in a particular direction and originate from two to three of the triplets. These processes impose a structural asymmetry on the basal body that has been related to the direction of the ciliary beat. The basal foot, which is composed of microfilaments that terminate in a dense bar, may be a focal point for the convergence of microtubules. *Satellites* or *pericentriolar bodies*, are electron-dense structures lying near the centriole that probably are nucleating sites for microtubules.

Ciliary Movement

Ciliary movement can be analysed easily by scraping the pharyngeal epithelium of a frog or toad with a spatula and placing the scrapings in a drop of physiologic salt solution between a slide and a coverglass. On the free surface of the epithelial cell, the rapid motion of the vibratile cilia can be seen. If a row of cilia is observed, the contraction is *metachronic* in the plane of the direction of motion; that is, it starts before or after the contraction of the next cilium. In this way true waves of contraction are formed. Now consider the other part that, in a plane perpendicular to the direction of motion, the contraction is *isochronic*; all the cilia are observed in the same phase of contraction at a given time.

The rhythmic contraction of cilia has been interpreted in different ways. A two-step process that involves *intraciliary* excitation followed by *interciliary* conduction has been proposed to explain the metachronic rhythm of the ciliary beat. This mechanism evidently does not depend on the nervous system, since it persists after the epithelium has been separated from the rest of the organism. However, cytoplasmic continuity is indispensable to its maintenance, for if a cut is made in the row of cilia the waves of contraction of the two isolated pieces becomes uncoordinated.

If a piece of epithelium is removed from the pharynx of a frog and implanted with a reversed orientation, the movement is maintained but in the opposite direction. Ciliary contractions are generally rapid (10 to 17 per second in the pharynx of the frog). Analysis of the motion has been facilitated greatly by stroboscopic and ultrarapid microcinematography. Ciliary movement may be pendulous, unciform (hooklike), infundibuliform, or undulant. The first two are carried out in a single plane. In the pendulous movement, typical of the ciliated Protozoa, the cilium is rigid and the motion is carried out by a flexion at its base.

On the other hand, in the unciform movement, the most common type in the Metazoa, the cilium is doubled upon contraction and takes the shape of a hook. In the infundibuliform movement, the cilium or flagellum rotates, describing a conical or funnel-shaped figure. In the undulant motion, characteristic of flagella contraction waves proceed from the site of implantation and pass to the free border. It was shown above that the plane of the beat of a cilium is perpendicular to that passing through the central microtubules.

Although the effective stroke is a single plane, the cilium moves away from the original plane during the recovery phase. This movement, which can be beautifully demonstrated by scanning electron microscopy,[23] implies that either the whole cilium rotates at its base or that the central microtubules rotate. Since the last century, ciliary and flagellar motion have been compared to that of muscle. However, the complex structure of the axoneme suggests that this motion may be more complex than the planar sliding of myofilaments in muscle.

The key process appears to be the sliding of the microtubular doublets over each other, with the associated making and breaking of the cross-bridges between adjacent doublets, represented by the dynein arms. The sliding of doublets during the bending of the cilium or flagellum has actually been observed and measured. This displacement of the doublets has been determined because the central microtubules extend to the very tip, whereas subfibres B terminate at a distance from the tip of 1 to 2 μm, and subfibres. A terminate slightly beyond subfibres B. This arrangement has also made it possible to demonstrate that the two subfibres do not move relative to one another within the doublet. In the diagram of the amount of sliding displacement is shown to depend on the distance separating the tubules and on the bend angle.

Experiments on ciliary motion has been aimed at demonstrating the involvement of ATP; it was mentioned above that the ATPase activity is associated with the protein *dynein* in the arms of subfibres. A Dynein constitutes about 8 per cent of the total ciliary protein. In glycerin-extracted cilia, flagella, and spermatozoan tails, the addition of ATP produces rhythmic activity that may persist for a few minutes or even hours. Interaction between tubulin and dynein seems to be essential in this type of cell motility. When studying myofilaments in muscle, we shall see that the actin-myosin system is the other process involved in contraction.

Several approaches are current by being used to study the molecular mechanisms in ciliary motion. One of them has been the use of sperm flagella in which the membranes have been removed by detergents. These sperm can be reactivated with ATP. Extraction of dynein arms by KC1 leads to a reduction in the frequency of beating, which can be restored to the original level by the addition of *dynein*. An anti-dynein antiserum inhibits the movement of spermatozoa without membranes, and ATP induces, it. The other observation is that, unlike muscle, cilia and flagella do not require Ca^{2+} for activation. In many cases, however, Ca^{2+} has a regulating effect on the wave form of the beating.

BASIC FEATURES OF CELL DIFFERENTIATION

The Differentiated State is Stable and persistence

One of the principal characteristics of cell differentiation of higher cells is that once established, the differentiated state is very stable and can persist throughout many cell generations. For example, a neuron will persist as such throughout the lifetime of a individual, and a cell committed to become a skin cell will gradually keratinize and eventually die. These persistent changes are very different from the type of regulation involved in enzyme induction and repression in bacteria, which is specially designed to respond rapidly to changes in the environments.

E, coli responds to the addition, of lactose to the culture medium by synthesizing B-galactosidase a few minutes later, and when lactose is exhausted, enzyme synthesis is stopped with equal speed. Cell differentiation of induced in the organism by various stimuli, but once it has been established it can persist even in the absence of the initial., it can persist even in the absence of the initial stimulus.

This is demonstrated, for example, by the fact that differentiated cloned cell lines, such as steroid-secreting cell lines and neuroblastoma cell lines (which differentiate into neurons), are able to grown indefinitely in vitro. Without this type of data it could have been argued that differentiation persists in the organism because the stimulus that has given rise to it also persists. In virtro, although such a stimule cannot be present, the differentiated state is maintained over many cell generations.

Determination

In many cases, before it is possible to recognize morphologically that a cell has differentiated, there is a period during which the cell is already committed to a particular change. In this interval, called *determination*, the cell has already focused its capacity for differentiation on a specific pathway. One of the classic examples of determination is provided by the *imaginal discs of Drosophila*. The discs are groups of cells which are present in the larva in an undifferentiated form, but which upon metamorphosis will give rise to legs, wings, antennae, and so forth. Each disc is predetermined to become a particular type of adult structure. This commitment is very stable and heritable, as shown by the disc transplantation experiments carried out by Hadorn. Discs transplanted into the abdomen of adult files remained undifferentiated but could be stimulated to differentiate by transplantation into a larva where the hormonal conditions would induce metamorphosis. Discs transplanted serially in the abdomens of adult flies for 9 years (1800 cell generations) were still able, after metamorphosis, to differentiate into structures belonging to the original disc.

Inter Dependency of Nucleus and cytoplasm

Nucleus and cytoplasm are interdependent; one cannot survive without the other. The cytoplasm provides most of the energy for the cell through oxidative phosphorylation (in mitochondria) and anaerobic glycolysis, and the cytoplasmic ribosomes contain most of the "machinery" for protein synthesis. On the other hand, the nucleus provides templates for specific synthesis (mRNA) and also supplies the other important RNA molecules (rRNA) and tRNA).

Any discussion of nucleocytoplasmic interrelations must consider:

- The mechanisms by which the genes contained in the chromosomes exert their control on the metabolic processes of the cytoplasm
- The mechanisms by which the cytoplasm influences gene activity.

Nuclear control of the cytoplasm was first considered during the last century by Balbiani in the so-called *merotomy* experiments. Such enucleated fragments are able to sustain most cellular activities: *e.g.*, they can form a cellulose membrane and carry on photosynthesis (plant cells)), react to stimuli and ingest food (amoebae), activate cilia (ciliated cells), undergo cytoplasmic streaming, and so forth. However, these cells generally survive only a short time and are incapable of growth and reproduction. Within 5 to 10 minutes after being enucleated by micromanipulation an ameba loses its surface tension, becomes spheroid, and develops numerous blunt pseudopodia. Its movements are slowed down and it can no longer digest foods. In this state it may survive for about 20 days. If a nucleus is successfully implanted within three days, the ameba becomes extended, starts to move normally, and digests foods. Finally, it may even divide and eventually produce a culture mass.

ACETABULARIA

The unicellular marine alga *Acetabularia* has been invaluable for the analysis of nucleocytoplasmic interactions. This giant cell has a stem between 3 and 5 cm long and a cap 1 cm in diameter. The unique nucleus is located in the basal or *rhizoid* end of the cell. The great experimental advantage of *Acetabularia* is that the nucleus can be removed simply by cutting off the rhizoid. The resulting enucleated cells are active in photosynthesis and can survive for many weeks.

Each *Acetabularia* species has a particular cap morphology. If a nucleus from *A. crenulata* is implanted into an enucleated *a mediterranea,* the resulting cell will eventually develop a cap of the *A. crenulata* type. If two nucleate cells of different species are grafted together, a hybrid having a cap with intermediate morphology is formed. This clearly shows that the shape of the cap is determined by the nucleus. Although dependent on the nucleus, the cytoplasm has some degree of functional autonomy. The information required to form the cap passes to the cytoplasm before cap formation, where it can be stored in an inactive form. *A. crenulata* does not form a cap in the darkness, only an elongated stem. If such a stem is nucleated and then exposed to light, a cap will develop. Therefore, the cap formation triggered by light is not dependent on new nuclear activity, and call the necessary information is present in the cytoplasm but is not converted to the final active form until a light stimulus is present.

The information for cap production is very stable; enucleated stalks maintained in the dark for weeks are still able to produce a cap upon illumination. This experimental result is probably due to the presence of stable mRNA are known to occur in other eukaryotic cells, for example, in the mammalian reticulocyte, which continues to produce hemoglobin for several days after the nucleus is eliminated in the course of normal erythropoiesis. The nucleus itself is dependent on the cytoplasm. *Acetabularia* has a giant nucleus in its rhizoid, but its volume decreases markedly when cytoplasmic energy production is impaired by darkness or by amputation of a large part of the cytoplasm.

Fusion of Cell

The fusion of cells by using inactivated *Sendai virus* (a member of the para-influenza viruses) provides the opportunity of placing a nucleus in a different cytoplasmic environment. Cell fusion can also be induced with other agents that affect membrane structure, such as polyethyleneglycol and lysolecithin. When different cells are fused, initially a *heterokaryon* is formed (*i.e.*, a single cell containing nuclei of two types). Eventually, both nuclei might enter mitosis synchronously, form a single metaphase plate, divide and produce a *hybrid cell line* (also known as a *synkaryon*). The cells of cell line have a single nucleus containing chromosomes of both parental nuclei.

In 1965 Henry Harris found that chick erythrocyte nuclei are reactivated when fused to HeLa cells (an undifferentiated human cell line so named because it was derived from the uterine carcinoma cells of a patient named *Henrietta Lacks*). These heterokaryons are of interest because the erythrocyte nucleus does not synthesize RNA or DNA. Chick erythrocytes are terminally differentiated cells which have a very condensed nucleus and which are destined to die (Mammalian erythrocytes normally eliminate their nuclei during red blood cell maturation, while the red cells of birds, amphibians, and reptiles retain the nucleus, which becomes inactivated). When fused to HeLa cells, the chick erythrocyte nucleus increases 20 times in volume, disperses its chromatin, resumes RNA synthesis, develops a nucleolus, and eventually replicates its DNA.

This reactivation process is accompanied by the uptake of large amounts of human nuclear proteins which are thought to reactivate the erythrocyte nucleus this was shown by experiments in which heterokaryons were stained by immunofluorescence with antisera obtained from patients suffering from *lupus erythematosus.* (Lupus is a human auto-immune disease characterized by the production of antibodies against one's own nuclear proteins.) After fusion with HeLa cells, the chick nuclei stained intensely with the human nuclear antibodies.

From the From these experiments it is that the synthesis of macromolecules in a nucleus is controlled by the cytoplasmic environment. Even though erythrocytes are terminally differentiated cells, they can reassume RNA and DNA synthesis. A similar result is obtained with macrophage-HoLa cell heterokaryons. Rabbit macrophases do not synthesize DNA but will do so if fused to HeLa cells[7]. Only the cell cytoplasm is required to reactivate the chick erythrocyte nucleus. This was established by fusing erythrocytes to enucleated HeLa cells. Populations of enucleated cultured cells can be obtained with the use of *cytochalasin* B, a drug that affects predominantly actin microfilaments. Cells exposed to this drug cease moving and extrude their nuclei, which remain attached to the rest of the cell by only a cytoplasmic stalk.

Cultured cells that have been grown attached to a plastic surface are centrifuged in the presence of cytochalasin B, and the nuclei form a pellet, while the cytoplasm of the cells remains attached. The detached nudeus is surrounded by the cell membrane and a small amount of cytoplasm and is called a *karyoplast.* The enucleated cytoplasm is called a *cytoplasm.* Enucleated cells are viable for at least two days after enucleation and perform many cell function is normally. For example, cell movements, pinocytosis, and contact inhibition are unaffected, and if cells are detached from the plastic surface by trypsinization, they can reattach and spread on a new Petri dish. Clearly, many cytoplasmic functions are independent of the cell nucleus. Since karyoplasts are still surrounded by the cell membrane it is possible to fuse then to a different cytoplasm by using Sendai virus.

Reconstituted cells, which arise from the fusion of a karyoplast and a cytoplast, can survive longer than enucleated cytoplasm, can synthesize RNA, and in some cases can undergo cell division. Karyoplasts and cytoplasts from L cells (a mouse cell line) were prepared by centrifugation in the presence of cytochalasin B and were fused with Sendia virus, To distinguish true reconstituted cells from cells that failed to enucleate (karyoplast preparations are always contaminated by small numbers of intact cells).

The uncleus was labeled with H-thymidine, and the cytoplasm, with latex beads of various diameters, (Cultured cells incubared in the presence of latex spares incorporate the beads by phagocytosis, thus providing a very useful cytoplasmic marker.) Cytochalasin enucleation and cell reconstitution are very powerful techniques because they provide the opportunity of placing a nucleus (containing very little of its own cytoplasm) in an entirely new cytoplasmic environment.

In a few cases it has been possible to activate latent genes coding for specific proteins by fusion to differentiated cells. For example, albumin-secreting rat hepatoma cells fused with mouse lymphocytes (which do not produce albumin) will occasionally give rise to clones of hybride cells (synkaryons) which are able to produce both rat and mouse albumin. Somatic cell hybridization experiments shown that the patterns of nucleic acid synthesis and gene expression by a nucleus can be modified by substances present in the cell cytoplasm. Similar conclusions can be drawn from work with nuclei transplanted into *Xenopus laevis* oöcytes.

Cell Fusion Provide Pure Antibodies of Medical Importance

When animals are injected with a macromolecule of a shape that is recognized ad foreign to that individual (*e.g.*, a protein or complex polysaccharide), *antibodies* appear in the serum several days later. Antibodies are protein molecules that are able to bind specifically to the foreign *antigen.* Many different antibodies appear in the serum of an immunized animal, each one recognizing a different part of the antigen's shape (or different antigens if more than one was injected). Furthermore, individuals of the same species have different immunological responses,, so that two antisera directed against the same an tigen can in fact be very different. This is a major problem in medicine because, for example, success of failure of an organ transplantation depends on whether the donor and the recipient patient have the correct *histocompatibility antigens* on the cell surface. Clearly, antibodies that can be used throughout the world as standardized diagnostic reagents are highly desirable. This has now been made possible by using cell fusion techniques to create cell lines that produce only a single type of antibody, which can then be obtained in unlimited amounts.

Antibodies are produced by the lymphocytes. (When producing large amounts of antibody, lymphocytes adopt a special morphology and are also called *plasma cells.) Each antibody-producing cell can synthesize only one type of antibody.* Lymphocytes do not multiply in culture, but C. Milstein has developed a technique where by a *single* antibody-producing cell can be propagated indefinitely in culture by hybridization with a tumor cell. Initially a mouse is immunized against the desired antigen. By using. Sendai virus, the lymphocytes from the spleen of this mouse are then fused to a mouse cell line derived from plasma cell tumors (called *myelomas* because they invade the bone marrow). The myeloma cell carries a mutation in the enzyme *hypoxanthing-guanine phosphorolysis transferase* and is therefore enable to grow in HAT medium. However, the resulting hybrid cells are able to grow in HAT medium and the clones that secrete the desired antibody have even identified and grown.

The fusion with the tumor cell "immortalizes" the spleen lymphocyle, which can then be grown indefinitely in culture or injected into mice, where the cells produce secreting tumour that can be maintained by serial transplantation. Since all of the cells of one clone are derived from a single lymphocyte, a *monoclonal antibody* of high purity is produced. Monoclonal antibodies are currently used as anti-viral antibodies (*e.g.*, for diagnosis of influenza virus type and for the treatment of rabies), as diagnostic reagents in clinical biochemistry, and for the classification of human histocompatibility antigens. Monoclonal antibodies provide a good example of the fact that the practical applications of science cannot be predicted beforehand. Cell fusion was initially developed to study the seemingly specialized cell biology problem of nucleocytoplasmic interactions but eventually produced immunological tools of practical importance to all mankind.

Xenopus Oöcytes

Oöcytes used for experiments in differentiation are growing egg cells originally located in be ovary of the female frogs. Each oocyte requires three months or more to grow into a very large cell 1:2 mm in diameter. During this period the oocyte accumulates proteins for use during early embryogenesis, and its nucleus (also called the *germinal vesicle)* is very active in RNA synthesis. This high transcriptional activity can be visualized morphologically in the form of lampbrush chromosomes that have lateral loops packed with nascent RNA molecules. During this growth period, the oocyte nucleus is arrested at the diplotene stage of meiotic prophase and therefore does not synthesize DNA, The stimulus to continue meiosis is provided by gonadotrophic hormones released from the frog's pituitary gland.

In the laboratory this stimulus can be mimicked by the injection of mammalian hormones (or urine from a pregnant woman, thus providing a test that was much used for the detection of pregnancy).

Hormone treatment rapidly induces maturation; the oöcytes are released from the ovary, progess through the first meiotic division while traversing the oviduct, and 15 hours later are laid as *eggs,* which are at meiotic metaphase II and ready to start development upon fertilization. If nuclei are transplanted into oöcytes, the pattern of nucleic acid synthesis closely resembles that of the host cell. Brain nuclei (which synthesize RNA but not DNA) were injected into oöcytes at different stages of their life cycle. If injected into growing oöcytes (which synthesize RNA but not DNA). The brain nuclei incorporate H-uridine but not ^{3}H-thymidine in radioautographic studies. When nuclei are injected into maturing oöcytes isolated from the oviduct after hormone stimulation (such oöcytes are undergoing meiotic division (such oöcytes are undergoing meiotic division I and therefore have condensed chromosomes), the chromosome become condensed and mitotic spindles are formed. If not synthesize RNA but actively synthesize DNA after penetration of sperm or the micro-injection needle), they do not synthesize RNA, but will start replicating their DNA

The oöcyte cytoplasm not only affects the pattern of macromolecular synthesis but is also able to *reprogram* the expression of individual genes in transplanted nuclei. *Xenopus* oöcytes that were injected with a suspension of about 200 HeLa nuclei. The nuclei survive for several weeks inside the oöcytes and in that period synthesize substantial amounts of RNA.

In the first few days after injection the somatic nuclei enlarge to about ten times their original volume. The injected material can also be introduced into the oocyte nucleus in this case the somatic nuclei swell to about 100 times their initial volume. These changes are accompanied by chromatin dispersion and a massive uptake of proteins from the surrouding cytoplasm. The injected nuclei resemble the oocyte's own nucleus morphologically, and occasionally the human chromosomes form structures reminiscent of lampbrush chromosomes

The resemblance is not only morphological. The oocyte cytoplasm reprograms the gene expression of the injected nuclei in such a way that only those genes which are normally active in oöcytes are expressed. This can be determined by taking advantage of the fact that the RNAs synthesized by the injected nuclei in the course of several days accumulate in the cytoplasm, where they coder for new proteins. De Robertis and Gurdon injected *Xenopus* cultured kidney cell nuclei into oöcytes of a different am phibian species (the salamander *Pleurodeles,* whose proteins can be readily distinguished from those of *Xenopus* by two-dimensional electrophoresis). Those genes that are normally expressed in kidney cells but not in oöcytes became inactive after injection into *Pleurodeles* oöcytes. More importantly, some oocyte-active genes (which were not expressed by the kidney cell nuclei) were activated by the oocyte cytoplasm. It is therefore clear that the oocyte contains components which can determine that a particular spectrum of protein-coding genes will be active and that others will be inative.

The work with cell fusion and *Xenopus* oöcytes indicates that the cytoplasm of all cells contains components which determine the state of activity of nuclear genes. We will see that cytoplasmic gene-controlling substances (or "determinants"), which sometimes becomes unequally distributed in the cytoplasm, play a crucial role in establishing the initial steps of cell differentiation in early embryos and adult tissues.

10

Immunology

Immunology is the study of the organs, cells, and chemical components of the immune system. The immune system creates both *innate* and *adaptive* immune responses. The innate response exists in many lower species, all the way up the evolutionary ladder to humans, and it acts through relatively crude means against large classes of pathogens. The adaptive response is unique to vertebrates, reacting to foreign invaders with specificity and selectivity.

The immune system must maintain a delicate balance, with potent defensive responses capable of destroying large numbers of foreign cells and viruses while refraining from undue destruction of the host's body. When the immune system cannot mount a sufficient defense of the host, there is an immune deficiency; this is seen in HIV infection and SCID. If, on the other hand, the immune system acts too vigorously and begins to attack the host, we have autoimmunity. This is a defiance of the integral immune system property of self/nonself recognition. That is, the immune system begins attacking or forming antibodies against the host's own body tissues. Examples of autoimmune diseases include Graves' disease, Hashimoto's thyroiditis, *myaesthenia gravis* and type I diabetes mellitus. The human immune system recognizes non-self entities and mounts an effector response to neutralize the organism. A faster and stronger memory response may occur upon later exposure. The memory response has been used throughout history to confer immunity upon several populations, even previous to our understanding of the physiological basis for such a response. Thucydides wrote in his History of the Peloponnesian War that persons who had been exposed to plague previously could care for the sick without danger. In the 19th century, variolation was commonplace; this was the removal of smallpox (variola virus) skin pustules which were subsequently put into small cuts in the skin of healthy people. This was itself a crude form of vaccination, with the crusty dry pustules acting as an incubator of attenuated virus.

Edward Jenner would later use the cowpox virus to vaccinate (from *vacca*, Latin for "cow") patients against smallpox, and Louis Pasteur attenuated rabies and injected it into a small boy, naming this substance a vaccine in honor of Jenner's earlier studies in the science of immunology.

As immunology progressed, many people began to question how these vaccines worked. Why should exposure to plague in Thucycides' time confer protection only against plague and not all disease? Why should cowpox, a similar disease to smallpox but clearly a less severe virus, give milk maids sufficient immunity to resist full smallpox infection? In short, what has caused this memory response to be relatively (yet not absolutely) specific as well as selective?

APPLICATION OF IMMUNOLOGY

Vaccination is the most widespread application of immunology. It aims to artificially induce immunological memory to protect against primary infections by established pathogens. In designing a vaccine, it is important to take the pathogen's route of infection and its mechanism of pathogenicity into account. Symptoms of diphtheria and tetanus are caused exclusively by bacterial toxins. Consequently, it is of prime importance that these toxins be inactivated; going after the bacteria themselves is secondary. The polyvalent vaccine for babies starting at the age of two months therefore contains inactivated toxins, so-called toxoids, to generate neutralizing antibodies. In case of infection, no harm is done, as the antibody-coated toxins cannot bind to their cellular receptors. For polio, it is necessary to catch the virus before it enters the CNS. The virus enters the body via the enteral pathway, first replicating in intestinal epithelial cells, then spreading via the blood. Two ways of immunization have been successfully developed.

In the vaccine developed by Jonas Salk, inactivated polio virus is injected, inducing neutralizing antibodies which prevent the virus from reaching the CNS. In contrast, oral live-attenuated polio vaccine, developed a few years later by Albert Sabin, induces local immunity in the gut. Secretory IgA in the intestinal lumen prevents the virus from infecting enteric cells. Following immunization, attenuated virus can be spread to contacts of the vaccinee, resulting in protection of additional individuals. Live-attenuated polio vaccine is easy to administer (no needles required, etc.) and cheap to produce, but it has one drawback: one of the three mutated virus strains included very rarely reverts to a pathogenic virus, causing vaccine-associated paralytic poliomyelitis (VAPP) in about one in a million newly vaccinated persons. In many countries having eradicated indigenous polio, this risk is higher than that from imported wild polio infections.

Therefore most developed countries, including Austria, switched back to inactivated vaccine a few years ago. Oral vaccination remains the method of choice in countries with ongoing wild poliovirus circulation and lower vaccination rates. Here, the larger reduction in numbers of poliomyelitis obtained by the simpler immunization protocol outweigh the small number of VAPP cases. Inevitably, vaccinations entail certain risks and therefore tend to cause highly emotional debates, especially among parents nervous about their kids' wellbeing. In deciding for or against a specific vaccination, it is imperative to quantify the risks of either decision. Let's consider measles as an example. 'Measles are a harmless children's illness', some parents argue, and were easily overcome before the advent of vaccination.

Therefore, the risk of vaccinating a healthy child would outweigh the benefit of preventing measles. From an individual adult's experience, this sounds plausible. But do the data support this view? Measles vaccine, part of MMR (measles, mumps, rubella) is a live-attenuated vaccine with two goals: induction of CD8+ T-memory cells and induction of neutralizing IgG. A small risk exists: many children develop a fever, some develop postvaccinal measles, a markedly attenuated form with a slight skin rash. A complication of encephalitis in extremely rare cases cannot be completely excluded, but if it exists, it certainly affects fewer than one in a million vaccinees (with events of this rarity, it is hard to establish causality).

On the other hand, risks are much higher if a non-vaccinated person attracts measles. The most frequent complication is *otitis media*, which is very painful. Viral pneumonia, for which no causal therapy exists, occurs in one of 200 cases. Encephalitis due to the virus occurs with a frequency between 1:1000 and 1:5000. Mortality associated with encephalitis is about 15%, and lasting neurologic defects are common in those who overcome the disease. An extremely rare complication, subacute sclerosing panencephalitis, which manifests itself many years after the acute infection, is universally fatal. As a concrete illustration of the risks incurred by non-vaccinated persons serves a limited outbreak documented in the Netherlands in 1999: About 2300 cases (almost exclusively in families opposing vaccination) were brought to the attention of authorities. Three children died.

Fifty-three had to be admitted to the hospital, of which 30 suffered from pneumonia, 4 from encephalitis and 19 from other complications. 130 people were treated for pneumonia at home, 152 for *otitis media*. Ethically not unproblematic, as long as only few people refuse to be vaccinated from fear of side effects, these individuals profit in two ways: they don't incur any potential risk from vaccination, but are nevertheless protected by herd immunity due to the vaccination of the vast majority of their contacts. As soon as the fraction of people opposing vaccination exceeds a certain threshold, however, relative risk is reversed, as exemplified in the Dutch measles outbreak. Back to methodological aspects of immunization. Vaccination against hepatitis B virus, of paramount importance for medical personnel, relies on induction of neutralizing antibodies, which prevent the virus from entering liver cells. In this case, the vaccine consists exclusively of the envelope protein of the virus, HBs-antigen, that is produced by recombinant DNA technology and self-assembles to empty viral envelopes. This excludes any potential for the emergence of a pathogenic virus by mutation. The majority of vaccines against viral diseases contain attenuated live viruses. If we intend to induce CD8+ T memory cells, against, e.g., the measles virus, we will not succeed by using inactivated material: for a strong CD8+ T cell response, viral peptides have to be produced in the cell and presented on MHC-I. The virus used for immunization has to be able to replicate, yet must not be dangerous. Attenuation of a human virus can be achieved by repeatedly passaging the original virus on animal cells, e.g. ape cells. By mutation and selection, the virus over time adapts to the monkey cells. For example, it may change its surface protein to better interact with its receptor on the ape cell. When used back in humans as a vaccine, it will have trouble replicating in human cells and only do so very slowly. Today, using recombinant DNA-technology, attenuating mutations can be generated directly without lengthy animal passaging. For vaccines against bacterial infections, where the goal is to induce antibodies, the tendency is to reduce complexity.

In the early days of vaccination, there was no other way than to use easy to produce, but biologically messy preparations of inactivated bacteria. Inevitably, these caused high rates of unwanted effects, which were partly responsible for the lingering reservations about vaccination in collective memory. To minimize the potential for side effects and complications, modern vaccines aim to use the lowest number of defined antigen molecules possible. These vaccine molecules sometimes look strange to the untrained eye. Remember T cell help, required to "release the safety catch" from B cells to prevent production of unnecessary and potentially dangerous antibodies. Many pathogenic bacteria, e.g., *Haemophilus influenzae* and *Streptococcus pneumoniae* are capsulated, making them invisible to neutrophils. Vaccine-induced antibodies against capsular polysaccharides can solve this problem by efficient opsonization. But when the bacterial polysaccharides are purified and injected, especially in children, inadequate amounts of antibody are produced. Why? T cell help is missing! T cell help is contingent on the B cell presenting an antigen-derived peptide on MHC-II, and in the case of pure polysaccharide antigens, the B cell has nothing to present.

The solution is a trick: take the polysaccharide and couple some polypeptide to it. This protein doesn't even have to be derived from *H. influenzae*. In contrast, it may be even better to use a protein that for sure has already generated TH2 cells in the children, for example tetanus toxoid. So, the "conjugated" vaccine against *H. influenzae* consists of a polysaccharide from *H. influenzae* coupled to the toxoid from *Clostridium tetani*. The B cell with a B cell receptor binding the polysaccharide will internalize the whole conjugate into its endosomal pathway and present peptides from the toxoid on MHC II. Lots of TH2 cells recognizing these peptides are around, since the child has been repeatedly immunized with tetanus toxoid as part of the polyvalent vaccination against diphtheria, pertussis, tetanus etc. So, the anti-tetanus TH2 cell gives help to the anti-*Haemophilus* B cell. Strange?- It works! Still, a vaccine molecule optimized along these lies by itself frequently fails to elicit a satisfactory immune response. Another aspect to consider in immunizations is the requirement for costimulation. Dendritic cells activate naive T cells only when expressing B7. This, in turn, depends on dendritic cell activation via pattern recognition receptors (PRRs) by pathogen-associated molecular patterns (PAMPs). Old immunization protocols used additives, termed adjuvants, that included bacterial PAMPs. The classical adjuvant for immunization of laboratory animals was *complete Freund's adjuvant*, consisting of inactivated *Mycobacterium tuberculosis* in mineral oil. While frequently causing side effects, the old human vaccines containing inactivated bacteria included their own adjuvants, as many of their components activated PRRs. Modern human vaccines contain adjuvants that are less prone to cause side effects. The most frequent is "alum", particles of aluminum salts, which directly activate NOD-like receptors and the inflammasome, thereby inducing expression of costimulatory molecules.

How about tuberculosis, the bacterium hiding in macrophages, where antibodies are of little use? The French vaccine pioneers A. Calmette and C. Guérin grew*Mycobacterium bovis* for years on glycerin-bile-potato media (BCG-Bacille Calmette Guérin). This treatment finally attenuated the mycobacterium to an extent that it could be used as an anti-bacterial live vaccine. It promotes an early and more intense immune response against *M. tuberculosis* by TH1 memory cells. However, a protective effect is found only in 60-80% of those immunized, is only relative, and is of limited duration. In addition, it prevents easy diagnosis of tuberculosis by the Mendel-Mantoux skin test, as the immunized test positive. Therefore, most countries have discontinued BCG immunization. An idea that sounds brilliant but hasn't yet resulted in a working routine protocol is DNA immunization. Plasmids encoding antigens could be injected right into the muscle, which then expresses the encoded proteins. It would even be possible to do that without a needle, shooting the DNA right into the muscle with a "gene gun", a modified air gun. Yet, in spite of positive proof of concept, actual immune responses so far have been too weak to switch to this method.

STRUCTURE AND FUNCTION OF IMMUNOLOGY

Immunology is a science that examines the structure and function of the immune system. It originates from medicine and early studies on the causes of immunity to disease. The earliest known reference to immunity was during the plague of Athens in 430 BC. Thucydides noted that people who had recovered from a previous bout of the disease could nurse the sick without contracting the illness a second time. In the 18th century, Pierre-Louis Moreau de Maupertuis made experiments with scorpion venom and observed that certain dogs and mice were immune to this venom. This and other observations of acquired immunity were later exploited by Louis Pasteur in his development of vaccination and his proposed germ theory of disease.

Pasteur's theory was in direct opposition to contemporary theories of disease, such as the miasma theory. It was not until Robert Koch's 1891 proofs, for which he was awarded aNobel Prize in 1905, that microorganisms were confirmed as the cause of infectious disease. Viruses were confirmed as human pathogens in 1901, with the discovery of the yellow fever virus by Walter Reed.

Immunology made a great advance towards the end of the 19th century, through rapid developments, in the study of humoral immunity and cellular immunity. Particularly important was the work of Paul Ehrlich, who proposed the side-chain theory to explain the specificity of the antigen-antibody reaction; his contributions to the understanding of humoral immunity were recognized by the award of a Nobel Prize in 1908, which was jointly awarded to the founder of cellular immunology, Elie Metchnikoff.

LAYERED DEFENCE

The immune system protects organisms from infection with layered defenses of increasing specificity. In simple terms, physical barriers prevent pathogens such as bacteria and viruses from entering the organism. If a pathogen breaches these barriers, the innate immune system provides an immediate, but non-specific response. Innate immune systems are found in all plants and animals. If pathogens successfully evade the innate response, vertebrates possess a second layer of protection, the adaptive immune system, which is activated by the innate response.

Here, the immune system adapts its response during an infection to improve its recognition of the pathogen. This improved response is then retained after the pathogen has been eliminated, in the form of an immunological memory, and allows the adaptive immune system to mount faster and stronger attacks each time this pathogen is encountered. Both innate and adaptive immunity depend on the ability of the immune system to distinguish between self and non-self molecules. In immunology, *self* molecules are those components of an organism's body that can be distinguished from foreign substances by the immune system. Conversely, *non-self* molecules are those recognized as foreign molecules. One class of non-self molecules are called antigens (short for *anti*body *gen*erators) and are defined as substances that bind to specific immune receptors and elicit an immune response.

Surface Barriers

Several barriers protect organisms from infection, including mechanical, chemical, and biological barriers. The waxy cuticle of many leaves, the exoskeleton of insects, the shells and membranes of externally deposited eggs, and skin are examples of mechanical barriers that are the first line of defence against infection. However, as organisms cannot be completely sealed against their environments, other systems act to protect body openings such as the lungs, intestines, and the genitourinary tract. In the lungs, coughing and sneezing mechanically eject pathogens and otherirritants from the respiratory tract. The flushing action of tears and urine also mechanically expels pathogens, while mucus secreted by the respiratory and gastrointestinal tract serves to trap and entangle microorganisms.

Chemical barriers also protect against infection. The skin and respiratory tract secrete antimicrobial peptides such as the β-defensins. Enzymes such as lysozyme and phospholipase A2 insaliva, tears, and breast milk are also antibacterials. Vaginal secretions serve as a chemical barrier following menarche, when they become slightly acidic, while semen contains defensins and zinc to kill pathogens. In the stomach, gastric acid and proteases serve as powerful chemical defenses against ingested pathogens.

Within the genitourinary and gastrointestinal tracts, commensal flora serve as biological barriers by competing with pathogenic bacteria for food and space and, in some cases, by changing the conditions in their environment, such as pH or available iron. This reduces the probability that pathogens will reach sufficient numbers to cause illness. However, since most antibiotics non-specifically target bacteria and do not affect fungi, oral antibiotics can lead to an "overgrowth" of fungi and cause conditions such as a vaginal candidiasis (a yeast infection). There is good evidence that re-introduction of probiotic flora, such as pure cultures of the lactobacilli normally found in unpasteurized yogurt, helps restore a healthy balance of microbial populations in intestinal infections in children and encouraging preliminary data in studies on bacterial gastroenteritis, inflammatory bowel diseases, urinary tract infection and post-surgical infections.

IMMUNOLOGICAL MEMORY

When B cells and T cells are activated and begin to replicate, some of their offspring become long-lived memory cells. Throughout the lifetime of an animal, these memory cells remember each specific pathogen encountered and can mount a strong response if the pathogen is detected again. This is "adaptive" because it occurs during the lifetime of an individual as an adaptation to infection with that pathogen and prepares the immune system for future challenges. Immunological memory can be in the form of either passive short-term memory or active long-term memory.

Passive Memory

Newborn infants have no prior exposure to microbes and are particularly vulnerable to infection. Several layers of passive protection are provided by the mother. During pregnancy, a particular type of antibody, called IgG, is transported from mother to baby directly across the placenta, so human babies have high levels of antibodies even at birth, with the same range of antigen specificities as their mother. Breast milk or colostrum also contains antibodies that are transferred to the gut of the infant and protect against bacterial infections until the newborn can synthesize its own antibodies. This is passive immunity because the fetus does not actually make any memory cells or antibodies—it only borrows them. This passive immunity is usually short-term, lasting from a few days up to several months. In medicine, protective passive immunity can also be transferred artificially from one individual to another via antibody-rich serum.

Active Memory and Immunization

Long-term *active* memory is acquired following infection by activation of B and T cells. Active immunity can also be generated artificially, through vaccination. The principle behind vaccination (also called immunization) is to introduce an antigen from a pathogen in order to stimulate the immune system and develop specific immunity against that particular pathogen without causing disease associated with that organism. This deliberate induction of an immune response is successful because it exploits the natural specificity of the immune system, as well as its inducibility. With infectious disease remaining one of the leading causes of death in the human population, vaccination represents the most effective manipulation of the immune system mankind has developed.

Most viral vaccines are based on live attenuated viruses, while many bacterial vaccines are based on acellular components of micro-organisms, including harmless toxin components. Since many antigens derived from acellular vaccines do not strongly induce the adaptive response, most bacterial vaccines are provided with additional adjuvants that activate the antigen-presenting cells of the innate immune system and maximize immunogenicity.

Disorders of Human Immunity

The immune system is a remarkably effective structure that incorporates specificity, inducibility and adaptation. Failures of host defence do occur, however, and fall into three broad categories: immunodeficiencies, autoimmunity, and hypersensitivities.

Immunodeficiencies

Immunodeficiencies occur when one or more of the components of the immune system are inactive. The ability of the immune system to respond to pathogens is diminished in both the young and the elderly, with immune responses beginning to decline at around 50 years of age due to immunosenescence. In developed countries, obesity, alcoholism, and drug use are common causes of poor immune function. However, malnutrition is the most common cause of immunodeficiency in developing countries. Diets lacking sufficient protein are associated with impaired cell-mediated immunity, complement activity, phagocyte function, IgA antibody concentrations, and cytokine production. Additionally, the loss of the thymus at an early age through genetic mutation or surgical removal results in severe immunodeficiency and a high susceptibility to infection.

Immunodeficiencies can also be inherited or 'acquired'. Chronic granulomatous disease, where phagocytes have a reduced ability to destroy pathogens, is an example of an inherited, orcongenital, immunodeficiency. AIDS and some types of cancer cause acquired immunodeficiency.

Autoimmunity

Overactive immune responses comprise the other end of immune dysfunction, particularly the autoimmune disorders. Here, the immune system fails to properly distinguish between self and non-self, and attacks part of the body. Under normal circumstances, many T cells and antibodies react with "self" peptides. One of the functions of specialized cells (located in the thymus andbone marrow) is to present young lymphocytes with self antigens produced throughout the body and to eliminate those cells that recognize self-antigens, preventing autoimmunity.

Hypersensitivity

Hypersensitivity is an immune response that damages the body's own tissues. They are divided into four classes (Type I – IV) based on the mechanisms involved and the time course of the hypersensitive reaction. Type I hypersensitivity is an immediate or anaphylactic reaction, often associated with allergy. Symptoms can range from mild discomfort to death. Type I hypersensitivity is mediated by IgE, which triggers degranulation of mast cells and basophils when cross-linked by antigen. Type II hypersensitivity occurs when antibodies bind to antigens on the patient's own cells, marking them for destruction. This is also called antibody-dependent (or cytotoxic) hypersensitivity, and is mediated by IgG and IgM antibodies. Immune complexes (aggregations of antigens, complement proteins, and IgG and IgM antibodies) deposited in various tissues trigger Type III hypersensitivity reactions. Type IV hypersensitivity (also known as cell-mediated or*delayed type hypersensitivity*) usually takes between two and three days to develop. Type IV reactions are involved in many autoimmune and infectious diseases, but may also involve *contact dermatitis* (poison ivy). These reactions are mediated by T cells, monocytes, and macrophages.

Other Mechanisms

It is likely that a multicomponent, adaptive immune system arose with the first vertebrates, as invertebrates do not generate lymphocytes or an antibody-based humoral response. Many species, however, utilize mechanisms that appear to be precursors of these aspects of vertebrate immunity. Immune systems appear even in the structurally most simple forms of life, with bacteria using a unique defence mechanism, called the restriction modification system to protect themselves from viral pathogens, called bacteriophages. Prokaryotes also possess acquired immunity, through a system that uses CRISPR sequences to retain fragments of the genomes of phage that they have come into contact with in the past, which allows them to block virus replication through a form of RNA interference.

Pattern recognition receptors are proteins used by nearly all organisms to identify molecules associated with pathogens. Antimicrobial peptides called defensins are an evolutionarily conserved component of the innate immune response found in all animals and plants, and represent the main form of invertebrate systemic immunity. The complement system and phagocytic cells are also used by most forms of invertebrate life. Ribonucleases and the RNA interference pathway are conserved across all eukaryotes, and are thought to play a role in the immune response to viruses.

Unlike animals, plants lack phagocytic cells, but many plant immune responses involve systemic chemical signals that are sent through a plant. Individual plant cells respond to molecules associated with pathogens known as Pathogen-associated molecular patterns or PAMPs. When a part of a plant becomes infected, the plant produces a localized hypersensitive response, whereby cells at the site of infection undergo rapid apoptosis to prevent the spread of the disease to other parts of the plant. Systemic acquired resistance (SAR) is a type of defensive response used by plants that renders the entire plant resistant to a particular infectious agent. RNA silencing mechanisms are particularly important in this systemic response as they can block virus replication.

TUMOR IMMUNOLOGY

Another important role of the immune system is to identify and eliminate tumors. The *transformed cells* of tumors express antigens that are not found on normal cells. To the immune system, these antigens appear foreign, and their presence causes immune cells to attack the transformed tumor cells. The antigens expressed by tumors have several sources; some are derived from oncogenic viruses like human papillomavirus, which causes cervical cancer, while others are the organism's own proteins that occur at low levels in normal cells but reach high levels in tumor cells. One example is an enzyme called tyrosinase that, when expressed at high levels, transforms certain skin cells (e.g. melanocytes) into tumors called melanomas. A third possible source of tumor antigens are proteins normally important for regulating cell growth and survival, that commonly mutate into cancer inducing molecules called oncogenes. The main response of the immune system to tumors is to destroy the abnormal cells using killer T cells, sometimes with the assistance of helper T cells. Tumor antigens are presented on MHC class I molecules in a similar way to viral antigens. This allows killer T cells to recognize the tumor cell as abnormal. NK cells also kill tumorous cells in a similar way, especially if the tumor cells have fewer MHC class I molecules on their surface than normal; this is a common phenomenon with tumors. Sometimes antibodies are generated against tumor cells allowing for their destruction by the complement system.

Clearly, some tumors evade the immune system and go on to become cancers. Tumor cells often have a reduced number of MHC class I molecules on their surface, thus avoiding detection by killer T cells. Some tumor cells also release products that inhibit the immune response; for example by secreting the cytokine TGF-β, which suppresses the activity of macrophages and lymphocytes. In addition, immunological tolerance may develop against tumor antigens, so the immune system no longer attacks the tumor cells. Paradoxically, macrophages can promote tumor growth when tumor cells send out cytokines that attract macrophages, which then generate cytokines and growth factors that nurture tumor development. In addition, a combination of hypoxia in the tumor and a cytokine produced by macrophages induces tumor cells to decrease production of a protein that blocks metastasis and thereby assists spread of cancer cells.

Physiological Regulation

Hormones can act as immunomodulators, altering the sensitivity of the immune system. For example, female sex hormones are known immunostimulators of both adaptive and innate immune responses. Some autoimmune diseases such as lupus erythematosus strike women preferentially, and their onset often coincides with puberty. By contrast, male sex hormones such astestosterone seem to be immunosuppressive. Other hormones appear to regulate the immune system as well, most notably prolactin, growth hormone and vitamin D.

When a T-cell encounters a foreign pathogen, it extends a vitamin D receptor. This is essentially a signaling device that allows the T-cell to bind to the active form of vitamin D, the steroid hormonecalcitriol. T-cells have a symbiotic relationship with vitamin D. Not only does the T-cell extend a vitamin D receptor, in essence asking to bind to the steroid hormone version of vitamin D, calcitriol, but the T-cell expresses the gene CYP27B1, which is the gene responsible for converting the pre-hormone version of vitamin D, calcidiol into the steroid hormone version, calcitriol. Only after binding to calcitriol can T-cells perform their intended function. Other immune system cells that are known to express CYP27B1 and thus activate vitamin D calcidiol, are dendritic cells,keratinocytes and macrophages. It is conjectured that a progressive decline in hormone levels with age is partially responsible for weakened immune responses in aging individuals. Conversely, some hormones are regulated by the immune system, notably thyroid hormone activity. The age-related decline in immune function is also related to dropping vitamin D levels in the elderly. As people age, two things happen that negatively affect their vitamin D levels. First, they stay indoors more due to decreased activity levels. This means that they get less sun and therefore produce less cholecalciferol viaUVB radiation. Second, as a person ages the skin becomes less adept at producing vitamin D.

The immune system is affected by sleep and rest, and sleep deprivation is detrimental to immune function. Complex feedback loops involving cytokines, such as interleukin-1 and tumor necrosis factor-α produced in response to infection, appear to also play a role in the regulation of non-rapid eye movement (REM) sleep. Thus the immune response to infection may result in changes to the sleep cycle, including an increase in slow-wave sleep relative to REM sleep.

Nutrition and Diet

Overnutrition is associated with diseases such as diabetes and obesity, which are known to affect immune function. More moderate malnutrition, as well as certain specific trace mineral and nutrient deficiencies, can also compromise the immune response. Foods rich in certain fatty acids may foster a healthy immune system. Likewise, fetal undernourishment can cause a lifelong impairment of the immune system.

Manipulation In medicine

The immune response can be manipulated to suppress unwanted responses resulting from autoimmunity, allergy, and transplant rejection, and to stimulate protective responses against pathogens that largely elude the immune system. Immunosuppressive drugs are used to control autoimmune disorders or inflammation when excessive tissue damage occurs, and to prevent transplant rejection after an organ transplant. Anti-inflammatory drugs are often used to control the effects of inflammation. Glucocorticoids are the most powerful of these drugs; however, these drugs can have many undesirable side effects, such as central obesity, hyperglycemia, osteoporosis, and their use must be tightly controlled. Lower doses of anti-inflammatory drugs are often used in conjunction with cytotoxic or immunosuppressive drugs such as methotrexate or azathioprine. Cytotoxic drugsinhibit the immune response by killing dividing cells such as activated T cells. However, the killing is indiscriminate and other constantly dividing cells and their organs are affected, which causes toxic side effects. Immunosuppressive drugs such as ciclosporin prevent T cells from responding to signals correctly by inhibiting signal transduction pathways.

Larger drugs (>500 Da) can provoke a neutralizing immune response, particularly if the drugs are administered repeatedly, or in larger doses. This limits the effectiveness of drugs based on larger peptides and proteins (which are typically larger than 6000 Da). In some cases, the drug itself is not immunogenic, but may be co-administered with an immunogenic compound, as is sometimes the case for Taxol. Computational methods have been developed to predict the immunogenicity of peptides and proteins, which are particularly useful in designing therapeutic antibodies, assessing likely virulence of mutations in viral coat particles, and validation of proposed peptide-based drug treatments. Early techniques relied mainly on the observation that hydrophilic amino acids are overrepresented in epitope regions than hydrophobic amino acids;however, more recent developments rely on machine learning techniques using databases of existing known epitopes, usually on well-studied virus proteins, as a training set. A publicly accessible database has been established for the cataloguing of epitopes from pathogens known to be recognizable by B cells. The emerging field of bioinformatics-based studies of immunogenicity is referred to as *immunoinformatics*. Immunoproteomics is the study of large sets of proteins (proteomics) involved in the immune response.

ADAPTIVE IMMUNE SYSTEM

The adaptive immune system evolved in early vertebrates and allows for a stronger immune response as well as immunological memory, where each pathogen is "remembered" by a signature antigen. The adaptive immune response is antigen-specific and requires the recognition of specific "non-self" antigens during a process called antigen presentation. Antigen specificity allows for the generation of responses that are tailored to specific pathogens or pathogen-infected cells. The ability to mount these tailored responses is maintained in the body by "memory cells". Should a pathogen infect the body more than once, these specific memory cells are used to quickly eliminate it.

Lymphocytes

The cells of the adaptive immune system are special types of leukocytes, called lymphocytes. B cells and T cells are the major types of lymphocytes and are derived from hematopoietic stem cells in the bone marrow. B cells are involved in the humoral immune response, whereas T cells are involved in cell-mediated immune response. Both B cells and T cells carry receptor molecules that recognize specific targets. T cells recognize a "non-self" target, such as a pathogen, only after antigens (small fragments of the pathogen) have been processed and presented in combination with a "self" receptor called a major histocompatibility complex (MHC) molecule. There are two major subtypes of T cells: the killer T cell and the helper T cell. Killer T cells only recognize antigens coupled to Class I MHC molecules, while helper T cells only recognize antigens coupled to Class II MHC molecules. These two mechanisms of antigen presentation reflect the different roles of the two types of T cell. A third, minor subtype are the $\gamma\delta$ T cells that recognize intact antigens that are not bound to MHC receptors. In contrast, the B cell antigen-specific receptor is an antibody molecule on the B cell surface, and recognizes whole pathogens without any need for antigen processing. Each lineage of B cell expresses a different antibody, so the complete set of B cell antigen receptors represent all the antibodies that the body can manufacture.

Killer T Cells

Killer T cells are a sub-group of T cells that kill cells that are infected with viruses (and other pathogens), or are otherwise damaged or dysfunctional. As with B cells, each type of T cell recognizes a different antigen. Killer T cells are activated when their T cell receptor (TCR) binds to this specific antigen in a complex with the MHC Class I receptor of another cell. Recognition of this MHC:antigen complex is aided by a co-receptor on the T cell, called CD8. The T cell then travels throughout the body in search of cells where the MHC I receptors bear this antigen. When an activated T cell contacts such cells, it releases cytotoxins, such as perforin, which form pores in the target cell's plasma membrane, allowing ions, water and toxins to enter. The entry of another toxin called granulysin (a protease) induces the target cell to undergo apoptosis. T cell killing of host cells is particularly important in preventing the replication of viruses. T cell activation is tightly controlled and generally requires a very strong MHC/antigen activation signal, or additional activation signals provided by "helper" T cells.

Helper T cells

Helper T cells regulate both the innate and adaptive immune responses and help determine which immune responses the body makes to a particular pathogen. These cells have no cytotoxic activity and do not kill infected cells or clear pathogens directly. They instead control the immune response by directing other cells to perform these tasks. Helper T cells express T cell receptors (TCR) that recognize antigen bound to Class II MHC molecules. The MHC:antigen complex is also recognized by the helper cell's CD4 co-receptor, which recruits molecules inside the T cell (e.g., Lck) that are responsible for the T cell's activation.

Helper T cells have a weaker association with the MHC:antigen complex than observed for killer T cells, meaning many receptors (around 200–300) on the helper T cell must be bound by an MHC:antigen in order to activate the helper cell, while killer T cells can be activated by engagement of a single MHC:antigen molecule. Helper T cell activation also requires longer duration of engagement with an antigen-presenting cell. The activation of a resting helper T cell causes it to release cytokines that influence the activity of many cell types. Cytokine signals produced by helper T cells enhance the microbicidal function of macrophages and the activity of killer T cells. In addition, helper T cell activation causes an upregulation of molecules expressed on the T cell's surface, such as CD40 ligand (also called CD154), which provide extra stimulatory signals typically required to activate antibody-producing B cells.

B lymphocytes and antibodies

A B cell identifies pathogens when antibodies on its surface bind to a specific foreign antigen. This antigen/antibody complex is taken up by the B cell and processed by proteolysis into peptides. The B cell then displays these antigenic peptides on its surface MHC class II molecules. This combination of MHC and antigen attracts a matching helper T cell, which releases lymphokines and activates the B cell. As the activated B cell then begins to divide, its offspring (plasma cells) secrete millions of copies of the antibody that recognizes this antigen. These antibodies circulate in blood plasma and lymph, bind to pathogens expressing the antigen and mark them for destruction by complement activation or for uptake and destruction by phagocytes. Antibodies can also neutralize challenges directly, by binding to bacterial toxins or by interfering with the receptors that viruses and bacteria use to infect cells.

HUMORAL AND CHEMICAL BARRIERS

Complement System

The complement system is a biochemical cascade that attacks the surfaces of foreign cells. It contains over 20 different proteins and is named for its ability to "complement" the killing of pathogens by antibodies. Complement is the major humoral component of the innate immune response. Many species have complement systems, including non-mammals like plants, fish, and some invertebrates. In humans, this response is activated by complement binding to antibodies that have attached to these microbes or the binding of complement proteins to carbohydrates on the surfaces of microbes. This recognition signal triggers a rapid killing response. The speed of the response is a result of signal amplification that occurs following sequential proteolytic activation of complement molecules, which are also proteases. After complement proteins initially bind to the microbe, they activate their protease activity, which in turn activates other complement proteases, and so on. This produces a catalytic cascade that amplifies the initial signal by controlled positive feedback. The cascade results in the production of peptides that attract immune cells, increase vascular permeability, and opsonize (coat) the surface of a pathogen, marking it for destruction. This deposition of complement can also kill cells directly by disrupting their plasma membrane.

Inflammation

Inflammation is one of the first responses of the immune system to infection. The symptoms of inflammation are redness, swelling, heat, and pain, which are caused by increased blood flow into tissue. Inflammation is produced by eicosanoids and cytokines, which are released by injured or infected cells. Eicosanoids include prostaglandins that produce fever and the dilation of blood vessels associated with inflammation, and leukotrienes that attract certain white blood cells (leukocytes). Common cytokines include interleukins that are responsible for communication between white blood cells; chemokines that promote chemotaxis; and interferons that have anti-viral effects, such as shutting down protein synthesis in the host cell. Growth factors and cytotoxic factors may also be released. These cytokines and other chemicals recruit immune cells to the site of infection and promote healing of any damaged tissue following the removal of pathogens.

Cellular barriers

Leukocytes (white blood cells) act like independent, single-celled organisms and are the second arm of the innate immune system. The innate leukocytes include the phagocytes (macrophages, neutrophils, and dendritic cells), mast cells, eosinophils, basophils, and natural killer cells. These cells identify and eliminate pathogens, either by attacking larger pathogens through contact or by engulfing and then killing microorganisms. Innate cells are also important mediators in the activation of the adaptive immune system.

Phagocytosis is an important feature of cellular innate immunity performed by cells called 'phagocytes' that engulf, or eat, pathogens or particles. Phagocytes generally patrol the body searching for pathogens, but can be called to specific locations by cytokines. Once a pathogen has been engulfed by a phagocyte, it becomes trapped in an intracellular vesicle called a phagosome, which subsequently fuses with another vesicle called alysosome to form a phagolysosome. The pathogen is killed by the activity of digestive enzymes or following a respiratory burst that releases free radicals into the phagolysosome. Phagocytosis evolved as a means of acquiring nutrients, but this role was extended in phagocytes to include engulfment of pathogens as a defence mechanism. Phagocytosis probably represents the oldest form of host defence, as phagocytes have been identified in both vertebrate and invertebrate animals.

Neutrophils and macrophages are phagocytes that travel throughout the body in pursuit of invading pathogens. Neutrophils are normally found in the blood stream and are the most abundant type of phagocyte, normally representing 50% to 60% of the total circulating leukocytes. During the acute phase of inflammation, particularly as a result of bacterial infection, neutrophils migrate toward the site of inflammation in a process called chemotaxis, and are usually the first cells to arrive at the scene of infection. Macrophages are versatile cells that reside within tissues and produce a wide array of chemicals including enzymes, complement proteins, and regulatory factors such as interleukin 1. Macrophages also act as scavengers, ridding the body of worn-out cells and other debris, and as antigen-presenting cells that activate the adaptive immune system. Dendritic cells (DC) are phagocytes in tissues that are in contact with the external environment; therefore, they are located mainly in the skin, nose, lungs, stomach, and intestines. They are named for their resemblance to neuronal dendrites, as both have many spine-like projections, but dendritic cells are in no way connected to the nervous system.

Dendritic cells serve as a link between the bodily tissues and the innate and adaptive immune systems, as they present antigen to T cells, one of the key cell types of the adaptive immune system. Mast cells reside in connective tissues and mucous membranes, and regulate the inflammatory response. They are most often associated with allergy and anaphylaxis. Basophils and eosinophils are related to neutrophils. They secrete chemical mediators that are involved in defending against parasites and play a role in allergic reactions, such as asthma. Natural killer (NK cells) cells are leukocytes that attack and destroy tumor cells, or cells that have been infected by viruses.

HUMORAL IMMUNITY

β -CELL ACTIVATION

Native B cells are fully differentiated B cells that have not yet encountered a complementary antigen; they bear membrane–bound IgM and IgD on their outer surface. Clonal selection begins when native B cells bind to their complementary antibodies: Then they multiply; produce soluble IgM and IgD, and undergo a high rate of mutation in their DNA-encoding region, resulting in affinity maturation.

Later, class switching occurs: Various activated B cells begin to make all classes of antibodies. Some members of the clone become short-lived effectors (plasma) cells, which make large quantities of antibody; others become long-lived memory cells, which fight future infections by the same pathogen.

Lymphocytes, including interleukins-1,-2,-4,-5, and -6, produced by NK cells help B-cell activation stimulated by most antigens. Some T-independent antigens stimulate B-cell activation without help. B cells are selected to respond to specific antigens in accordance with the particular antibody present on the B cell membrane prior to encountering an antigen.

When a B cell detects an antigen with which it can react, it binds to the antigen (sensitized or activates) and divides many times to produce a clone of many plasma cells and some memory cells. Many B cells require helper T cells to proliferate and differentiate into both plasma and memory B cells. Plasma cells synthesize and release large numbers of antibodies. Memory cells remain in lymphoid tissues ready to respond to subsequent exposure to the same antigen.

Properties of Antibodies (Immunoglobulins)

Antibodies are glycoproteins because carbohydrate groups are attached to them.Structually, antibodies consist of two heavy and two light polypeptide chains, joined by disulfide bonds to form a Y shape. The upper ends of the Y, consists of variable region in both the light and heavy chains, differ from antibody to antibody.

These variable regions form the two antigen-binding sites (part of the Fab fragment), which are responsible for the specificity of the antibody. The remaining part of the molecule consists of constant regions (Fc) that are similar in all antibodies of a particular class. The constant region binds to and activates phagocytes and complement.

Class of Immunoglobins

Five classes of immunoglobulins have been identified. Each class has a particular kind of constant region which gives that class its distinguishing properties. IgD, IgE, and IgG are secreted as monomers, IgM as pentamers, and IgA as monomers or dimmers.

IgG

The main class of antibodies found in blood, accounts 20 percent of all plasma proteins. IgG is produced in large quantities during a secondary response. The antigen-binding sites of IgG attach to antigens on microorganisms and their tissue-binding sites attach to receptors on phagocytic cells. IgG is the only immunoglobulin that can cross the placenta from mother to fetus and provide antibody protection.

IgG – monomer, predominant type, specific immune response, transmammary, transplacental.

IgA

Occurs in small amounts in blood and in larger amounts in body secretions such as tears, milk, saliva, mucus and linings of GI tracts. Secretory IgA, which consists of two monomer units held together by a J chain, has an attached secretory component, which protects the IgA from proteolytic enzymes and facilitates its transport.

Mucosal surfaces such as in the respiratory , GI tracts are major sites for invasion by pathogens. The main function of IgA is to bind antigens on microorganisms before they invade. It also activates complement, which helps to kill the microorganisms.

IgA – monomer (plasma), dimer (mucus/saliva/tears/milk/intestinal secretions), inhibit microbe attachment to epithelia

IgM

IgM is found as a monomer on the surface of B cells and is secreted as a pentamer by plasma cells. IgM consists of five units connected by their tails to a J chain and has 10 peripheral antigen-binding sites. As IgM binds to antigens, it also activates complement and causes microorganisms to clump together. High levels of IgM indicate recent infection or expose to antigen. IgM – monomers (antigen receptor) penerates (plasma), nonspecific immune response, blood typing

IgE

IgE has a special affinity for receptors of basophils in blood or mast cells in tissue. It binds to these cells by tissue binding sites, leaving antigen-binding sites free to bind antigens which can develop allergy. When IgE binds antigens, the basophils and mast cells secrete histamine which produce allergy symptoms. IgE are elevated in patients with allergies and worm parasites. IgE – monomer, tonsils, skin, mucus membranes. Stimulates basophils to release histamine

IgD

IgD, monomer, is found mainly on B-cell membranes and function is unknown.Only trace amount.

Primary and Secondary Responses

Primary responses are the immune system's first encounter with foreign antigens. First produce IgM, then IgG. Memory cells remain in lymphoid tissue, ready to respond to subsequent exposure to the same antigen. The primary response of B cells can occur by two mechanisms.

(a). T-independent antigens - B cells can be activated by binding antigen, proliferating and forming plasma cells.

(b). T-dependent antigens – B cells becomes an antigen presenting cell, and activate the T helper cell (T_H). The activated T_H cells then secretes lymphokines that further activate the B cell causing it to differentiate and proliferate producing B memory cells and T_H memory cells.

Secondary responses bring fast and efficient destruction of antigens recognized by B and T memory cells. IgM is produced in a smaller quantities and IgG is produced in much larger quantities than in primary response.

Kinds of Antigen-Antibody Reactions

Humoral immunity depends upon the production of antibodies by B cells. Antibodies protect against invading microbes by neutralization, opsonization, and activating complement. Humoral immunity is most effective against bacteria, antigen-antibody reaction result in agglutination (clumping).

(1). Neutralization of pathogens and toxins by IgA or IgG.

(2). Opsonization of bacteria by IgG, which is activated by antigen-bound IgM or IgG, produces complement C_3b ,a powerful opsonin, they coat microbes so that they can be phagocytized.

(3). Cell lysed by complement after or directly by IgMs or IgG immune complexes.

THE IMMUNE SYSTEM

We are constantly being exposed to infectious agents and yet, in most cases, we are able to resist these infections. It is our immune system that enables us to resist infections. The immune system is composed of two major subdivisions, the innate or non-specific immune system and the adaptive or specific immune system. The innate immune system is our first line of defence against invading organisms while the adaptive immune system acts as a second line of defence and also affords protection against re-exposure to the same pathogen. Each of the major subdivisions of the immune system has both cellular and humoral components by which they carry out their protective function. In addition, the innate immune system also has anatomical features that function as barriers to infection. Although these two arms of the immune system have distinct functions, there is interplay between these systems (i.e., components of the innate immune system influence the adaptive immune system and vice versa).

Although the innate and adaptive immune systems both function to protect against invading organisms, they differ in a number of ways. The adaptive immune system requires some time to react to an invading organism, whereas the innate immune system includes defenses that, for the most part, are constitutively present and ready to be mobilized upon infection. Second, the adaptive immune system is antigen specific and reacts only with the organism that induced the response. In contrast, the innate system is not antigen specific and reacts equally well to a variety of organisms. Finally, the adaptive immune system demonstrates immunological memory. It "remembers" that it has encountered an invading organism and reacts more rapidly on subsequent exposure to the same organism. In contrast, the innate immune system does not demonstrate immunological memory.

All cells of the immune system have their origin in the bone marrow and they include myeloid (neutrophils, basophils, eosinpophils, macrophages and dendritic cells) and lymphoid (B lymphocyte, T lymphocyte and Natural Killer) cells, which differentiate along distinct pathways. The myeloid progenitor (stem) cell in the bone marrow gives rise to erythrocytes, platelets, neutrophils, monocytes/macrophages and dendritic cells whereas the lymphoid progenitor (stem) cell gives rise to the NK, T cells and B cells. For T cell development the precursor T cells must migrate to the thymus where they undergo differentiation into two distinct types of T cells, the CD4+ T helper cell and the CD8+ pre-cytotoxic T cell. Two types of T helper cells are produced in the thymus the TH1 cells, which help the CD8+ pre-cytotoxic cells to differentiate into cytotoxic T cells, and TH2 cells, which help B cells, differentiate into plasma cells, which secrete antibodies. The main function of the immune system is self/non-self discrimination. This ability to distinguish between self and non-self is necessary to protect the organism from invading pathogens and to eliminate modified or altered cells (e.g. malignant cells). Since pathogens may replicate intracellularly (viruses and some bacteria and parasites) or extracellularly (most bacteria, fungi and parasites), different components of the immune system have evolved to protect against these different types of pathogens.

It is important to remember that infection with an organism does not necessarily mean diseases, since the immune system in most cases will be able to eliminate the infection before disease occurs. Disease occurs only when the bolus of infection is high, when the virulence of the invading organism is great or when immunity is compromised. Although the immune system, for the most part, has beneficial effects, there can be detrimental effects as well. During inflammation, which is the response to an invading organism, there may be local discomfort and collateral damage to healthy tissue as a result of the toxic products produced by the immune response. In addition, in some cases the immune response can be directed toward self tissues resulting in autoimmune disease.

ANATOMICAL BARRIERS TO INFECTIONS

Mechanical Factors

The epithelial surfaces form a physical barrier that is very impermeable to most infectious agents. Thus, the skin acts as our first line of defence against invading organisms. The desquamation of skin epithelium also helps remove bacteria and other infectious agents that have adhered to the epithelial surfaces.

Movement due to cilia or peristalsis helps to keep air passages and the gastrointestinal tract free from microorganisms. The flushing action of tears and saliva helps prevent infection of the eyes and mouth. The trapping effect of mucus that lines the respiratory and gastrointestinal tract helps protect the lungs and digestive systems from infection.

Chemical factors

Fatty acids in sweat inhibit the growth of bacteria. Lysozyme and phospholipase found in tears, saliva and nasal secretions can breakdown the cell wall of bacteria and destabilize bacterial membranes. The low pH of sweat and gastric secretions prevents growth of bacteria. Defensins (low molecular weight proteins) found in the lung and gastrointestinal tract have antimicrobial activity. Surfactants in the lung act as opsonins (substances that promote phagocytosis of particles by phagocytic cells).

Biological Factors

The normal flora of the skin and in the gastrointestinal tract can prevent the colonization of pathogenic bacteria by secreting toxic substances or by competing with pathogenic bacteria for nutrients or attachment to cell surfaces.

Humoral Barriers to Infection

The anatomical barriers are very effective in preventing colonization of tissues by microorganisms. However, when there is damage to tissues the anatomical barriers are breached and infection may occur. Once infectious agents have penetrated tissues, another innate defence mechanism comes into play, namely acute inflammation. Humoral factors play an important role in inflammation, which is characterized by edema and the recruitment of phagocytic cells. These humoral factors are found in serum or they are formed at the site of infection.

1. Complement system – The complement system is the major humoral non-specific defence mechanism. Once activated complement can lead to increased vascular permeability, recruitment of phagocytic cells, and lysis and opsonization of bacteria.
2. Coagulation system – Depending on the severity of the tissue injury, the coagulation system may or may not be activated. Some products of the coagulation system can contribute to the non-specific defenses because of their ability to increase vascular permeability and act as chemotactic agents for phagocytic cells. In addition, some of the products of the coagulation system are directly antimicrobial. For example, beta-lysin, a protein produced by platelets during coagulation can lyse many Gram positive bacteria by acting as a cationic detergent.
3. Lactoferrin and transferrin – By binding iron, an essential nutrient for bacteria, these proteins limit bacterial growth.
4. Interferons – Interferons are proteins that can limit virus replication in cells.
5. Lysozyme – Lysozyme breaks down the cell wall of bacteria.
6. Interleukin-1 – Il-1 induces fever and the production of acute phase proteins, some of which are antimicrobial because they can opsonize bacteria.

Cellular Barriers to Infection

Part of the inflammatory response is the recruitment of polymorphonuclear eosinophiles and macrophages to sites of infection. These cells are the main line of defence in the non-specific immune system.

1. Neutrophils – Polymorphonuclear cells are recruited to the site of infection where they phagocytose invading organisms and kill them intracellularly. In addition, PMNs contribute to collateral tissue damage that occurs during inflammation.

2. Macrophages – Tissue macrophages and newly recruited monocytes, which differentiate into macrophages, also function in phagocytosis and intracellular killing of microorganisms. In addition, macrophages are capable of extracellular killing of infected or altered self target cells. Furthermore, macrophages contribute to tissue repair and act as antigen-presenting cells, which are required for the induction of specific immune responses.
3. Natural killer (NK) and lymphokine activated killer (LAK) cells – NK and LAK cells can nonspecifically kill virus infected and tumor cells. These cells are not part of the inflammatory response but they are important in nonspecific immunity to viral infections and tumor surveillance.
4. Eosinophils – Eosinophils have proteins in granules that are effective in killing certain parasites.

THE IMMUNE SYSTEM'S ROLE IN PROTECTION

Germs are everywhere: in air, soil, rocks, and water; in plants and animals; and, of course, in our own bodies. Some thrive in intense heat, while others require extreme subzero temperatures. Many have a remarkable ability to replicate themselves rapidly—reproducing every few hours in some cases—an evolutionary trick that makes them highly adaptable to changing environments and to the medicines we use to light them. By studying germs and how our body responds to them, scientists have unraveled many of the battle secrets—and vulnerabilities—of the immune defence system, revealing how it normally operates in its daily war on microbes and how its malfunction can contribute to disease.

ENEMY INVADERS

Scientists now know that most biological bad guys fall into four basic categories: viruses, fungi, bacteria, and parasites. Each type of predator has its own invade-and-infect strategy.

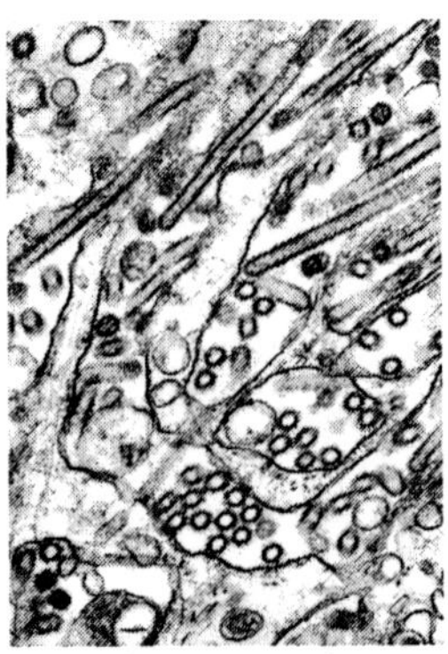

Fig. Bird flu, or avian influenza, refers to a group of related viral diseases that attack birds.

Viruses

Viruses are not whole cells but consist of one or more molecules of DNA or RNA. DNA, or deoxyribonucleic acid, carries a cell's genetic information and synthesizes RNA, or ribonucleic acid, which carries information from the nucleus to the body of the cell to assemble proteins. The viruses may be shaped like rods or spheres or may be multisided. In some ways, viruses are like biological pirates: they invade cells, hijack the internal machinery, and start reproducing, unleashing thousands of duplicate viruses (known as clones) that invade and take over other cells. Viruses also change their appearance, or mutate, often, and they can jump from one species to another.

Fungi

Take a walk in the woods or stroll along the aisles of a grocery store and you'll see one type of fungus—a mushroom. Fungi are actually a primitive vegetable and are found in air, soil, plants, and water. Thousands of types of fungi have been identified, including yeast, mold, and mildew, and there are surely many more types, about half of which cause disease in humans. Disease-causing fungi tend to infect moist areas of the body.

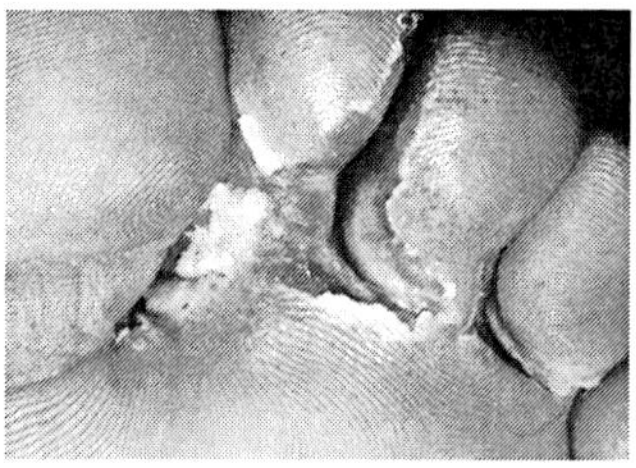

Fig. Athlete's foot attacks between the toes, eroding the skin.

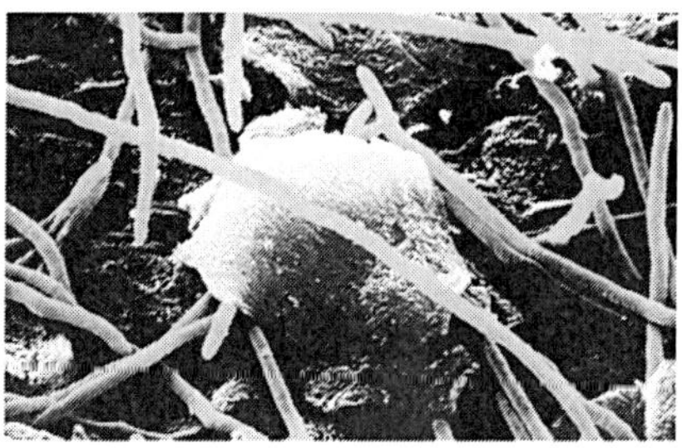

Fig. Pictured are the rounded, single-cell spores and threads characteristic of the athlete's foot fungus.

Athlete's foot, for example, is a common rash that results from fungi that often lurk in a shared shower (as at the gym) and then thrive and reproduce when your feet sweat. But not all fungi cause disease; some actually help fight it. Penicillin, the antibiotic that transformed the treatment of infectious diseases, was discovered when a laboratory scientist noticed that a petri dish with mold growing on it was free of bacteria.

Bacteria

Bacteria are single-celled organisms that may be shaped like balls, rods, or spirals. Larger and more complicated than viruses, bacteria don't need to hijack your cells to wreak havoc—they can do it all on their own. Fortunately, most bacteria are helpful. Some live in the stomach and intestine and help digest food. But about one in eight bacteria can make you sick. Harmful bacteria can cause problems in three ways: some invade and attack a specific part of the body; others produce chemical poisons that cause illness; still others multiply so much that they obstruct blood vessels or prevent the heart from functioning normally. Some examples of bacterial infections and the way in which they occur:

- Food poisoning, caused by eating contaminated food;
- Anthrax poisoning, for which authorities carefully watch, caused by inhaling or swallowing contaminated powder; and
- Lyme disease, which is transmitted by infected ticks and causes muscle and joint pain and sometimes damages the nervous system.

E. coli is a bacterium that exists naturally in the lower intestine of humans. It is also used as a research organism in labs. A toxin it produces can cause illness.

A computer model shows the structure of the anthrax bacterial toxin, which is a bioterrorism concern. Cutaneous anthrax (anthrax infection on the skin) causes painful blisters and swelling, while inhalation anthrax.

Parasites

Parasites comprise single-celled protozoa and multicellular animals, such as nematodes and helminths (worms), that require a moist environment to survive and usually cause disease in humans. A person can become infected after drinking contaminated water or eating infected food, or eating with unwashed hands. One example of a parasitic infection is malaria, which causes recurring bouts of fever and chills after someone is bitten by an infected mosquito. Although the immune system responds, parasites are capable of mounting follow-up attacks: they can go into hiding in the cells for a while and then, start replicating again when the immune defence has diminished.

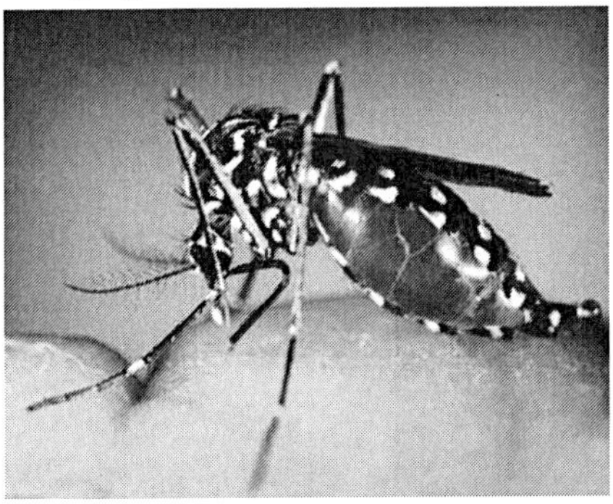

Fig. Mosquitoes can spread diseases, such as West Nile virus and malaria when they bite.

Fighting Back

Considering this daily barrage from all directions, it's a wonder that we aren't sick all the time. Fortunately, the human immune system has evolved to handle this onslaught, operating like a highly efficient killing machine to fend off germs wherever and whenever they appear. Although the immune system employs a variety of weapons and strategies based on the specific threat, in general your immune system mounts a three-step defensive process.

Sounding the Alarm

The first defenders on the scene of a germ attack are components of your innate immune system. The white blood cells that make up the innate immune system circulate throughout the body constantly, much like police on patrol, always on the lookout for biological suspects. These patrolling white blood cells belong to the phagocyte family, but they have many subtypes (just as a police force consists of detectives, sergeants, captains, and patrol officers). Phagocytes consist of macrophages, dendritic cells, and granulocytes. In various ways, the different types of phagocytes identify, engulf, and ingest germs and other invaders. Phagocytes lead the way in many critical innate reactions.

Knowing that where one harmful virus or germ lurks more may be hiding, phagocytes sound chemical alarms to bring the more specialized immune cells to the scene. First, dendritic cells display an antigen—a chemical that identifies the invader—so that the appropriate immune system specialist cells are able to recognize the culprit. The specialists, which consist of B and T cells, are known collectively as lymphocytes, and they make up the body's adaptive immune system. As its name implies, the adaptive immune system adapts and adjusts to specific threats as the need arises, whereas the innate immune system is pre-existing and less specific. Although B cells can recognize and respond to antigens without much assistance, T cells require a second "danger" signal in the form of a biological flag, known as an MHC molecule, which an antigen-presenting cell (such as a macrophage or dendritic cell) uses to clearly designate that an invader is foreign.

This flow chart describes the steps of the body's immune response to a foreign organism or substance. A foreign substance triggers first the innate and subsequently the adaptive immune systems. In a successful response the pathogens are killed and antibodies are produced to more efficiently respond to future attacks.

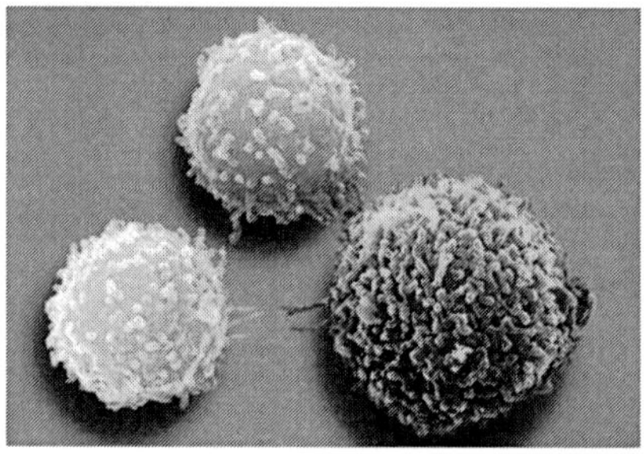

A scanning electron micrograph image shows a macrophage (right) and two lymphocytes (left). Lymphocytes and macrophages are types of white blood cells.

These danger signals rouse the T cells, which quickly multiply and rush to the scene. The elapsed time before these defenses arrive may be days to weeks, but repeated exposure to an invader will teach T cells to respond quicker. Vaccines, which will be discussed later, take advantage of this immune system memory.

The Battle Escalates

Lymphocytes mount a two-pronged attack, one directed at infected cells and the other at hostile microbes circulating in the blood. The cell-targeting attack is directed by T cells. Killer T cells directly kill infected cells that have been marked for destruction by the phagocytes, while helper T cells coordinate the attack and send for reinforcements as needed. Meanwhile, B cells produce antibodies that bind to free-floating microbes circulating in the blood so that they cannot infect other cells. Phagocytes then engulf and destroy the antibody-studded invaders. Antibodies also activate complement proteins, which destroy microbes by punching holes in them. As the battle rages at the microscopic level, you may start to be aware that something is amiss. If you've been infected with a cold virus, for example, your throat will become sore, your eyes watery, and your sinuses congested. These are the physical signs of inflammation, the buildup of fluid and cells that occurs as the immune system fights a hostile invader.

Remember and Recover

Once all invaders and infected cells have been destroyed, the immune system soldiers that once multiplied so quickly decline in number. Inflammation subsides, and symptoms gradually disappear. But certain memory B and T cells remain, to remember how to attack the invader if it returns.

Even with all these physiological weapons at our disposal, microbes occasionally manage to outsmart our immune sentries or elude detection. One reason is that microbes evolve rapidly and humans do not, giving the germs an advantage over our immune defenses. Moreover, if our defence system is impaired in any way—either by an inherited condition or because we've been exposed to certain environmental toxins or medical treatments that suppress immunity—we are more susceptible to infection.

Failing to Protect: Immune Dysfunction Spells Trouble

For all its highly evolved mechanisms for identifying and fighting off the daily onslaught of germs, the immune system sometimes fails to provide the protection we need. Because the system is so complex, relying on a battalion of specialized units (in the form of T cells, B cells, macrophages, and an arsenal of antibodies and other biochemicals) to perform specific defence tasks and to coordinate with one another, dysfunction in any one of these units can render the entire system inadequate.

One might compare the immune system to the U.S. Department of Defence. When waging a war, the Defence Department relies on all of its forces—Army, Navy, Air Force, Marines, as well as specialized intelligence units, paramilitary forces, antiterrorist teams, and so on—to work together in a coordinated fashion to defeat the enemy. It also relies heavily on its weapons—bombs, missiles, guns, tanks, battleships, warplanes, and so forth. As with the immune system, the effectiveness of the war machine as a whole depends largely on the ability of each component to efficiently fulfill its individual role. If a reconnaissance plane spies an enemy battalion approaching, but it can't get the message to troops on the ground, or if the troops don't have the ammunition or armor they need to fend off the attack, the enemy is free to wreak havoc.

In the same way, if any one component of the immune system is not up to par, germs can quickly get the upper hand. This is the case in immune deficiency diseases (IDDs), which result when one or more parts of the immune system are missing or defective.

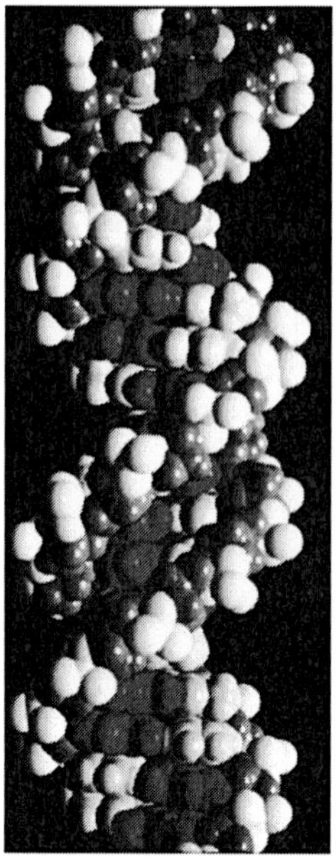

A computer model shows the two intertwined strands of nucleotides that form the "double-helix" structure of DNA, which contains the body's genetic code.

An IDD can be inherited or acquired through an infection (such as HIV) or illness, or it can result as a side effect of certain immunosuppressive medical treatments, or treatments that suppress natural immune responses. Acquired immunodeficiency syndrome (AIDS) is the best-known immune deficiency disease ("Special Focus: AIDS," page 32), but there are many other less well-known and less prevalent examples. Some cancer treatments, including chemotherapy drugs, radiation, and high doses of a group of medicines called steroids, can weaken the immune system and render a person more vulnerable to infection.

Genes Problem

Primary immune deficiency (PID) diseases are the result of an inherited genetic defect that interferes with the immune system's normal development in one way or another. More than eighty different PID diseases have been identified, each one producing a constellation of symptoms depending on which piece or pieces of the defensive system are faulty. Symptoms of individual diseases can range from mild or nonexistent to devastatingly severe. About 25,000 to 50,000 Americans have been diagnosed with a severe form of primary immune deficiency disease, but there are probably many thousands more who have milder forms. One of the most common of these diseases, affecting about 1 in 600 Americans, is selective immunoglobulin A (IgA) deficiency. IgA is a particular type of antibody that defends against infections at mucous membranes that line the mouth, airways, and digestive tract. IgA deficiency results when B cells do not mature properly and fail to produce

IgA antibodies at the levels required. Many people with the deficiency remain healthy; others may suffer recurrent infections of the ear, sinuses, or lungs.

Living in a Bubble

Far less common is severe combined immunodeficiency disease (SCID), sometimes called "bubble boy" disease, which affects about one child in a million and is usually fatal. Infants born with this inherited condition have dramatic abnormalities in both innate and adaptive immunity, leaving them utterly defenseless against infections of any type and necessitating that they remain in a germ-free environment (the plastic "bubble" in which such children often must live acts as a barrier to infectious microbes).

At least a quarter of SCID cases are linked to a defect in a gene that specifies the genetic code for a particular enzyme, adenosine deaminase (ADA). The absence of ADA interferes with metabolic processes within the cell, setting off a cascade of molecular events that are particularly lethal to T and B cells. Scientists have tested controversial gene therapy approaches to treating SCID, with the goal of replacing the missing gene and restoring production of ADA. While this approach has proven successful in some cases, treated children died as a result of complications from the therapy, prompting an end to clinical trials until scientists conduct further laboratory research.

Targeting Tumor Cells

In addition to defending us against germs, the immune system identifies and attacks cancerous tumor cells. When a normal cell turns cancerous, many properties of the cell change, and these changes, some of which occur on the surface of the cell, are recognized by the immune system as antigens. The dendritic cells, for example, recognize dying cancer cells and process them to make their antigens more conspicuous to T cells. Under normal circumstances, the immune system fights back to contain and neutralize the cancer. However, if the immune system is defective in some way, or tumor cells are so widespread or so fast-growing that the immune defenses are overwhelmed, cancer cells can develop and thrive.

NORMAL AND CANCER CELLS
Structure

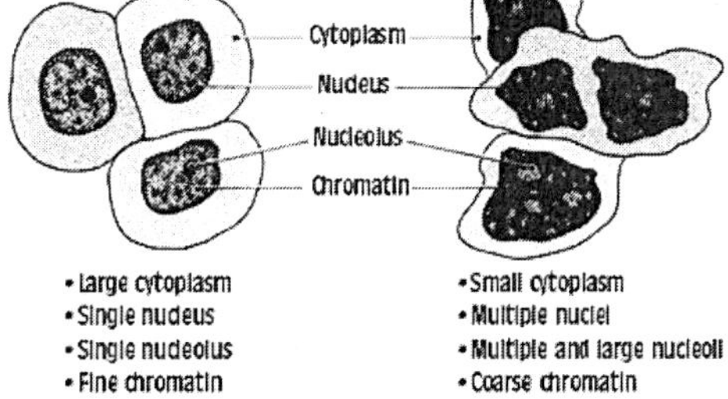

NCI

A great deal of research is now directed at developing immune therapy for cancer. Sometimes referred to as cancer vaccines, these approaches seek to bolster the immune system's response to tumor cells. A number of research groups are using various strategies to "pump up" tumor-killing immune cells or to increase levels of specific types of antibodies that can seek out and destroy tumor cells. The National Cancer Institute, part of the federal government's National Institutes of Health, is the main agency responsible for coordinating and funding clinical trials testing cancer immune therapy.

Disease in the Immune System

Like tragic cases of "friendly ?re" on the battlefield of war, sometimes the immune system mistakenly attacks "self" tissue. This self-attack can cause symptoms that range from the annoyance of a runny nose, as in allergies, to the devastating progressive degeneration of joints and organs associated with rheumatoid arthritis.

A disease in which the immune system attacks the body is called an autoimmune disorder. Autoimmune disorders comprise at least 80 different conditions that together affect 5 to 9 percent of Americans, or between 14 million and 22 million people. The misdirected attack on self tissue may target one or several body parts, depending on the disease. For example, in type 1 diabetes the immune system zeroes in on the pancreas, damaging cells that secrete the hormone insulin and rendering the body inefficient at processing glucose (a type of sugar). In multiple sclerosis, the objects of attack are cells in the central nervous system (the brain and spinal cord), particularly those in myelin, the fatty substance that sheathes nerve fibres and enhances the transmission of nerve signals. In psoriasis, skin cells are targeted. Systemic lupus erythematosus, on the other hand, is an example of a non-organ-specific autoimmune disease: organs and tissues throughout the body may be affected, and different people experience different sets of symptoms.

No Tolerance for Self

Each autoimmune disorder is associated with a unique combination of health problems. All of them seem to result from the malfunctioning of immune tolerance mechanisms. "Tolerance" refers to the process by which the developing immune system normally eliminates any immune cells that are autoreactive—that is, immune cells that mount a response to the body's own tissue.

Tolerance is a multilayered system, with a series of checks and balances built in to prevent self-attack from either innate or adaptive immune cells. In some cases, the self-reactive cells are removed or deleted. In other cases, they are silenced by immune cells called regulatory or suppressor cells. Unfortunately, the process is somewhat leaky, and each of us ends up having a subpopulation of autoreactive immune cells floating around our body. If at some point in our life we are exposed to a microbe that happens to carry antigens resembling those on a particular organ—a situation immunologists call molecular mimicry—then the self-reactive cells that slipped through the checkpoints of tolerance mistakenly take aim at the body, striking out against the very tissue that they are supposed to be defending. Some scientists reason that such a scenario may trigger multiple sclerosis, for example.

A combination of many factors, inherited and environmental, increases one's susceptibility to autoimmune disease. On the genetic side, scientists are trying to identify the specific genes involved in various disorders, but that search has proved difficult because these disorders do not appear to result from single genetic defects. Rather, it is likely that a combination of genes, each of which may increase one's vulnerability, interact with environmental and lifestyle factors to produce disease. Environmental risks for autoimmune disease may include exposure to certain toxins or chemicals that are in the air, the ground, or the food we eat, or to certain viral infections.

For reasons that are not clear, autoimmune disorders strike women more than men. Some scientists believe the female hormone estrogen may contribute to this increased incidence among women, but estrogen's role in immune function is complex and still being sorted out. In 2002 the National Institutes of Health committed $51 million to a five-year research initiative aimed at unraveling the puzzle of autoimmune disorders. The plan, which established nine Autoimmune Centers of Excellence at large academic research institutions, funds both basic research on the underlying biology of these disorders and clinical trials to test potential treatments. The goal is to speed the translation of scientific discoveries about the basic biology of autoimmunity into new therapies that will benefit patients.

Sneezing and Wheezing

Though they are not technically autoimmune disorders, allergies and asthma also result from an inappropriate immune response. An allergic reaction occurs when normally harmless substances, such as pollen or chemicals in pet dander, trigger an immune response. In people prone to allergies, the response produces an overabundance of the antibody called IgE, or immunoglobulin E, which in turn triggers plasma cells in the blood and mast cells in the skin, tongue, lungs, nose, and intestinal tract to release histamine, a biochemical that produces common allergy symptoms.

Asthma results when mast cells in the lungs and airways are provoked into producing histamine as a result of contact with an allergen, or because of some other precipitating factor such as exercise. The airways can become constricted, causing the difficult breathing, wheezing, and coughing that are typical of an asthma attack. Without treatment, asthma can be deadly. Although the overall number of deaths from asthma is low, at least 17 million

Americans have asthma, and its prevalence appears to be increasing. Researchers are trying to understand why.

the Immune System Can't Beat Every Enemy

You may be wondering why, if the immune system is such an ef?cient germ-killing machine, so many people are sick with the flu at certain times of the year. Or why, if our immune system can fight off cancer cells and other life-threatening illnesses, it can't protect us from the common cold. Why do we hear so much in the news these days about "new" viral diseases (such as West Nile, SARS, and bird flu) and resurging "old diseases" (such as tuberculosis, malaria, and whooping cough)?

Despite the remarkable efficiency of the normally functioning immune system, people who manage to escape unscathed from every infectious microbe they encounter are rare. The highly adaptable nature of germs means that new threats are always emerging as microbes reinvent themselves, learn to outsmart our medicines, and jump from other species to humans.

Adapting to Change

Perhaps the biggest reason our immune systems can't seem to keep up with every "bug" out there waiting to infect us is a simple one: microbes are more adaptable to changing environments than we are. Bacteria, parasites, and viruses replicate rapidly in response to environmental pressures and, while doing so, alter their structures in subtle ways that make them undetectable to the immune system.

Flu viruses are among the most changeable of microbes; the genes that compose them are in a continual state of flux and reassortment. This is why last year's ?u vaccine is unlikely to protect fully against this year's dominant strain (or strains) of ?u, and why the ?u vaccine is reformulated every year. It is also why, without immunization, we are more likely to catch the ?u: since the virus has morphed into a new composition, our immune system doesn't recognize it, and we end up suffering the classic symptoms: fever, chills, muscle aches, and fatigue. These symptoms are the result of our immune system trying to rid the body of the virus. Annually, many Americans come down with the ?u and an average of 20,000 to 40,000 people die from it. The elderly, young children, and people whose immune systems are compromised are most at risk for serious complications from the ?u.

The common cold, like the ?u, is the result of a viral infection, usually called a rhinovirus. Scientists have identified more than 110 different strains of rhinovirus and at least another hundred viruses that cause colds in humans. Nearly half of all adult colds are of unknown origin, which makes the sought-after "cure for the common cold" a distant goal. School-age children get colds most often, probably because they are in such close proximity to one another. Children have an average of six to ten colds a year, compared with adults' two to four a year, on average. People over 60 have even fewer colds—less than one a year, on average—which may simply reflect the fact that many older people have less day-to-day contact with other people who could infect them with a cold virus.

Pandemics and Emerging Microbes

Every few decades a particularly virulent strain of the flu emerges and produces a pandemic, an epidemic that occurs across a wide geographic area. The last flu pandemic occurred in 1968, and many scientists believe we are long overdue for the next. Global health experts are keeping a close watch on bird flu (also known as avian flu), which many fear might produce a pandemic. First identified in chickens, this flu strain began causing alarm because it was found in humans for the first time in 1997, and by 2005, many dozens of human cases had been documented in several countries.

One of the biggest subjects in contemporary microbiological research (the study of microbes) is that of "emerging microbes." In the last two decades alone, more than thirty newly recognized infectious diseases have emerged, including AIDS, toxic shock syndrome, Lyme disease, hantavirus, hepatitis C, and SARS. In most cases, these diseases are not really new, but rather their incidence is rising more rapidly than ever, making them an ever greater threat to humans. Also of concern are old diseases that are now re-emerging, even though they were mostly wiped out thanks to improved public health measures and widespread immunization. Tuberculosis is one example. Driving the spread of these diseases is increasing globalization, with the growing world population traveling greater distances than ever before, carrying disease-causing microbes with them. Climate changes and the widespread logging of Earth's forests are also factors in the spread of infectious diseases to new geographic areas and populations.

Index